Lecture Notes in Mathematics

Edited by A. Dold and B. Eckmann

Subseries: Fondazione C.I.M.E., Firenze
Adviser: Roberto Conti

1385

A. Anile Y. Choquet-Bruhat (Eds.)

Relativistic Fluid Dynamics

Lectures given at the 1st 1987 Session of the
Centro Internazionale Matematico Estivo (C.I.M.E.)
held at Noto, Italy, May 25–June 3, 1987

Springer-Verlag
Berlin Heidelberg New York London Paris Tokyo Hong Kong

Editors

Angelo M. Anile
Dipartimento di Matematica, Città Universitaria
Viale A. Doria 6, 95125 Catania, Italy

Yvonne Choquet-Bruhat
Département de Mécanique, Université P. et M. Curie
Tour 66, 4, Place Jussieu
75252 Paris Cedex, France

Mathematics Subject Classification (1980): 83, 83C, 83E

ISBN 3-540-51466-X Springer-Verlag Berlin Heidelberg New York
ISBN 0-387-51466-X Springer-Verlag New York Berlin Heidelberg

Printing and binding: Druckhaus Beltz, Hemsbach/Bergstr.
2146/3140-543210 – Printed on acid-free paper

CB-SHh
SCIMON

PREFACE

In the field of Relativistic Fluid Dynamics, there has been only one previous conference (the C.I.M.E. course of 1970, held in Bressanone with the late professor Cattaneo as Director) and the only other book on the subject is the excellent monograph by professor Lichnerowicz, dated 1967, entitled **Relativistic Hydrodynamics and Magnetohydrodynamics** and published by Benjamin. Therefore it is no surprise that after 17 years the proceedings of a course on this subject should amount to a rather substantial book. In 17 years the subject has developed greatly, mainly with regard to applications which previously would never have been imagined.

In particular there has been a tremendous development in the field of plasma physics (relativistic fluids are a good model for high-energy astrophysical plasmas) and nuclear physics (relativistic fluids are currently used in the analysis of the heavy ion reactions). Therefore relativistic fluid dynamics is a working tool in vastly different areas such as astrophysical plasmas and nuclear physics.

This is the explanation for the fact that, since 1970, there has been no other general course on the subject. In fact there have been sessions on relativistic fluids in conferences on plasma physics and on nuclear physics separately. However this tended to obscure the underlying mathematical structure of the subject and made more difficult to transfer results and techniques from one area to another.

Having realized this, we thought that a course on this subject could bring expertise and interest from several areas (astrophysics, plasma physics, nuclear physics, mathematical methods) and provide an appropriate arena for fruitful discussions and exchanges of ideas

The main lecture courses had the objective of introducing the most significant aspects of relativistic fluid dynamics. Their topics were: **covariant theory of conductivity in ideal fluid and solid media; covariant fluid mechanics and thermodynamics: an Introduction; hamilton techniques for relativistic fluid dynamics;** and **stability theory, relativistic plasmas.**

The lectures were delivered by leading scientists in these areas (B. Carter, W. Israel, D. Holm, H. Weitzner) and constitute an up-to-date and thorough treatment of these topics.

They were also several interesting contributions from the seminars on specialized topics. Not all of them, for reasons of space, have been included in this volume. In particular, the seminars by Dudyński and Ekiel-Jezewska, Granik, Hiscock and Lindblom, Deb Ray, Boillat, were omitted. The important topics treated by these authors are covered, however, in other publications.

About fifty people (including research students and senior scientists) participated actively in the course.

We thank all the lecturers and the participants for their invaluable contribution to the success of the course. We thank also the C.I.M.E. foundation and its Director, Professor Conti and secretary, Professor Zecca, for having sponsored the Course and for their constant help and encouragement. Thanks are also due to the City of Noto (world famous for its beautiful beaches and splendid baroque architecture) for its support of the conference and the lavish hospitality. Finally we are grateful to the local organizing committee (Dr. Muscato, Professors Miceli, Fianchino and Fortuna) for their support and dedication to the success of the meeting.

TABLE OF CONTENTS

COVARIANT THEORY OF CONDUCTIVITY IN IDEAL FLUID OR SOLID MEDIA.

Brandon Carter

Institute for Theoretical Physics, U.C.S.B.,
Santa Barbara, California 93106,
and
Groupe d'Astrophysique Relativiste – D.A.R.C.,
C.N.R.S. – Observatoire de Paris, 92195 Meudon, France.

Contents:

Abstract

After a preparatory account of the established theory of non-conducting perfect fluid media, with emphasis on the important but traditionally neglected concept of the 4-momentum 1-form associated with each chemically independent constituent, it is shown how to generalise the theory to allow for conductivity by extending the variational formalism in terms of independent displacements of the world-lines.

Attention is concentrated initially on the simplest possible conducting model, in which appart from a single conserved particle current the only other constituent is the entropy-current whose flow world-lines are displaced independently of those of the conserved particles in the variational formulation, resistive dissipation being included by allowing the variationally defined force density acting between the particle and entropy currents to be non-zero. The model so obtained is fully determined by the specification of the resistivity coefficient and the traditional thermodynamic variables of the corresponding non-conducting thermal equilibrium state if it is restricted by postulating that it satisfies a "regularity ansatz" to the effect that the separate 4-momenta associated with the (non-conserved) entropy and the (conserved) particles are respectively directed allong the corresponding flow directions. It is shown that this regularity ansatz is consistent with good hyperbolic causal behaviour, unlike a previous ansatz proposed by Landau and Lifshitz, which is interpretable as a degeneracy requirement to the effect that the separate 4-momenta have the same direction as each other, and which results in (inevitably superluminal) parabolic behaviour. Another ansatz, proposed much earlier by Eckart, is shown to be effectively equivalent to the mixed-up requirement that the 4-momentum associated with the entropy to be directed not along its own flow direction but along that of the particles, and (as recently shown by Hiscock and Lindblom) results in even worse (quasi-elliptic) behavior.

After this analysis of the simplest possible well behaved thermally conducting model, it is shown how the principles by which it was constructed can be extended to allow for multiple (including electrically charged) currents, in solid as well as fluid media.

Introduction.

On of the main objectives of this course will be to demonstrate the availability of a simple and natural way of treating thermal conductivity in relativistic hydrodynamics using an effectively unique "off the peg" model to be designated by the qualification "regular", which singles it out within a wider class of in general "anomalous" (albeit mathematically well behaved and for some purposes physically well adapted) models of a somewhat more complicated type, and which distinguishes it also from the older and better known models due to Eckart[1] and to Landau and Lifshitz [2] whose mathematical behaviour has long been known to be blatently pathological, due essentialy to their failure to make proper allowance for the inertial delay time that should normally occur between the application of any external driving force (in this case the effect of a thermal gradient) and the build up of the corresponding response (in this case a proportional heat flux)[3,4].

Following lines originally developped in the non-relativistic domain by Muller[5], a considerable body of more recent work, mainly due to Israel and Stewart[6,7,8] has shown how the causal pathology in the more primitive earlier models can be satisfactorily overcome within a larger and much more elaborate class of "second order" models containing many adjustable parameters and functions that allow a model within this class to be "tailored" to fit particular physical contexts with considerable accuracy, using as a test case the much studied example of a monoatomic Boltzman gas[9]. In many practical situations, however, the cost in time (or in money, which in numerical computing and many other contexts often amounts to the same thing) of high accuracy tailoring is effectively prohibitive. Moreover lack of detailed knowledge of the subject to be fitted may render accurate tailoring impossible in any case, even if cost is no object. (Anyone with experience of shopping for clothes as a surprise present for someone else will be familiar with this problem.) It is therefore useful to have the option of using an inexpensive "off the peg model" that is guaranteed to be intrinsically trouble free as well as being reasonably well adapted to the most commonly ocurring situations, even if it cannot claim the high accuracy (at the expense of restrictive specialisation) of more elaborate models.

The "regular" model[10,11] to be described here is intended to fulfill such a need. Like the similarly motivated but unsuccessful earlier attempts by Eckart and by Landau and Lifshitz, this regular model can be considered as a limiting special case within the more general and complicated Israel-Stewart class. The mathematical properties of this entire class of models has recently become much better understood due to the work of Hiscock and Lindblom [12,13,14] who have carried out much more thorough analyses of causality and local stability properties than were available before. In particular they have cleared up the confusion that existed in the litterature on the question of whether the newer Landau-Lifshitz model was essentially distinct from the earlier Eckart model or whether it was merely the same theory (at least modulo unimportant higher order corrections) presented in terms of a different reference system. In a recent study of the special subclass of "first order" models within the general "second order" Muller-Stewart-Israel category, Hiscock and Lindblom have shown[13] that while the Landau-Lifshitz model is a a partial differential system that (like the ancient non-relativistic Fourier heat conduction model) exhibits parabolic (instead of causally desirable hyperbolic) behaviour as had been generally realised before, on the other hand the Eckart model is even worse (with the corollary that it is an essentially distinct theory) in that it actually displays quasi-elliptic behaviour! The regular model whose use is being advocated here has not yet been subjected to a thorough Lindblom-Hiscock type analysis, but its manner of construction ensures in advance that – subject to inequalities such as as must be imposed on the equation of state even for a simple non-conducting perfect fluid model – it will be entirely free of such flaws.

The approach that lead directly to the derivation of the regular thermal conduction model presented here differed from the the traditional approach (by which its existence had been overlooked in favour of causally unsatisfactory alternatives) in that the traditional approach was primarily based on analysis of the *stress-momentum-energy tensor*, with components $T^{\mu\nu}$ say, whereas the alternative approach, as developped in the present course, attatches greater importance to the (traditionally neglected) concept of the *momentum-energy covector* with components π_μ say instead. Except for a system consisting of very weakly coupled parts, the stress-momentum-energy tensor is in general fundamentally well defined only for the system as a whole (whence the futility of the historic Abraham-Minkowski controversy about how to distinguish its "material" and "electromagnetic" contributions in a polarised and thus elecromagnetically interacting medium). On the other hand in the kind of system to be considered here, even strongly interacting currents have corresponding *separately* well defined momentum-energy covectors.

In the simplest kind of thermally conducting model (including those of Eckart and Landau-Lifshitz type) there are just two dynamically independent current vectors, namely an (in general non-conserved) entropy current, with components s^μ say, together with a single (conserved) particle number current, with components n^μ say, and there will therfore be two distinct corresponding momentum energy covectors, with components ${\pi^0}_\mu = \Theta_\mu$, and ${\pi^1}_\mu = \chi_\mu$ say (where the choice of symbols Θ and χ is intended as a reminder of the respective relationships with *temperature* and *chemical potential* that will be described in due course). In the most general models to be described, the momenta may be independent both of each other and of the corresponding currents, but in order to obtain a simple general purpose "off the peg" model in which, appart trom the specification of a resistivity scalar, all thermodynamic function of state are determined uniquely by their analogues (as presumed to be known a priori) in thermal equilibrium, then some restrictive simplifying ansatz is required. The Landau-lifshitz type models may be accounted for in this approach as being implicitly based on an ansatz to the effect that the momentum covectors Θ_μ and χ_μ in question should not be vectorially independent, but that they should be proportional to each other (thereby determining a unit covector which turns out to be the timelike eigenvector of the full stress-momentum-energy tensor): having imposed such a degeneracy condition on the momenta, it is not surprising that one obtains the degeneracy property of parabolicity for the characteristics of the corresponding system. The Eckart type models can analogously be accounted for as being based on an ansatz that is even more obviously inappropriate, namely that the *thermal* momentum-energy covector, Θ_μ should be proportional to the covariantly modified version, n_μ of the *particle* current, whose own momentum-energy covector χ_μ is thereby forced to have the "anomalous" property of being directed elsewhere: again, it is scarcely surprising that such a mix-up leads to the flagrantly pathological property of quasi-elliptic behaviour.

Without going to the trouble of carrying out a causality analysis, it is obvious that neither of the prescriptions just described is compatible with the elementary common sense requirement of *consistency with the weakly-coupled limit* (as exemplified by the astrophysically familiar kind of situation in which the entropy is almost entirely carried by a "black-body radiation" gas of photons and perhaps electon-positron pairs, in comparatively weak interaction with a conserved background of heavy non-relativistic particles) in which the system may be approximated by two independent simple perfect fluids in which each of the momenta will necessarily have the same direction as the covariantly modified version of *is own* corresponding current, i.e Θ_μ will be proportional to s_μ, and not to n_μ which instead must be proportional to χ_μ. The "regular" model, as reccommended for "off the peg" use, is simply based on the postulate that the foregoing property of proportionality between each momentum-energy covector and the covariant version of the *corresponding* current should be preserved even when the effective coupling is strong. Since the decoupled limit is clearly well behaved in the sense of compatibility (subject to the usual inequalities) with normal causality, this good behaviour will evidently carry over into the wider class of coupled models characterised by the same "regularity ansatz".

The development of the subject in the present course will be based on a policy of adhering as closely as possible to a variational formulation at each stage, introducing dissipative effects in terms of the variationally defined "external" forces that would be required to vanish in the conservative strictly variational case. As well as showing the appropriate way to define the momentum-energy covectors that play the key role in our discussion, the variational approach has the advantage of taking care automatically of many of the mathematical self-consistency requirements that would otherwise have to be imposed on a piecemeal basis and which would end by going most of the way towards imposition of a variational structure in any case. (Any minor residual loss of generality is to be considered as acceptable according to the spirit of this course, whose purpose is to set up the simplest workable general purpose models for a treating broad classes of physical phenomena, rather than seeking to build the most elaborate and accurate models for specialised application.) The final, and most obvious (though for our main purpose accessory) bonus of the variational approach is that in ideal limit when the relevant dissipation coefficients (in our case the one of central interest being the thermal resistivity) are set equal to zero, one obtains a conservative system with the type of special properties whose implications and systematic exploitation are described in the accompanying lecture notes of Holm. Appart from the physical distinction that we shall be essentially concerned here with the inclusion of dissipative effects, a basic mathematical distinction between the approach to be developed here and the approach developped in the accompanying course of Holm is that the latter is based the use of a "(3+1)-decomposition" with respect to some specially chosen time-cordinate that is introduced so as to allow the direct adaptation to relativistic systems of methods (of generalised Hamiltonian type) originally developped in the context of Newtonian mechanics, whereas our present approach will be based on the contrary principle (with complementary advantages and disadvantages) of adhering to a fully covariant treatment at all stages.

As compared with the accompanying course on the full class of "second order" models by Israel, the main physical restriction that will be imposed as a simplification throughout the present course is that we shall take no account of viscous effects. Although there is no reason in principle why they should not be dealt with in within the mathematical framework of the variational approach used here, the inclusion of viscous effects will be postponed for a future occasion since it would nevertheless involve technical complications that would risk obscuring some of the very simple, but until now generally overlooked, points that I hope to put over here. This course does however go beyond the accompanying courses in a different direction by allowing for "chemical" (in the general sense, including nuclear) interactions, which were not included in the previously cited work, but which are more important than viscous effects in many astrophysical contexts, and which are comparatively simple to deal with because their description can mainly be carried out in terms of scalars, as compared with the vectors and covectors needed for describing conduction effects and the tensors needed for describing viscous effects. The final section (which is included as an optional extra) also contains as its main content a description of the way to allow for the possiblity that the thermal conductivity under consideration may be occurring in a elastic-solid (as opposed to fluid) background (as would apply in the case of a neutron star crust). Although it would be mathematically simpler, allowance for viscous stress would involve a further step away from the strictly variational structure, and its description would involve further physically independent and therfore debateable postulates. On the other hand, although the technical machinery needed for dealing with (shear dependent) elastic stress is more elaborate mathematically than would be required for the inclusion of (shear-rate dependent) viscous stress, the fact that it involves no additional mechanism of dissipation makes treatment of elastic stress particularly simple from the point of view of the amount of physical input required.

I. NON-CONDUCTING MULTICONSTITUENT FLUIDS.

1.1 Mathematical requisites: Cartan derivatives and Lie derivatives.

Before describing the first of the physical models with which we shall be concerned, we shall start by explaining some of the basic mathematical machinery and terminology that will be used throughout this course. We shall work in terms of a background manifold, $\mathcal{M}$ say, with local coordinates x^μ, $\mu = 1, ..., n$, where the dimension will of course just be $n = 4$ in the ordinary space-time applications that will be considered. Familiarity with the usual Riemannian covariant differentiation operation ∇, with local coordinate representation ∇_μ, will be taken for granted. However although such a differentiation operation is generally covariant in the sense of being defined independently of any preferred linear structure, it does depend on the specification of a fundamental (pseudo-) metric with components $g_{\mu\rho}$ say satisfying $\nabla_\nu g_{\mu\rho} = 0$. Since we shall find it profitable to work as far as possible with concepts and relationships that are covariant in the stronger sense of being independent even of the metric, we shall prefer, whenever it is feasable, to use the exterior differentiation scheme of Cartan which we now recapitulate briefly, both to fix the terminology and notation conventions, (which vary considerably throughout the physics litterature) and because its advantages in fluid mechanics (as opposed e.g. to electromagnetic theory), although coming to be more widely recognised (see e.g. the work of Schutz[15]), is not yet as widely known as it deserves to be.

The basic Cartan exterior calculus scheme is specialised in that it applies only to *covariant* tensors (which we shall distinguish from contravariant and mixed tensors by underlining) that are fully *antisymmetric*, i.e. to p-forms, ($p \leq n$) as defined in terms of tensor components $\omega_{\mu_1\ldots\mu_p}$ satisfying

$$\omega_{\mu_1\ldots\mu_p} = \omega_{[\mu_1\ldots\mu_p]} \tag{1.1}$$

(where square brackets denote antisymmetrised averaging), but the severity of this restriction is mitigated by the fact that such tensors, $\underline{\omega}$ are the only ones for which integration over a p-surface S is well defined in the absence of any previously specified (e.g. linear) structure on the manifold, since one can construct an (unambiguously additive) scalar by contracting such a p-form with the surface element p-vector (meaning a fully antisymmetric *contravariant* tensor, which we shall distinguish by an overhead arrow) with components $dS^{\mu_1\ldots\mu_p}$ given in terms of a tangent space basis consisting of infinitesimal displacements $dx^\mu_{(1)}, dx^\mu_{(2)}, \ldots, dx^\mu_{(p)}$ by

$$d\vec{S} = d\vec{x}_{(1)} \wedge d\vec{x}_{(2)} \wedge \ldots \wedge d\vec{x}_{(p)} \tag{1.2}$$

where the (associative though not commutative) exterior product operation is defined in accordance with the normalisation convention introduced by Cartan (though not followed by all subsequent authors) by

$$(\underline{\omega} \wedge \underline{\Omega})_{\mu_1\ldots\mu_p\mu_{p+1}\ldots\mu_{p+q}} = \frac{(p+q)!}{p!q!}\omega_{[\mu_1\ldots\mu_p}\Omega_{\mu_{p+1}\ldots\mu_{p+q}]} \tag{1.3}$$

for any p-form $\underline{\omega}$ and q-form $\underline{\Omega}$. Using the notation $\rfloor$ for *inner* multiplication as defined by contraction with the normalisation convention

$$\vec{S}\rfloor\underline{\omega} = \frac{1}{p!}\,\omega_{\mu_1\ldots\mu_p} dS^{\mu_1\ldots\mu_p} \tag{1.4}$$

one can define the integral of $\underline{\omega}$ over S by a limit process as the surface elements are made infinitesimally small of the corresponding sum:

$$\int_S d\vec{S}\rfloor\underline{\omega} = \mathop{\mathcal{L}t}_{dS\to 0} \sum_{dS} d\vec{S}\rfloor\underline{\omega}\,. \tag{1.5}$$

In order to avoid confusion with the traditional physicist's use of the symbol "d" to indicate "infinitesimal variations" (i.e tangent space elements) as above, we shall not follow the newer mathematician's custom of using "d" as an abbreviation for the exterior differentiation operation definable in the more explicit notation as "$\partial\wedge$" , where ∂ denotes the elementary partial differentiation operation with coordinate representation given simply as

$$\partial_\mu = \frac{\partial}{\partial x^\mu} \tag{1.6}$$

Thus we distinguish between the infinitesimal variation $d\phi$ of a scalar field ϕ due to an infinitesimal displacement $d\vec{x}$ on the one hand, and the corresponding gradient 1-form which we denote by $\partial\phi$ (but which in customary mathematicians shorthand would be indiscriminately denoted by the same symbol as the image displacement $d\phi$) on the other hand, the relation between them being given by

$$d\phi = \partial\phi \rfloor d\vec{x} = (\partial\phi)\cdot d\vec{x} = (\partial_\mu\phi)dx^\mu\ . \tag{1.7}$$

where we introduce the traditional use of a simple dot, $\cdot$, to indicate contraction of just one pair of adjacent indices, as distinct from the contraction of all possible indices that is indicated by the symbol $\rfloor$ (the result being of course the same in this particular case). In this purely scalar example the antisymmetrised product symbol $\wedge$ is quite redundant. For a form $\underline{\omega}$ of higher order, $p \geq 1$, the antisymmetrisation indicated by the $\wedge$ symbol in the exterior product $\partial\wedge\underline{\omega}$ is a substantive requirement for general covariance, but for this very reason can in many (though by no means all) contexts, including the present work, be taken to be understood implicitly, without danger of ambiguity, even when the wedge symbol is tacitly dropped in the interest of brevity as we shall do from now on, writing $\partial\underline{\omega}$ for $\partial\wedge\underline{\omega}$ with coordinate components given by

$$(\partial\underline{\omega})_{\mu_1\mu_2\ldots\mu_{p+1}} = (p+1)\partial_{[\mu_1}\omega_{\mu_2\ldots\mu_{p+1}]} \tag{1.8}$$

The exterior differentiation operation as so defined has the well known cohomology property associated with the name of Poincaré, to the effect that for an arbitrary p-form $\underline{\omega}$

$$\partial\partial\underline{\omega} = 0 \tag{1.9}$$

and that at a local (but not necessarily global) level one has, conversely

$$\partial\underline{\Omega} = 0 \ \Rightarrow\ \exists\underline{\omega} \ :\ \underline{\Omega} = \partial\underline{\omega}\ . \tag{1.10}$$

One also has the associated Stoke's theorem property to the effect that the integral over a closed p-surface $\partial\Sigma$ bounding a $(p+1)$-volume Σ say will be given by

$$\oint_{\partial\Sigma} d\vec{S}\rfloor\underline{\omega} = \int_\Sigma d\vec{\Sigma}\rfloor\partial\underline{\omega}\ . \tag{1.11}$$

The development of the antisymmetric differential calculus can be taken considerably further so as to apply to contravariant tensors whenever a preferred volume measure n-form $\underline{\epsilon}$ is specified, since it may be used (even in the absence of any corresponding metric tensor) for relating p-forms to *dual* $(n-p)$-vectors and vice versa. Thus if $\vec{\beta}$ is a q-vector (i.e. an antisymmetric contravariant tensor of order q) then we can construct its dual $(n-q)$-form ${}_*\vec{\beta}$ according to the formula ${}_*\vec{\beta} = \vec{\beta}\rfloor\underline{\epsilon}$, i.e.

$$ {}_*\beta_{\mu_1\ldots\mu_{n-q}} = \frac{1}{q!}\, \beta^{\rho_1\ldots\rho_q}\, \epsilon_{\rho_1\ldots\rho_q\mu_1\ldots\mu_{n-q}}\, . \tag{1.12}$$

Using an upper star prefix for the inverse mapping from (covariant) p-forms to (contravariant) $(n-p)$-vectors, as defined by

$$ {}_*({}^*\underline{\omega}) = \underline{\omega}\,, \quad {}^*({}_*\vec{\beta}) = \vec{\beta} \tag{1.13}$$

the interior product of a p-form $\underline{\omega}$ and a q-vector $\vec{\beta}$ can be expressed (depending on whether p is larger or smaller than q) in terms of outer (Cartan) multiplication in one or other of the forms

$$ \vec{\beta}\rfloor\underline{\omega} = {}_*(({}^*\underline{\omega})\wedge\vec{\beta}) \quad \text{if } p\geq q\,,$$

$$ \vec{\beta}\lfloor\underline{\omega} = {}^*(\underline{\omega}\wedge({}_*\vec{\beta})) \quad \text{if } q\geq p \tag{1.14}$$

This suggests the convenience of defining the *inner* derivative, or "divergence" of a q-vector $\vec{\beta}$ to be

$$ \mathrm{div}\vec{\beta} = {}^*(\partial({}_*\vec{\beta})) \tag{1.15}$$

In order for this to be well defined the only prerequisite structure that has to be given on the manifold is the measure $\underline{\epsilon}$, the specification (by a choice of affine connection) of a general purpose covariant differentiation operation ∇ being unnecessary. However whenever a covariant differentiation operator actually is given, subject of course to consistency with the measure in the sense that $\nabla\underline{\epsilon} = 0$, the divergence operation defined by (1.15) will be expressible directly in coordinate or condensed notation as

$$ (\mathrm{div}\vec{\beta})^{\mu_1\ldots\mu_{q-1}} = \nabla_\lambda\, \beta^{\mu_1\ldots\mu_{q-1}\lambda}, \quad \mathrm{div}\vec{\beta} = (-1)^{q-1}\nabla\cdot\vec{\beta}\, . \tag{1.16}$$

We may use the generalised divergence relation defined by (1.15) to express the Stokes theorem (1.11) in the dual *Green Theorem* form commonly preferred by physicists:

$$ \oint_{\partial\Sigma} \underline{dS}\rfloor\vec{\beta} = (-1)^n \int_\Sigma \underline{d\Sigma}\rfloor\mathrm{div}\vec{\beta}\, . \tag{1.17}$$

where the abbreviation

$$ \underline{dS} = *d\vec{S} \tag{1.18}$$

has been used for the dual surface element.

Another important kind of differentiation operation, which shares with exterior differentiation the property of bing well defined and generally covariant independently of any background linear or Riemannian structure or even of any measure that may be present is *Lie differentiation* with respect to any smooth vector field $\vec{\xi}$ say, which we shall denote by the symbol $\vec{\xi}\mathcal{L}$. It is definable for any kind of field X (not just tensors, but also densities, affine connections, et cetera) that is geometric in the sense of being bijectively mappable by any non-singular differentiable automorphism $f: \quad x \mapsto fx$ of the support manifold onto a well defined naturally induced retraction image fX,

$$f: \ X(fx) \mapsto fX(x)\ . \tag{1.19}$$

Letting $f(t)$ denote the one-parameter family of diffeomorphisms constructed by dragging the manifold a parameter distance t along the integral curves of

$$\frac{dx^\mu}{dt} = \xi^\mu \tag{1.20}$$

the corresponding Lie derivative is definable as

$$\vec{\xi}\mathcal{L}X = \frac{d}{dt}(f(t)X)\Big|_{t=0} \quad . \tag{1.21}$$

In the case of a quantity that is *tensorial* with mixed indices $T_{\mu\ldots}^{\ \nu\rho\ldots}$ say, the Lie derivative is given explicitly by the general formula

$$\begin{aligned}(\vec{\xi}\mathcal{L}T)_{\mu\ldots}^{\ \nu\rho\ldots} = \xi^\lambda\partial_\lambda T_{\mu\ldots}^{\ \nu\rho\ldots} + T_{\lambda\ldots}^{\ \nu\rho\ldots}\partial_\mu\xi^\lambda + \ldots \\ - T_{\mu\ldots}^{\ \lambda\rho\ldots}\partial_\lambda\xi^\nu - \ldots\end{aligned} \tag{1.22}$$

with an additional term for each further index, the most familiar special case being that of the Lie derivative of another vector field, $\vec{\eta}$ say for which one obtains the simple Lie commutator:

$$\begin{aligned}\vec{\xi}\mathcal{L}\vec{\eta} = [\vec{\xi},\vec{\eta}] = -\vec{\eta}\mathcal{L}\vec{\xi}\,, \\ [\vec{\xi},\vec{\eta}]^\mu = \xi^\lambda\partial_\lambda\eta^\mu - \eta^\lambda\partial_\lambda\xi^\mu\,.\end{aligned} \tag{1.23}$$

Another familiar special case concerns the spacetime metric $g_{\mu\rho}$ used for specifying the covariant differentiation operator ∇ by the requirement that it should give $\nabla_\lambda g_{\mu\rho} = 0$ for which one obtains

$$(\vec{\xi}\mathcal{L}g)_{\mu\rho} = 2\nabla_{(\mu}\xi_{\rho)} \tag{1.24}$$

(with the standard convention that round bracket on indices indicates symmetrised averaging over permutations) which vanishes when ξ is the generator of a one-parameter isometry group.

Of particular importance for our present purposes is the case of the "differential forms", i.e. covariant fully antisymmetric tensors to which the Cartan exterior differential calculus described above applies: for any p-form $\underline{\omega}$, the Lie derivative is expressible concisely in the above notation scheme by Cartan's formula

$$\vec{\xi}\mathcal{L}\underline{\omega} = \vec{\xi}\cdot(\partial\underline{\omega}) + \partial(\vec{\xi}\cdot\underline{\omega}) \tag{1.25}$$

This formula may be used directly for the evaluation of the rate of change with respect to the parameter t introduced by (1.20) of the flux of $\underline{\omega}$ over any p-surface $\Sigma(t)$ that is obtained from an initial p-surface $\Sigma(0)$ by dragging along the integral curves of $\vec{\xi}$, which by definition of the Lie derivative is given by

$$\frac{d}{dt}\int_{\Sigma(t)} d\vec{\Sigma}\rfloor\underline{\omega} = \int_{\Sigma} d\vec{\Sigma}\rfloor\vec{\xi}\mathcal{L}\underline{\omega}\,. \tag{1.25}$$

Using the Cartan formula (1.25) and the Stokes theorem (1.11) one obtains the rate of variation of the flux integral in the form

$$\frac{d}{dt}\int_{\Sigma(t)} d\vec{\Sigma}\rfloor\underline{\omega} = \int_{\Sigma} d\vec{\Sigma}\rfloor\vec{\xi}\cdot(\partial\underline{\omega}) + \oint_{\partial\Sigma} d\vec{S}\rfloor\vec{\xi}\cdot\underline{\omega}\,. \tag{1.27}$$

One can use this important general identity to discuss various kinds of conservation theorem that may be relevant, depending on whether one is most concerned with the form $\underline{\omega}$, the surface Σ of the vector field $\vec{\xi}$. For example the form $\underline{\omega}$ itself may be said to be conserved independently of any reference to a particular surface Σ or vector field $\vec{\xi}$ if it is "closed" in the sense that $\partial\underline{\omega} = 0$ since then (directly by the Stokes theorem) the total flux through *any* surface Σ that is itself "closed" (compact, without boundary) will vanish. However such a closure property would not be sufficient to make the rate of variation (1.27) vanish for a surface Σ with non-empty boundary $\partial\Sigma$ unless the displacement vector ξ itself vanished on $\partial\Sigma$. What *would* be sufficient for the surface integral to be conserved under transport by $\vec{\xi}$ (even if this vector field were not zero on the boundary) would be to have $\vec{\xi}\mathcal{L}\underline{\omega} = 0$, in which we may say that "$\underline{\omega}$ is *weakly* conserved by the vector field $\vec{\xi}$". We reserve the statement that "$\underline{\omega}$ is *strongly* conserved by the vector field $\vec{\xi}$" to mean that the flux integral is conserved under arbitrary displacements along the integral curves of $\vec{\xi}$, i.e. to mean that $\underline{\omega}$ is weakly conserved by $\alpha\vec{\xi}$ for an arbitrarily variable scalar renormalisation factor α. It is apparent from (1.27) that in order to have conservation by $\vec{\xi}$ in this *strong* sense (meaning in effect that $\int_\Sigma d\vec{\Sigma}\rfloor\underline{\omega}$ only depends on which integral curves of $\vec{\xi}$ are intercepted by Σ) one needs that $\underline{\omega}$ should simultaneously satisfy $\vec{\xi}\cdot(\partial\underline{\omega}) = 0$ and $\vec{\xi}\cdot\underline{\omega} = 0$. (Thus if $\underline{\omega}$ is intrinsically conserved in the sense of being a "closed" form, the only remaining requirement for strong conservation by any vector field $\vec{\xi}$ is that $\vec{\xi}$ should be a nullvector of $\underline{\omega}$ in the sense that its contraction with $\underline{\omega}$ vanishes.) It follows from the general commutation relation for Lie differentiation operations,

$$[\vec{\xi}\mathcal{L}, \vec{\eta}\mathcal{L}] = [\vec{\xi}, \vec{\eta}]\mathcal{L}\,. \tag{1.28}$$

that if *both* $\vec{\xi}$ and $\vec{\eta}$ conserve the flux $\underline{\omega}$ in either the strong or the weak sense, then so does the commutator $[\vec{\xi}, \vec{\eta}]$. This means that the set of *all* fields $\vec{\xi}$ that conserve $\underline{\omega}$ in respectively the weak or the strong sense will mesh together consistently to form a well defined congruence of (weakly or strongly) flux conserving surfaces.

The classic physical example of the application of the foregoing considerations is that of the Maxwellian electromagnetic field tensor with components $F_{\mu\rho}$ in ordinary 4-dimensional space-time. Since this field must always satisfy the Faraday-Maxwell equation $\partial\underline{F} = 0$ (which is the well known integrability condition for the existence of a 4-potential 1-form $\underline{A}$ such that $\underline{F} = \partial\underline{A}$) the nullvector condition $\xi^\mu F_{\mu\rho} = 0$ is necessary even for weak conservation but sufficient even for strong conservation of $\underline{F}$ by $\vec{\xi}$. For a general electromagnetic field $\underline{F}$ with non-degenerate component matrix there will be no solutions at all for any such flux conserving fields $\vec{\xi}$, but in the degenerate "force-free" case characterised by the condition ${}^*\underline{F}\rfloor\underline{F} = 0$ there will be a not just one- but *two*- parameter family of nullvectors $\vec{\xi}$ at each point, which by the above reasonning will then mesh together to form a well defined congruence of "magnetic" 2-surfaces over which the flux will be strongly conserved. One of the purposes of the following lectures is to show that perfect fluids provide an example which, although less widely known than the pure electromagnetic case just described, is actually an even richer illustration of the Cartan differentiation and integration theory that has just been summarised.

1.2 Canonical form and associated conservation laws for ideal fluid systems.

The concept of what I shall refer to as the *canonical formulation* of the equations of motion for a congruence of space-time world-lines representing a fluid flow with *canonically normalised* tangent vector

$$\beta^\mu = \frac{dx^\mu}{d\tau} , \tag{1.29}$$

as specified with respect to a suitable *canonical*-(not necessarily proper- or coordinate-) time parametrisation τ, is derived from ordinary single-particle Lagrangian mechanics for individual particles following the flow world-lines. The most familiar form of the Lagrangian equations for stationarity with respect to infinitesimal world-line displacements of the canonically parametrised world-line integral of some given Lagrangian scalar function L say of the local space-time position coordinates x^μ and of the canonical velocity coordinates β^μ is just

$$\frac{d\pi_\mu}{d\tau} = \frac{\partial L}{\partial x^\mu} \tag{1.30}$$

where the generalised momentum components are determined from the Lagrangian function $L(x^\mu, \beta^\nu)$ as the partial derivatives

$$\pi_\mu = \frac{\partial L}{\partial \beta^\mu} \tag{1.31}$$

However this familiar form (1.30) of the single-particle Lagrangian equations has the inconvenient property that (unlike (1.31) which defines a well behaved momentum covector) it relates quanties that are non-tensorial in nature, since the quantity $d\underline{\pi}/d\tau$ on the left is not a covariant but just an ordinary derivative. Whenever we are dealing with a congruence of solutions of the Lagrangian equations and not just one particular particle world line, the awkward coordinate transformation properties of (1.30) may be remedied[16] by taking advantage of the fact that since the canonical velocity will be well defined as a scalar field over the space-time manifold $\mathcal{M}$ the Lagrangian function $L(x, \vec{\beta})$ will determine a corresponding scalar field L over $\mathcal{M}$. The gradient ∂L of this scalar field will of course (like the generalised momentum $\underline{\pi}$) be a well behaved covector, whose relationship to the non-covectorial partial derivative on the right hand side of (1.30) will be expressible as

$$\partial_\mu L = \frac{\partial L}{\partial x^\mu} + \pi_\rho \partial_\mu \beta^\rho \ . \tag{1.32}$$

This shows that (1.30) can be converted into tensorial, or to be more explicit covectorial, form by adding $\pi_\rho \partial_\mu \beta^\rho$ to each side. When this is done, the result is expressible in the notation of the preceeding section as

$$\vec{\beta}\mathcal{L}\underline{\pi} = \partial L \ , \tag{1.33}$$

i.e. the Lie derivative of the canonical momentum covector with respect to the canonical velocity vector equals the gradient of the Lagrangian scalar field. This tensorially well behaved equation (1.33) is what I refer to as the *canonical* formulation of the equations of motion of the flow. It will be shown explicitly in the following sections how this canonical formulation can be derived in a physically natural manner for a wide extensive classes of idealised fluid systems, starting with the case of the simple barytropic perfect fluid model in both its original Eulerian (Newtonian mechanical) version and its generalised ("special" or "general") relativistic version. In so far as the present section is concerned we shall postpone discussion of the physical derivations and interpretations of (1.33) but will instead give a brief account of some of its most immediate mathematical implications.

The first property of the canonical formulation (1.30) that needs to be emphasised is the existence of an alternative *Hamiltonian* as opposed to Lagrangian version. Not only in the general case, for which the velocities can be eliminated as functions of momenta, but even in the degenerate case for which a proper Hamiltonian function of position and momenta does not exist, it will always be possible to use the Legendre prescription for H in the form $\pi_\mu \beta^\mu - L$ to define an induced Hamiltonian scalar field

$$H = \underline{\pi} \cdot \vec{\beta} - L \tag{1.34}$$

over $\mathcal{M}$, and to use it to convert the Lagrangian version (1.33) of the canonical equation into the equivalent Hamiltonian version

$$2\beta^\rho \partial_{[\rho}\pi_{\mu]} = -\partial_\mu H \qquad (1.35)$$

This latter version could alternatively have been derived directly from the standard form of the single particle Hamiltonian equations (as formulated in terms of the canonically preferred time parametrisation τ) by a procedure analogous to the derivation of (1.33) from (1.30) whenever a well behaved Hamiltonian formulation exists, but it has the advantage of being derivable by the less direct Lagrangian route in any case. The Hamiltonian version of the canonical formulation can be expressed more concisely in terms of the Cartan type notation scheme described in the preceeding section using the abbreviation

$$\underline{\underline{w}} = \partial \underline{\pi} \qquad (1.36)$$

for what is appropriately describable as the canonical *vorticity 2-form* of the flow, with coordinate components $w_{\mu\rho}$, and

$$\underline{\mathcal{F}} = -\partial H \qquad (1.37)$$

for what is appropriately describable as the canonical *force 1-form* with components $\mathcal{F}_\mu$. In this terminology the Hamiltionian version of the canonical formulation is simply

$$\vec{\beta} \cdot \underline{\underline{w}} = \underline{\mathcal{F}} \ . \qquad (1.38)$$

One of the most obvious consequences of this latter version is the orthogonality property of the force,

$$\vec{\beta} \cdot \underline{\mathcal{F}} = 0 \qquad (1.39)$$

(which follows directly from the antisymmetry of the 2-form $\underline{\underline{w}}$). This is equivalent to the most fundamental of all variational conservation laws, to the effect that the value of the Hamiltonian scalar H itself is *always* constant *along* the flow lines

$$\vec{\beta} \mathcal{L} H = 0 \ . \qquad (1.40)$$

We shall see however that it occurs very commonly (the case of an simple "barytropic" perfect fluid being the simplest example) that H is restrained to be *uniform* meaning that it is constant not just along but also across the flow lines. In such a *force-free* case, as characterised by $\partial H = 0$ we shall refer to the system as being *uniformly canonical* . Since by its construction the vorticity 2-form is always closed,

$$\partial \underline{\underline{w}} = 0 \qquad (1.41)$$

it can be seen that the *uniformly canonical equation of motion*, $\vec{\beta} \cdot \underline{\underline{w}} = 0$, is just the condition for a 2-surface flux integral of the vorticity $\underline{\underline{w}}$ to be *strongly* conserved by the flow lines with tangent vector field $\vec{\beta}$ in the sense explained at the end of the preceeding section, meaning that $\int_S d\vec{S} \rfloor \underline{\underline{w}}$ depends only on which particular flow world-lines are intercepted by the 2-surface S; moreover since the antisymmetric component matrix must have even rank, and since the existence of the null eigenvector $\vec{\beta}$ implied by the uniformly canonical equation of motion rules out the possibility that it have the maximal rank, namely 4, that would ingeneral be possible in ordinary space-time, it follows that (except in the zero rank irrotational case with $\underline{w} = 0$) the vorticity must have rank 2, so that the the 1-dimensional flow lines will be included within a congruence of 2-dimensional strongly flux conserving "vorticity surfaces". In the general, non-uniformly canonical case we shall not have vorticity conservation in this strong sense, but nevertheless we shall always have

$$\vec{\beta}\mathcal{L}\underline{w} = 0 \tag{1.43}$$

which means that although a vorticity flux integral $\int_S d\vec{S}\rfloor \underline{w}$ will in general be affected by arbitrary displacements along the flow world lines it will however be conserved by displacement by a *uniform* distance as measured by the *canonical* parameter τ fixed by the normalisation of $\vec{\beta}$ in accordance with (1.29). By Stokes theorem (or directly from the basic Lagrangian form (1.30) of the canonical equation of motion) it can be seen that the same consideration applies to the *circulation integral* round a circuit C say bounding such a 2-surface S, i.e. we shall always have

$$\frac{d}{d\tau}\oint_C dx^\mu \pi_\mu = 0\,. \tag{1.43}$$

In the Newtonian example of the ordinary barytropic Eulerian fluid, the time independence of any comoving circulation integral has been well known since the time of Kelvin, over a century ago, but as far as I know it has not been explicitly pointed out before now that this result can be generalised to the non-barytropic (two-constituent) case (whose canonical formulation is described in the next section) in both the relativistic and the Newtonian cases provided that one is prepared to think not in terms of a standard linear time displacement measure (such as is given uniquely in Newtonian theory and modulo Lorentz transformations in Special though not General relativistic theory) but with the appropriate (in general improper) *canonical* time.

The applications to be considered in the following sections will not only include cases where the flow characterised by $\vec{\beta}$ is that of a "free", i.e. effectively decoupled ideal fluid consituent, but also cases of perfect conductivity in which the fluid in question may be strongly coupled to its environment. The canonical equation of motion (which only has three independent tensor components, in view of the identity (1.39)) will never constitute a complete set of equations of motion for the system, but must be supplemented by appropriate equations of state and additional dynamic equations (which may or may not have cononical form) for any other constituents that may be coupled in, as well as purely kinematic equations, which in the simplest case will just reduce to a conservation law for a number flux associated with the flow. At the level of its most fundamental geometric expression, such a number flux is to be represented by an $(n-1)$-form $\underline{N}$ say (which in 4-dimensional space-time will therefore have tensor components of the form $N_{\mu\nu\rho}$) satisfying the orthogonality condition

$$\vec{\beta}\cdot\underline{N}=0\,. \tag{1.44}$$

whose integral $\int_\Sigma d\vec{\Sigma}\rfloor\underline{N}$ over a hypersurface Σ is to be interpreted as the number of particles of the flow crossing the hypersurface. In view of (1.44) the necessary and sufficient condition for such a number flux to be strongly conserved in the sense of the preceeding section is that the form $\underline{N}$ should be closed, i.e.

$$\partial\underline{N}=0\,. \tag{1.45}$$

Of course if one admits the use of a volume measure $\underline{\epsilon}$ (which in 4-dimensional spacetime will have components $\epsilon_{\lambda\mu\nu\rho}$) then one may use the dual alternative language most commonly employed by physicists, whereby instead of $\underline{N}$ one works with its dual, $^*\underline{N}=\vec{n}$ say, which is just an ordinary vector tangent to the flow, like $\vec{\beta}$, to which $\vec{n}$ is necessarily parallel by (1.44). In this dual language, the conservation condition (1.45) is expressible in the less fundamental but more widely familiar form $\nabla\cdot\vec{n}=0$.

Independently of the existence of any such conserved particle number flux, the canonical equations of motion enable us to construct a "screw current" 3-form $\underline{Z}$, with components $Z_{\mu\nu\rho}$ that always has the analogous property of being strongly conserved by the flow, according to the prescription

$$\underline{Z}=\underline{\mathcal{F}}\wedge\underline{w}\,. \tag{1.46}$$

which is such that *underlineZ* is automatically closed, being in fact exact:

$$\partial\underline{Z}=0\,,\quad \underline{Z}=\partial(\underline{\pi}\wedge\underline{\mathcal{F}})=-\partial(H\underline{w})\,. \tag{1.47}$$

It therefore evident that it always satisfies the conditions

$$\vec{\beta}\cdot\underline{Z}=0\,,\quad \vec{\beta}\cdot(\partial\underline{Z})=0 \tag{1.48}$$

that are necessary and sufficient to guarantee conservation by the flow generated by $\vec{\beta}$ in the strong sense as defined in the preceeding section, i.e.

$$(\alpha\vec{\beta})\mathcal{L}\underline{Z}=0 \tag{1.49}$$

for an arbirarily variable scalar field α. In physicist's dual language this "screw current" would be representable in ordinary 4-dimensional space-time (though not on higher dimensions) by a simple vector, $^*\underline{Z}=\vec{z}$ say, which (by the first equation of (1.48)) will be parallel to the flow tangent vector $\vec{\beta}$, and which (by (1.47)) will be conserved in the usual sense that $\nabla\cdot\vec{z}=0$.

In the ordinary 4-dimensional case, it can be seen from the formal identity between the properties (1.44) and (1.45) that were postulated for $\underline{N}$ and those that have just been established for $\underline{Z}$, that the latter must be related to the former (whenever it is specified) by a simple proportionality factor ς say,

$$\underline{Z} = \varsigma \underline{N} \tag{1.50}$$

and that the particle number conservation condition (1.45) will be equivalent to the condition that ς (like H) be constant allong the world lines

$$\vec{\beta}\mathcal{L}\varsigma = 0 \tag{1.51}$$

The existence of such a constant of the motion was originally noticed in the special case of an Eulerian perfect fluid in Newtonian theory by Ertel[15,17,18]. The generalisation of this quantity (for which the term "enstrophy" has since been coined) to the case of a relativistic perfect fluid been pointed out much more recently by Friedman[19] and by Katz[20].

Another example of a physically significant conservation law that is virtually obvious in the present approach, but which long eluded discovery by the unweildy methods traditionally used in Newtonian hydrodynamics, is that of *helicity* which is defineable as the 3-form $\underline{\Omega}$ say (with components $\Omega_{\mu\nu\rho}$) given by

$$\underline{\Omega} = \underline{\pi} \wedge \underline{w} \; . \tag{1.52}$$

Unlike the screw current $\underline{Z}$ (which is closed in the general case and which vanishes trivially in the special uniformly canonical case characterised by $\underline{\mathcal{I}} = 0$) the helicity 3-form $\underline{\Omega}$ is not closed in the general case but obeys a (non-trivial) closure conservation law in the uniformly canonical case, its exterior derivative being given in general by

$$\partial\underline{\Omega} = \underline{w} \wedge \underline{w} = -2\underline{Z} \wedge \partial\tau \tag{1.53}$$

where the last expression involves the a scalar field τ defined to be any solution of the canonical time evolution equation (1.29) (the orthognoality property expressed by the first equation of (1.48) ensures that the gauge ambiguity corresponding to an arbitrary choice of initial hypersurface $\tau = 0$ will not affect exterior product of $\partial\tau$ with $\underline{Z}$). Either from these expressions or from the alternative versions

$$(\vec{\beta} \cdot \underline{\pi})\partial\underline{\Omega} = 2\underline{\Omega} \wedge \underline{\mathcal{I}} = 2\underline{\pi} \wedge \underline{Z} \tag{1.54}$$

it can be seen that $\underline{\Omega}$ will be closed whenever $\underline{Z}$ vanishes on an initial hypersurface (and hence, by (1.49), everywhere else) and more particularly that $\underline{\Omega}$ will necessarily be closed in the uniformly canonical case characterised by $\underline{\mathcal{I}} = 0$, for which $\underline{Z}$ must vanish identically. Quite generally, the helicity 3-form $\underline{\Omega}$ shares with the momentum 1-form $\underline{\pi}$ the property of obeying a weak non-local conservation law in the sense that its Lie derivative with respect to the canonical velocity vector β although not zero is nevertheless exact:

$$\vec{\beta}\mathcal{L}\Omega = \partial(\vec{\beta} \cdot \underline{\Omega})$$

$$= \partial(\underline{\pi} \wedge \partial L) = \partial(L\underline{w}) \tag{1.55}$$

which not only tells us that the helicity will remain zero on any flow world-line on which is is zero initially, but also, more generally that the helicity integral $\int_\Sigma d\vec{\Sigma} \rfloor \underline{\Omega}$ will always be conserved in the *weak* sense (i.e. by a *canonically* parametrised displacement along the flow lines) for any 3-volume Σ with zero helicity (initially and hence at all later times) on its boundary, $\partial\Sigma$. This boundary condition requirement will for example be satisfied necessarily in the case for an integral over the *entire* volume of an isolated (e.g. self gravitating) body of fluid surrounded by empty space. It will also hold for a non-isolated but bounded core volume whenever it is surrounded by an irrotational outer envelope. Such results are of interest in the context of dynamo theory, the purely magnetic limit case having been the first in which helicity as defined here was studied explictly[21]. The concept of helicity in the case of ordinary Eulerian fluid mechanic was developed even more recently[22], and the relativistic generalisation was not found until the present approach was adopted[16].

1.3 Standard formulation for (chemically active) perfect fluid systems.

After dealing so far only with mathematical abstractions, we now come to consider a concrete physical model exemplifying the general concepts that have been presented. As a preliminary to the derivation (in the following section) of the canonical formulation for the conservative case with not more than two independent constituents, we shall start by considering the rather more general case of a non-conducting perfect fluid with an *arbitrary* number of chemically independent and perhaps chemically interacting constituents (where as far as physical applications are concerned, the term "chemical" is to be interpreted in the broad sense as including whatever - nuclear or other - interactions may be relevant), our purpose being to show that the generally accepted model for such a fluid has a dynamic equation of motion that can be converted to what we shall refer to as the *standard formulation*, involving only exterior (not covariant) differentiation, and therefore particularly convenient as a starting point for the methods of analysis described in the preceeding sections. The use of such a formulation was pionneered by Lichnerowicz[24] but has not been as widely adopted as it deserves to be (being in fact totally ignored in many more recent textbooks, including even the best and most comprehensive) as a result perhaps of the unfortunate nomenclature in which it was originally presented, whereby the physical interpretation was unnecessarily obscured: in particular, the crucially important 1-forms that are unambiguously identifiable as *momenta* in the traditional dynamical sense of the word (as used in section 1.2) were referred to[24] as "currents" (and correspondingly denoted by the symbol $\underline{C}$ instead of the more traditionally appropriate Greek or Latin letters $\underline{\pi}$ or $\underline{p}$) which is highly misleading for physicists (if not perhaps for mathematicians) who are accustomed to reserve the term current for *contra*variant vector fields, or their space-time duals which are not 1-forms but 3-forms.

What we refer to as the standard (as opposed to canonical) formulation of the basic dynamical equation of motion is expressed in terms of a flow tangent vector $\vec{u}$ with standard normalisation (which in general will not be the same as the canonical normalisation characterising the parallel tangent vector $\vec{\beta}$ whose relationship to $\vec{u}$ will be described in the next section) and a set of number densities n_X say (as measured in the rest frame determined by $\vec{u}$) and corresponding 4-momentum 1-forms $\underline{\pi}^X$ with space-time components π^X_μ where the generalised chemical index X ranges over values labelling the relevant independent constituents which might consist of electrons and ions, chemical molecules in the strict traditional sense, nuclear species et cetera, of which one of the most important in any thermodynamically non-trivial situation will be the *entropy* , i.e the relevant information that is suppressed in the averaging process involved in going from an underlying "microscopic" model to a the continuum level usually described as "macroscopic" (These terms are of course purely relative: the "microscopic" level might involve individual stars in a continuum model for a a globular cluster or galactic nucleus, while the "macroscopic" level might involve sub-nuclear length scales in a continuum model for heavy ion ion collisions in a high energy accelerator.)

In view of its central thermodynamic role, we shall reserve the index value $X = 0$ for the entropy and will introduce the special symbols $s = n_0$, for the entropy density and $\underline{\Theta} = \underline{\pi}^0$, for the associated thermal 4-momentum covector, the latter notation being motivated by the intimate relationship that will be seen to hold between the momentum covector of the entropy and the ordinary thermodynamic temperature which we shall denote by Θ (the symbol T being reserved here for the stress-energy-momentum tensor with components $T_\mu{}^\nu$). In the particular case of a pure thermal gas of "black body radiation" the entropy is the only constituent that needs to be considered, but more generally we shall wish to allow for an arbitrary number of particle species with number densities n_1, n_2, ... and associated 4-momenta $\underline{\pi}^1$, $\underline{\pi}^2$, ... When later on we shall need to use index values running only over these other (strictly positive) chemical index values, so as to exclude the zeroth value labelling the entropy which is included in the full range $\{0, 1, 2, ...\}$ of X, we shall shall use *early* capital index symbols A, B which are thus to be understood as having the restricted range $\{1, 2, ...\}$ (in agreement with the notation convention used in the accompanying lecture notes of Israel, with which I have tried to be consistent as far as possible).

In terms of the quantities that have just been listed, and with the understanding that X is subject to the usual summation convention with respect to its entire range, including the zero value corresponding to the entropy whenever it is effectively present, the standard formulation of the equation of motion is given simply as

$$n_X \vec{u} \cdot (\partial \underline{\pi}^X) = 0 \tag{1.56}$$

In the relativistic case, for which the standard velocity normalisation determined by the space-time metric tensor $g_{\mu\rho}$ that is used for index raising and lowering is given simply as

$$u_\mu u^\mu = -c^2 \tag{1.57}$$

(where c is the speed of light) the derivation of (1.56) is quite straightforward as will be shown explicitly below. In the Newtonian case, which from the point of view of the space-time metric is a *degenerate* limit, the equations of motion can still be converted into the standard form (1.56), but the right way to proceed is not quite so obvious and rather more work is required, for which the interested reader is referred elsewhere[25].

The generally accepted defining property of a relativistic "perfect" fluid is that it should have a energy-momentum-stress tensor expressible in terms of the (positive indefinite) orthogonal space-projection given (subject to the metric signature convention $(\,+\,+\,+\,-\,)$ implicit in (1.57)) by

$$\gamma_{\mu\rho} = g_{\mu\rho} + c^{-2}\, u_\mu u_\rho \tag{1.58}$$

in the form

$$T^{\mu\rho} = \rho u^\mu u^\rho + P\gamma^{\mu\rho} \tag{1.59}$$

where ρ and P are to be interpreted as the total mass density (including the relativistic energy contribution) and the pressure respectively, whose values are to be specified in terms of the relevant number densities n^X by appropriate equations of state. The basic dynamic equation of motion is then conventionally postulated to have the form of a covariant "conservation" law (which is only a conservation law in the strict sense in the Minkowski space limit from which gravitational effects have been excluded) of the form

$$\nabla_\mu T^\mu{}_\rho = f^{\text{ext}}_\rho \tag{1.60}$$

where $\underline{f}^{\text{ext}}$ is a long range external force contribution which (unless one wishes to allow for such exotic possibilities as the chromohydrodynamics mentionned in the accompanying course of Holm) can only be of electromagnetic origin (since gravitation is implicitly taken care of in the covariant derivative in (1.60)) which means that it will be given in terms of the Maxwell field, with components $F_{\mu\rho}$ expressible in terms of 4-potential components A_μ in the form

$$\underline{F} = \partial \underline{A}\,, \tag{1.61}$$

by an electric force law of the form

$$\underline{f}^{\text{ext}} = \underline{F} \cdot \vec{j}\,, \qquad \vec{j} = e^{\text{X}} n_{\text{X}} \vec{u} \tag{1.62}$$

where $\vec{j}$ is the electric 4-current vector, as specified in terms of constants e^{X} representing the (fixed) electric charge per particle associated with the corresponding species (and where that of the entropy may be presumed to be zero, $e^0 = 0$).

The fundamental equation of state is that for the mass density ρ, or equivalently, if one prefers, for the energy density ε as defined by a relation of the form

$$\rho = m^{\text{X}} n_{\text{X}} + \varepsilon \tag{1.63}$$

where each of the parameters m^{X} represents a *residual mass*, (commonly referred to loosely as "the rest mass") per particle. As far as the general formalism is concerned, the introduction of the concept of residual masses is an unnecessary complication, but it may be may be useful for some specific purposes (particularly the analysis of Newtonian limits). Such a residual mass will conventionally be specified in terms of some lowest attainable energy state (which will always be zero for the entropy, $m^0 = 0$) whose definition is in principle quite arbitrary in so far as the meaning of "attainable" may be chosen in any way that seems convenient. Ultimately (if necessary by compression into microscopic black holes as a last resort) the energy is *all* extractable, and there is thus no physical loss of generality if one simply sets each of the masses m^{X} to zero at the outset. The formalism of the theory will in any case be invariant under arbitrary energy-origin adjustments of the form

$$m^{\text{X}} \mapsto m^{\text{X}} - c^{\text{X}} \tag{1.64}$$

where the c^{X} are arbitrary constants. By varying the fundamental equation of state for ρ as a function of the relevant number densities n_{X} one defines corresponding (variable) "effective mass" functions μ^{X} (which, unlike the "rest masses", are relativistically well defined) to be the partial derivative coefficients in the expression

$$d\rho = \mu^{\text{X}} dn_{\text{X}}\,. \tag{1.65}$$

These "effective mass" functions may, if one wishes, be expressed in the form

$$\mu^{\text{X}} = m^{\text{X}} + \chi^{\text{X}} c^{-2} \tag{1.66}$$

in terms of "non-relativistic chemical potentials" χ^{X}, which like the m^{X} are energy-origin dependent in the sense that under (1.64) they must evidently transform according to

$$\chi^{\text{X}} \mapsto \chi^{\text{X}} + c^{\text{X}} c^2\,. \tag{1.67}$$

They may be considered as being defined (subject to a convention for the choice of the m^{X}) by the energy variation expansion

$$d\varepsilon = \chi^X dn_X = \Theta ds + \chi^A dn_A \tag{1.68}$$

where the latter version is obtained by introducing the explicit notation $s = n_0$ for the entropy density and $\Theta = \chi^0$ for the thermal chemical potential i.e. for the *temperature* as ordinarily defined, in units such that Boltzman's constant k is equal to one. Thus introducing the term "entropon" as an abbreviation for "one unit of entropy" in such a ($k = 1$) unit system, with the understanding that the "entropon rest mass" m^0 is zero, we see that that μ^0, which is interpretable as *the effective mass per entropon* will be given in terms of the temperature simply by

$$\mu^0 = \Theta c^{-2} , \tag{1.69}$$

(from which it is evident that this thermal mass μ^0 will vanish in the Newtonian limit $c \to \infty$.)

It is perhaps worthwhile to digress at this point to mention that the introduction of the concept of temperature via (1.68) does not mean that such a model model is only applicable to situations of strict thermal equilibrium: in contexts of sufficiently weak coupling, e.g. between internal molecular excitation levels and external kinetic energy levels, the quantity Θ so defined could be an average over distinct subsystem temperatures; nevetheless the way in which such an average is taken (which depends on the way in which the subsystems are accounted for in terms of a set of population number densities n^X which for an an adequate out-of-equilibrium desription might need to be rather large) will not in general be defined in an unambiguously natural manner except in the equilibrium limit, for which all reasonable definitions of temperature should agree.

To complete the specification of the perfect fluid model, one postulates that the internal mechanics are such as to ensure that the energy componente of the stress-energy momentum "conservation" equation (1.60), i.e. the contraction

$$u^\rho \nabla_\mu T^\mu{}_\rho = 0 \tag{1.70}$$

should be satisfied as an *identity* so that only three of the four components of (1.63) should be dynamically independent, since otherwise the system would be overdetermined since the flow world-lines each have only three degrees of freedom. Substituting from (1.68) and using (1.65) one sees that the required energy identity will be expressible more explicitly as

$$(\rho + Pc^{-2} - \mu^X n_X)\theta = \mu^X r_X \tag{1.71}$$

where θ is the flow divergence as defined by

$$\theta = \nabla \cdot \vec{u} \tag{1.72}$$

and the r_X are the creation rates per unit space-time volume of the corresponding species, as defined by

$$r_X = \nabla \cdot \vec{n}_X \,, \qquad \vec{n}_X = n_X \vec{u} \,. \tag{1.73}$$

While it is of course possible to imagine more complicated models (involving effects such as bulk hysteresis) the simplest way of achieving this, and the way that is normally understood to be implicit in the use of the term "perfect fluid", is to require that each side of (1.71) separately should vanish identically. As far as the left hand side is concerned, this means that the equation of state for P can not be independent of that for ρ but should be given in terms of it by

$$P = c^2(\mu^X n_X - \rho) \tag{1.74}$$

The consequent vanishing of the right hand will be expressible as

$$\Theta \, \nabla \cdot \vec{s} = -c^2 \mu^A r_A \,, \qquad \vec{s} = s\vec{u} \,, \tag{1.75}$$

(where we recall our convention to the effect that summation should be taken over all positive values A, but not the zero value corresponding to the entropy, whose contribution is written out separately on the left). In a strictly conservative system for which all the separate creation rates r_A are postulated to be zero, this would ensure that the entropy creation rate $r_0 = \nabla \cdot \vec{s}$ would also be zero.

More generally, we can allow for situations in which a set of reactions, labelled by a bracketted chemical reaction symbol $[C]$ say, are occurring, according to "chemical reaction formulae" of the form $N^{[C]}_A (X^A) = 0$ where (X^A) is the symbol for the Ath species, and where the ratios in the interactions are usually rational so that the numbers $N^{[C]}_A$ may without loss of generality be taken to be integers, positive or negative according to whether the species in question is consumed or created in the interaction. The total creation rates will then be obtained by summing over the values of the reaction index $[C]$ in the form

$$r_A = -r_{[C]} N^{[C]}_A \tag{1.76}$$

where the quantities $r_{[C]}$ are the corresponding rates of the particular reactions involved. The simplest way of ensuring that (1.75) is consistent with the "second" law inequality

$$\nabla \cdot \vec{s} \geq 0 \tag{1.77}$$

is to make the usual postulate that the interactions are governed by a linear law which should have the form

$$r_{[C]} = -\kappa_{[C][D]} N^{[D]}_A \mu^A c^2 \tag{1.78}$$

for some *positive definite* and (by the Onsager principle) symmetric chemical interaction matrix $\kappa_{[C][D]}$ whose coefficients are functions of state, i.e. of the number densities n_X. If one is not interested in the detailed chemical pathways but only in the outcome, one may eliminate the separate reaction indices $[C]$ by working with a composite interaction matrix Ξ_{AB} constructed from the fundamental interaction matrix $\kappa_{[C][D]}$ by the transformation

$$\Xi_{AB} = c^2 N^{[C]}_{A} \kappa_{[C][D]} N^{[D]}_{B} \tag{1.79}$$

In terms of this composite interaction matrix, which will inherit from $\kappa_{[C][D]}$ the property of being symmetric and at least *non-negative* (though it need not be strictly positive definite), the total creation rates will be expressible more concisely as

$$\nabla \cdot \vec{n}_A = r_A = -\Xi_{AB}\mu^B \ , \tag{1.80}$$

so that the total rate of entropy creation per unit space-time volume will be given by

$$\nabla \cdot \vec{s} = r_0 = \frac{\mu^A \Xi_{AB} \mu^B c^2}{\Theta} \ . \tag{1.81}$$

Once one has specified that the entropy and other creation rates r_0 and r_A are governed respectively by (1.81) and (1.80), thereby determining the evolution of the quantities s and n_A that are needed in the primary equation of state for ρ and the secondary equation of state (as given by (1.74)) for P, the motion will be fully determined by the stress-momentum-energy "conservation" law (1.60), of which we have now ensured that only three components will be dynamically independent, so that there will be just one dynamical solution for arbitrary initial values of the full set of number densities n_X and of the components u^μ subject to the normalisation condition (1.57). In order to be able to replace the (three independent components of) the usual form (1.60) of the dynamic equations of motion by the (three independent components) of what we have dubbed the *standard* formulation, as given by (1.56), i.e. in explicit coordinate form

$$2n_X u^\mu \partial_{[\mu}\pi^X_{\ \rho]} = 0 \ , \tag{1.82}$$

all that remains is to specify the appropriate expressions for the components π^X_μ of the total 4-momentum 1-form i.e. the *energy-momentum covector* of each species (including the entropy). It can easily be seen that this is to be done in terms of the sum of an *extrinsic* (electromagnetic) energy-momentum contribution and an it intrinsic (chemical) energy-momentum contribution in the form

$$\underline{\pi}^X = \underline{\chi}^X + e^X \underline{A} \tag{1.83}$$

where the *chemical 4-momentum* contribution $\underline{\chi}^X$, and its rest frame (chemical energy) component (which reduces to χ^X if one adopts the convention that the residual mass m^X is zero) are given by

$$\chi^X_\mu = \mu^X u_\mu \ , \qquad -\vec{u}\cdot\underline{\chi}^X = \chi^X + c^{-2}m^X \ . \tag{1.84}$$

(The definition (1.84) implicitly involves an index lowering operation, whereby the set of kinematic 4-velocity components u^μ is mapped onto a covariant image set u_μ by contraction with the metric: $\vec{u} \mapsto \underline{u} = \underline{g}\cdot\vec{u}$. It is because this process does not have a straightforward analogue in Newtonian theory that the derivation of the standard formulation for a traditional Eulerian fluid model[25] is more delicate than the relativistic case described here.) In the particular case of the entropy, for which there is no associated electric charge, i.e. $e^0 = 0$, the total 4-momentum $\underline{\pi}^0$ will thus be given directly by

$$\underline{\pi}^0 = \underline{\Theta}\,, \qquad \Theta_\mu = c^{-2}\Theta u_\mu \tag{1.85}$$

in which it may be remarked that although the effective thermal mass $\mu^0 = c^{-2}\Theta$ (as given by (1.69)) will tend to zero in the Newtonian limit, the temperature form $\underline{\Theta}$ as so defined will nevertheless have a regular non-zero limit.

It is the equivalence (1.85) between the thermal momentum form $\underline{\pi}^0$ and the temperature form $\underline{\Theta}$ that justifies the interpretation of $c^{-2}\Theta$ as the effective mass, μ^0 of the "entropon" in the non-conducting perfect fluid case. Failure to recognise the role of the thermal inertia μ^0, (whose relevance, even in the familiar non-conducting case, has thus been clearly brought to light) was the root cause of the causality difficulties in earlier attempts [1,2] to set up more general thermally conducting models.

1.4 Reduction to Canonical form in (conservative) two-constituent case.

The standard formulation of the multiconstituent non-conducting perfect fluid model set up in the previous section has the convenient property of "chemical covariance" in so far as it is invariant under chemical basis transformations (e.g. from molecular to atomic accounting systems for the dissolved constituents in an aqueous solution) of the form $n_X \mapsto n'_X = N_X^{\,Y} n_Y$ provided the other quantities involved are subject to corresponding transformations of covariant or contravariant type as determined by the same constant-valued (usually rational or even integer valued) transformation matrix $N_X^{\,Y}$: in particular the 4-momenta will transform contravariantly, according to $\underline{\pi}^X \mapsto \underline{\pi}'^X$ with $\underline{\pi}^X = \pi'^Y N_Y^{\,X}$.

The investigation of the properties of this model will be continued in this and the following subsection subject to the restriction that the system will now be supposed to be strictly *conservative*, which means that the chemical interaction matrix Ξ_{AB} must be taken to be zero so that we shall have a separate conservation law of the form

$$\nabla \cdot (n_X \vec{u}) = 0 \tag{1.86}$$

for each of the species involved (including the entropy).

One general consequence of the conservation conditions (1.86) is conservation of extrinsic energy (or of angular or linear momentum) as defined with respect to any stationary (or axial or translational) isometry generator $\vec{k}$ that may be present, i.e. a solution of the Killing vector equations specified with respect to the metric g by

$$\vec{k}\mathcal{L}g = 0 \tag{1.87}$$

(with the left hand side as given by (1.24)) since by (1.82) one will obtain

$$\nabla \cdot (n_X \mathcal{E}^X \vec{u}) = 0\,, \qquad \mathcal{E}^X = -\vec{k} \cdot \underline{\pi}^X \tag{1.88}$$

which represents a relativisticly generalised version of the historic Bernouilli theorem, in which for each species the quantity $\mathcal{E}$ is interpretable (if the Killing vector is at least asymptotically timelike) as the corresponding (extrinsic) *energy per particle.*

Under the conservative circumstances characterised by (1.86) (and sometimes even more generally) it is often convenient to use some preferred choice, n_1 say, among the independent number densities for accounting purposes, and to specify the others in terms of it by ratios ν_X defined by

$$n_X = \nu_X n_1 \tag{1.89}$$

(which in particular means that we shall have $\nu_1 = 1$ as an identity). The use of such a preferred reference number density n_1 involves losing the full advantages of the manifest chemical covariance of the standard formulation as described in the previous paragraph, but working with the not-fully-chemically-covariant variables ν_X has the compensating advantage that their evolution equations are particularly simple, at least in the conservative case, since subject to (1.86) these ratios will simply be *constant* allong the flow world-lines, i.e.

$$\vec{u}\mathcal{L}\nu_X = 0\,. \tag{1.90}$$

Another such simplification can be made whenever a solution of the Killing equations (1.87) exists, since use of the preferred reference number density n_1 then enables the corresponding Bernouilli type conservation law (1.88) to be converted into a statement of the more familiar kind to the effect that the corresponding total energy per reference particle should also be constant along each flow world-line,

$$\vec{u}\mathcal{L}\mathcal{E} = 0\,, \qquad \mathcal{E} = \nu_X \mathcal{E}^X = -\vec{k} \cdot \underline{\pi}\,. \tag{1.91}$$

In this last expression we have introduced the use of the concept of the *total 4-momentum per (preferred) reference particle*, $\underline{\pi}$, as defined (in a chemically non-covariant way) by

$$\pi_\mu = \nu_X \pi^X_\mu = c^{-2} h u_\mu\,, \qquad h = \nu_X \mu^X c^2 = \frac{P + \rho c^2}{n_1}\,. \tag{1.92}$$

in which hc^{-2} plays the role of the effective combined inertial mass per reference particle, where h is interpretable as the relativistic enthalpy per reference particle. The total 4-momentum per reference particle, $\underline{\pi}$ defined in this way can be used in the conservative case characterised by (1.86) to convert the standard dynamical equation (1.56) to the simple form

$$\vec{u} \cdot (\partial \underline{\pi}) = -\lambda^{X} \partial \nu_{X} \tag{1.93}$$

where for each species we define what is interpretable as an associated distinct "Lagrangian scalar" field λ^{X} according to the prescription

$$\lambda^{X} = \vec{u} \cdot \underline{\pi}^{X} = -\mu^{X} c^{2} . \tag{1.94}$$

Now in view of the fact that $\partial \nu_{1}$ must vanish identically, the number of terms effectively present in the summation over X on the right of (1.93) will be *one less* than the total number of independent constituents involved. In particular if there are not more than two independent constituents, there will be not more than one term left on the right of (1.93), which according to the counting system introduced in the previous subsection would be designated by the index value $X = 0$ if the remaining constituent involved is to be interpreted as the entropy, and by the index value $X = 2$ if it is anything else.

For the remainder of this subsection, let us restrict our attention to such relatively simple situations, in which appart from the reference species (designated by $X = 1$) there is only one other species, characterised by the (only independent) number density ratio ν say, with

$$\nu^{X} = \nu , \tag{1.95}$$

where $X = 0$ or $X = 2$ as the case may be. Whichever value X has, we may use the corresponding "Lagrangian" scalar field λ^{X} as a rescaling factor to relate the standard 4-velocity vector $\vec{u}$ to a corresponding *canonical* velocity vector $\vec{\beta}$ according to the prescription

$$\vec{u} = -\lambda^{X} \vec{\beta} \tag{1.96}$$

which us enables us to reduce (1.93) (in this two-constituent case) to the form

$$\vec{\beta} \cdot (\partial \underline{\pi}) = \partial \nu \tag{1.97}$$

which can be seen to have the required canonical form, in its Hamiltonian version (1.35), provided the Hamiltonian scalar field is identified as being given by

$$H = -\nu . \tag{1.98}$$

In particular, when the independent variable is the entropy (i.e. in the X=0 case) it follows from (1.96) and (1.85) that the canonical velocity vector that is to be used in the resulting weak conservation laws for vorticity (1.43) and helicity (1.55) in the "non-baritropic" case will be given simply by

$$\vec{\beta} = \frac{1}{\Theta} \vec{u} , \tag{1.99}$$

i.e. the canonical renormalisation factor will just be the inverse of the temperature.

2 TWO-CONSTITUENT MODEL FOR CONDUCTING FLUID

2.1 World-line displacement variation principle.

The rather elegant mathematical properties that have just been demonstrated to hold for ordinary non-conducting perfect fluid models can to a large extent be comprehended as resulting directly from the existence of a *variational* formulation of the kind whose study was pioneered by Taub[26], using an action integral

$$I = \int_{\mathcal{M}} \underline{d\mathcal{M}} \rfloor \Lambda\,, \qquad \Lambda = -\rho c^2 \tag{2.1}$$

over a space-time 4-volume $\mathcal{M}$ of the scalar mass-energy density field ρ, which is to be considered as being expressed as a function of the relevant number densities n_X whose permissible infinitesimal variations dn_X are *not* allowed to be chosen arbitrarily, but are restricted to be *induced by infinitesimal displacements of the flow world-lines*, as determined by a freely chosen infinitesimal displacement vector field $\vec{\xi}$ say, as well as by any infinitesimal variations $d_E\mathbf{g}$ of the metric that one might wish to consider, where we introduce the use of the symbol d_E for variations of *Eulerian* type, as evaluated at a point that is "fixed" according to some geometric prescription (as distinct from the *Lagrangian* variations to be considered later, which are evaluated at points that are displaced with the world-lines). It is in fact instructive to examine the effect of metric variations[27] even when (as in the present course) one is not essentially concerned with the dynamics of the gravitational field (which would be allowed for by adding another action integral contribution, $I_{\rm grav}$ say, which for Einstein's theory would have a form analogous to (2.1) but with ρ replaced by a term proportional to the Ricci scalar), since even when the background is considered as being passive, as for example in the flat special relativistic case, consideration of such variations is needed for the purpose of evaluating the appropriate variational stress-energy-momentum tensor $\mathbf{T}$ whose "covariant conservation" will be ensured as an identity.

Allowance for the effect of external electromagnetic forces on charged constituents necessarily involves the (gauge-dependent) 4-potential introduced by (1.61), requiring the addition to the action integral of a coupling contribution of the familiar form

$$I_{\rm elec} = \int_{\mathcal{M}} \underline{d\mathcal{M}} \rfloor \Lambda_{\rm elec}\,, \qquad \Lambda_{\rm elec} = \vec{j}\cdot\underline{A} \tag{2.2}$$

where the current $\vec{j}$ is as given by (1.62) (while a further Maxwellian field contribution, $I_{\rm F}$ say – with integrand proportional to the contraction of the Maxwell field $\underline{F}$ with itself

– would be needed if we were concerned with the active dynamics of the electromagnetic field).

A more conventional kind of variational principle, involving only the variation of local scalar fields rather than world-lines, can also be constructed, but it requires the introduction of non-locally defined – gauge-dependent – Clebsch-type auxiliary fields[25,28,29]. As a preliminary to the application of the mathematically more elaborate but physically more directly meaningful kind of variation principle considered here, it is a straightforward geometric exercise[30] to verify that the effect of a flow world-line displacement vector field $\vec{\xi}$ and a space-time metric variation $d_E \mathrm{g}$ will be to cause a corresponding (Eulerian) infinitesimal change in each number density n_X that will be given by

$$d_E n_X = -\nabla_\mu (n_X \xi^\mu) - n_X c^{-2} u^\mu u^\rho \nabla_\mu \xi_\rho - \frac{1}{2} n_X \gamma^{\mu\rho} d_E g_{\mu\rho} \tag{2.3}$$

Taken together with the formula

$$d_E (\underline{d}\mathcal{M}) = (\underline{d}\mathcal{M}) \frac{1}{2} g^{\mu\rho} d_E g_{\mu\rho} \tag{2.4}$$

for the Eulerian variation of the 4-volume element, the formula (2.3), in conjunction with the basic density variation formula (1.65), is sufficient for the purpose of evaluating the allowed variations of (2.1). (The heavy use of the symbol d in expressions such as (2.4) provides part of the motivation for insisting on designating exterior differentiation instead by the less overworked and for that purpose more appropriate symbol ∂ .) To evaluate the variations of (2.2) one also needs to know the effect on the unit tangent vector to the flow world-lines, which can be seen[30] to be given by

$$d_E u^\mu = -\gamma^\mu_\nu [\vec{\xi}, \vec{u}]^\nu + \frac{c^{-2}}{2} u^\mu u^\nu u^\rho d_E g_{\nu\rho} \tag{2.5}$$

(where it is to be recalled that the square brackets indicate the Lie commutator as defined by (1.23)).

The effect of such a variation on the combined action integral

$$I + I_{\mathrm{elec}} = \int_{\mathcal{M}} \underline{d}\mathcal{M} \rfloor (\Lambda + \Lambda_{\mathrm{elec}}) \tag{2.6}$$

can now be evaluated in a very convenient manner using the terminology introduced in the preceeding section, which allows us to express the general Eulerian variation of the total action density scalar $\Lambda + \Lambda_{\mathrm{elec}}$ in the form

$$d_E (\Lambda + \Lambda_{\mathrm{elec}}) = \pi^X_\mu d_E n^\mu_X + \frac{1}{2} \mu^X n_X u^\mu u^\rho d_E g_{\mu\rho} + j^\mu d_E A_\mu \,. \tag{2.7}$$

This shows that, whereas from the single-particle point of view the momenta are dynamic conjugates of corresponding velocities, on the other hand from the continuum point of view *the 4-momenta $\underline{\pi}^{\mathrm{X}}$ are interpretable as the dynamic conjugates of the corresponding number currents $\vec{n}_{\mathrm{X}}$.*

The final result of applying (2.4) and (2.7) to (2.6) is an integral variation formula of the form

$$d(I+I_{\mathrm{elec}}) = \int_{\mathcal{M}} \underline{d}\mathcal{M} \rfloor (\tilde{f}_\mu \xi^\mu + \frac{1}{2} T^{\mu\nu} d_{_E} g_{\mu\nu} + j^\mu d_{_E} A_\mu)$$

$$+ \oint_{\partial\mathcal{M}} \underline{d}\Sigma \rfloor (\vec{u} \wedge \vec{\xi}) \cdot \underline{\Pi} \,. \tag{2.8}$$

where the quantity $\underline{\Pi}$ appearing in the final boundary hypersurface integral (whose contribution may be ignored as far as the variation principle is concerned) is the *total 4-momentum density,*

$$\Pi_\mu = n_{\mathrm{X}} \pi^{\mathrm{X}}_\mu = (\rho + P c^{-2}) u_\mu - c^{-2} u_\nu j^\nu A_\mu \tag{2.9}$$

with P as given by (1.74). (The chemically non-covariant total 4-momentum per preferred reference particle $\underline{\pi}$ that was introduced the previous subsection by (1.92) could have been alternatively defined by $\underline{\Pi} = n_{_1} \underline{\pi}$.) The variational stress-momentum-energy tensor can be read out from (2.8) as

$$T^\nu_\mu = c^2 (n_{\mathrm{X}} \mu^{\mathrm{X}} \gamma^\nu{}_\mu - \rho g^\nu{}_\mu)$$

$$= u^\nu \Pi_\mu - j^\nu A_\mu + P g^\nu{}_\mu \tag{2.10}$$

again with P as given by (1.74), from which it can be seen that this expression agrees with the originally postulated form (1.59). Finally the variational force density form can be read out as

$$\underline{\tilde{f}} = \vec{n}_{\mathrm{X}} \cdot (\partial \underline{\pi}^{\mathrm{X}}) + \underline{\pi}^{\mathrm{X}} \nabla \cdot \vec{n}_{\mathrm{X}}$$

$$= \underline{f} - \underline{f}^{\mathrm{ext}} \tag{2.11}$$

with the external force density $\underline{f}^{\mathrm{ext}}$ as given by (1.62), where the latter version is based on the postulate of charge conservation and the internal force balance identity that defines the effective material force density $\underline{f}$:

$$\nabla_\mu j^\nu = 0 \,, \qquad \nabla_\nu T^\nu{}_\mu = f_\mu \,,$$

$$\underline{f} = \vec{n}_{\mathrm{X}} \cdot (\partial \underline{\chi}^{\mathrm{X}}) + \underline{\chi}^{\mathrm{X}} \nabla \cdot \vec{n}_{\mathrm{X}} \,. \tag{2.12}$$

Appart from the invocation of the electric charge conservation postulate, we have so far been dealing in this section only with definitions and kinematic identities. We are now ready to interpret the physical statement expressed by the world-line variation principle, which is just the requirement that the equations of motion should be given by the vanishing of the variational force density $\underline{\tilde{f}}$, as given by (2.11). By the second version of (2.11) this is evidently equivalent to the condition that the external force, if any, should be equal to the effective material force defined by (2.12):

$$\underline{\tilde{f}} = 0 \qquad \Leftrightarrow \qquad \underline{f}^{\text{ext}} = \underline{f} \ , \tag{2.13}$$

which by (2.12) is in manifest agreement with the originally postulated form (1.60).

It can be seen directly from (2.11) that the 3-components of (2.13) orthogonal to the flow lines are precisely equivalent to the *standard form* (1.82) of the field equations as derived in section 1.3, while (again invoking charge conservation, which is manifestly necessary for the gauge invariance of (2.11)) the remaining component along the flow lines is just the internal (chemical) energy conservation law

$$\mu^X \nabla \cdot \vec{n}_X = 0 \tag{2.14}$$

that was expressed by (1.75).

2.2 Double world-line displacement variational model.

The outcome that the non-conducting perfect fluid theory should have the variational formulation described in the previous section is not an accident, but an almost inevitable consequence of the necessity (discussed in section 1.3) of satisfying suitable identities in order to avoid overdetermination of the system by the stress-momentum-energy conservation requirement (1.63): the only natural way of doing this was to impose consistency conditions that could subsequently be interpreted as Noether identities of the world-line variational formulation, whose nature is such that it automatically avoids overdetermination while at the same time guaranteeing conservation of the associated variational stress-momentum-energy tensor.

In seeking a natural generalisation from non-conducting to conducting fluids it would be possible to go step by step through the imposition of self consistency conditions in the traditional manner that was followed in section 1.3 (indeed historically this is what I actually did in my original derivation[10] of the model to be set up in this subsection) but the experience of the immediately preceeding subsection, suggests that the simplest (though of course not necessarily the most general) mathematical ansatz for ensuring the satisfaction of the relevant self-consistency requirements is to adopt a variational formulation at the outset.

In order to achieve the objective of obtaining the *simplest possible* thermally conducting model compatible with the general good behaviour requirements, including in particular that of *causality*, that are discussed in the accompanying lectures of Israel, we shall start by restricting our attention to cases for which (appart from the space-time metric **g** itself) the only algebraicly independent fields are a pair of current vectors $\vec{s}$ and $\vec{n}$ say, which for the application with which we are principally concerned, are to be respectively interpreted as representing the entropy current and one other conserved (e.g. baryonic) particle current. In so much they depend only on the *eight* algebraicly independent tensor components s^μ, n^μ such cases (which may be described as *pure two-constituent fluid models*) will be a special subclass within the wider class of *fourteen* component models whose study was pioniered by Muller[5] and developed by Israel[6] in the manner described in his accompanying lectures. The *six* components that have to be dropped in the reduction from fourteen to eight are interpretable as those corresponding to *viscous stress* which will be ignored in the present course. The use of the more elaborate fourteen component model allowing for viscosity will of course be necessary for many applications involving fluids for which the Prandtl number (a dimensionless ratio of viscous to thermal resistivity coefficients) is of order unity, as is the case for the simple Boltzman mono-atomic gas model whose macroscopic description provided one of the principal motivations for the development of the Muller-Israel type of theory. There are however many quite different practical contexts (involving physical systems ranging from low temperature quantum condensates, to which the two constituent superfluid model is applicable, to high temperature gases with photons providing the dominant pressure contribution) in which the role of viscosity is effectively insignificant for one reason or another, but in which thermal conductivity may be of major importance. It is for cases such as these that the model to be described here should be the most appropriate.

The obvious way of modifying the variational theory described in the previous subsection so as to allow for conductivity is simply to drop the restriction that the world-lines of the different constituents be locked together, i.e. to allow the flow world-lines of each constituent to vary independently[31]. In the two-constituent case with which the present section is concerned our starting point will be an action density scalar Λ say specified in a covariant way as a function of the two currents $\vec{s}$ and $\vec{n}$ under consideration, which in practise means that it depends only on their three independent scalar combinations, which may be taken to be

$$s = \frac{1}{c}(-s_\mu s^\mu)^{\frac{1}{2}}\,, \qquad x = \frac{1}{c}(-s_\mu n^\mu)^{\frac{1}{2}}\,, \qquad n = \frac{1}{c}(-n_\mu n^\mu)^{\frac{1}{2}}\,, \tag{2.15}$$

The function $\Lambda(s,x,n)$ generalises (within the geometrically allowed parameter range, $x^2 - sn \geq 0$) the action density function $\rho(s,n)$ that would apply to the limiting case of thermal equilibrium without conduction (as effectively characterised by the restraint $x^2 - sn = 0$). A more detailed discussion of the functional form that is appropriate for Λ in the conducting case will be postponed until the following subsections.

Whatever its detailed form, any such covariantly defined function of $\vec{s}$ and $\vec{n}$ must respond to infinitesimal (Eulerian) variations of these fields and of the background metric in accordance with a formula of the form

$$d_E \Lambda = \Theta_\mu d_E s^\mu + \chi_\mu d_E n^\mu + \frac{1}{2}(\Theta^\mu s^\rho + \chi^\mu n^\rho) d_E g_{\mu\rho} \tag{2.16}$$

where the partial derivative coefficients of the currents, which in our systematic accounting system for multiconstituent fluids would be designated respectively as

$$\pi^0{}_\mu = \Theta_\mu \,, \quad \pi^1{}_\mu = \chi_\mu \,, \tag{2.17}$$

are to be understood as *defining* what are appropriately interpretable as the *thermal 4-momentum* covector $\underline{\Theta}$ (representing the "energy-momentum per entropon")and the *chemical 4-momentum* covector $\underline{\chi}$ (representing the intrinsic energy-momentum per conserved particle) that are dynamically conjugate to the corresponding current vectors $\vec{s}$ and $\vec{n}$. The quoted form for the coefficients of the metric variation is fixed in terms of these quantites by the Noether principle32, which also implies that the momenta will satisfy the symmetry condition

$$\Theta^{[\mu} s^{\rho]} + \chi^{[\mu} n^{\rho]} = 0 \,. \tag{2.18}$$

We now proceed, in analogy with the work in the preceeding subsection, by considering the induced variation in the integral

$$I = \int_{\mathcal{M}} \underline{d}\mathcal{M} \rfloor \Lambda \tag{2.19}$$

the only essential difference being that instead of a congruence of world-lines with displacements specified by a single vector field $\vec{\xi}$ we now have two distinct congruences of world-lines, i.e the integral curves of $\vec{s}$ and $\vec{n}$, whose displacements will correspondingly be determined by two independent vector fields, $\vec{\xi}_0$ and $\vec{\xi}_1$ say. It can be seen (from the combination of (2.3) and (2.5)) that the the corresponding Eulerian variations induced in the currents will respectively be given by

$$d_E s^\mu = -s^\mu \nabla \cdot \vec{\xi}_0 + [\vec{s}, \vec{\xi}_0]^\mu - \frac{1}{2} s^\mu g^{\nu\rho} d_E g_{\nu\rho} \,,$$

$$d_E n^\mu = -n^\mu \nabla \cdot \vec{\xi}_1 + [\vec{n}, \vec{\xi}_1]^\mu - \frac{1}{2} s^\mu g^{\nu\rho} d_E g_{\nu\rho} \,, \tag{2.20}$$

By substituting these expressions in (2.16) we obtain the induced variation of the integral (2.18) in the form

$$dI = \int_{\mathcal{M}} \underline{d}\mathcal{M} \rfloor (f^0{}_\mu \xi_0{}^\mu + f^1{}_\mu \xi_1{}^\mu + \frac{1}{2} T^{\mu\nu} g_{\mu\nu})$$

$$+ 2 \oint_{\partial\mathcal{M}} (\underline{d}\Sigma)_\mu (s^{[\mu} \xi_0{}^{\rho]} \Theta_\rho + n^{[\mu} \xi_1{}^{\rho]} \chi_\rho) \,, \tag{2.21}$$

in which the variational stress-momentum-energy tensor can be read out in the explicit form

$$T_\mu^{\ \nu} = \Theta_{\ \mu} s^\nu + \chi_{\ \mu} n^\nu + \Psi g_\mu^\nu \tag{2.22}$$

(whose symmetry, though not manifest, is guaranteed by the Noether identity (2.18)) where Ψ (the relevant generalisation of the non-conducting fluid pressure P in the previous section) is given as

$$\Psi = \Lambda - \Theta_{\ \mu} s^\mu - \chi_{\ \mu} n^\mu \tag{2.23}$$

As usual the boundary hypersurface integral contribution may be ignored as far as the variation principle is concerned. The quantities that matter are the effective force density forms $\underline{f}^0$ and $\underline{f}^1$ which would be set equal to zero in the strictly conservative case when the variation principle is rigorously applied. They can be read out as

$$\underline{f}^0 = \underline{\Theta} \nabla \cdot \vec{s} + \vec{s} \cdot (\partial \underline{\Theta}) \ , \tag{2.24}$$

and

$$\underline{f}^1 = \underline{\chi} \nabla \cdot \vec{n} + \vec{n} \cdot (\partial \underline{\chi}) \ . \tag{2.25}$$

It can be seen that these force densities will automatically (as a Noetherian identity) satisfy the energy relation

$$\nabla_\rho T_\mu^{\ \rho} = f^0_{\ \mu} + f^1_{\ \mu} \ . \tag{2.26}$$

It therefore follows that in order to get a mechanically consistent model in which the **T** as defined by (2.21) is interpretable as the stress-energy-momentumin the usual sense, we must adopt a force law for which the particle and entropy contributions are "equal and opposite" in the sense of Newton's third law, i.e. we must have

$$\underline{f}^0 = -\underline{f}^1 \ . \tag{2.27}$$

In order for the particle number current to be conserved in the model, i.e. to have

$$\nabla \cdot \vec{n} = 0 \ , \tag{2.28}$$

it is also necessary to satisfy the orthogonality requirement

$$\vec{n} \cdot \underline{f}^1 = 0 \ . \tag{2.29}$$

If we wish to allow for resistive dissipation then the entropy current will not have to be conserved, but by (2.19) it must in any case satisfy

$$(\vec{s} \cdot \underline{\Theta}) \nabla \cdot \vec{s} = \vec{s} \cdot \underline{f}^0 \ . \tag{2.30}$$

Up to this stage, everything we have done has followed ineluctably from the principles stated at the beginning of this subsection. As our only additional postulate it is now obviously natural to add in an axiom of the usual kind to the effect that the resitive force density (as defined in this particular way) should, like Λ, be also a purely algebraic function of the two primary currents involved, i.e. $\vec{s}$ and $\vec{n}$. By the orthogonality property (2.29), it must therefore be proportional to the 1-form $\underline{\sigma}$ say representing the relative entropy transport with respect to the conserved particles as given in terms of the notation of (2.15) by

$$\sigma_\mu = s_\mu - \left(\frac{x}{n}\right)^2 n_\mu\,, \qquad n^\mu \sigma_\mu = 0\,. \tag{2.31}$$

To complete the determination of the system it thus remains only to specify the appropriate scalar proportionality factor, Z say, as a function of the independent scalar variables s, x, n, fo as to obtain the required force-density in the form

$$\underline{f}^{\text{o}} = Z(\vec{s}\cdot\underline{\Theta})\,\underline{\sigma} \tag{2.32}$$

where the normalisation of Z has been adjusted (by the inclusion of the factor $\vec{s}\cdot\underline{\Theta}$) in such a way that the entropy creation rate will be given simply by

$$\nabla\cdot\vec{s} = Z\,\vec{s}\cdot\underline{\sigma}\,, \tag{2.33}$$

in which the right hand side will automatically have the non-negativity property required by the second law of thermodynamics provided the scalar proportionality factor Z is itself non-negative.

When the theory has been completed in this way, the full set of equations of motion for the eight unknown components of the currents $\vec{s}$ and $\vec{n}$ may be taken as being given by the pair of (single component) divergence equations (2.28) and (2.33) together with the result of substitiuting (2.32) and (2.27) in (2.24) and (2.25) which gives the pair of (effectively triple component) dynamical equations

$$\vec{s}\cdot(\partial\underline{\Theta}) = Z\vec{s}\cdot(\underline{\Theta}\wedge\underline{\sigma}) \tag{2.34}$$

and

$$\vec{n}\cdot(\partial\underline{\chi}) = -Z(\vec{s}\cdot\underline{\Theta})\,\underline{\sigma}\,. \tag{2.35}$$

2.3 The Regularity Ansatz and its Pathological Predecessors.

The complete specification of a particular model of within the general class that was set up in the previous subsection requires the choice of just two scalar functions, namely Λ and Z, of the three independent scalar variables s, x, n that were defined by (2.15). Although its order of magnitude, and particularly the positivity requirement

$$Z(s,x,n) \geq 0 \tag{2.36}$$

are important for local stability, the detailed functional behavior of Z does not matter very much as far as the qualitative behavior of the system is concerned, so that it will often be an adequate approximation to set it to a fixed (average) value provided the relevant variables do not have too wide a range in the problem under study. On the other hand the behaviour of the system is very sensitively dependent on the form of the function $\Lambda(n,x,s)$. Unlike Z which enters into the differential equations of motion only in undifferentiated form, Λ enters in not just once but twice differentiated form, which means that the behaviour of the system will be qualitatively dependent on its second partial derivatives, and will therefore have an even more sensitive dependence on its first partial derivatives, which turn up directly in the expansion (2.16) that defines the 4-momenta. Explicitly, the respective components of the thermal and conserved-particle 4-momenta will be given in terms of these partial derivatives by

$$\Theta_\mu = \pi^0{}_\mu = \mathcal{C} s_\mu + \mathcal{A} n_\mu \,,$$
$$\chi_\mu = \pi^1{}_\mu = \mathcal{B} n_\mu + \mathcal{A} s_\mu \,. \tag{2.37}$$

in which the coefficients are given by

$$c^2\mathcal{C} = -2\frac{\partial \Lambda}{\partial s^2} = \frac{-1}{s}\frac{\partial \Lambda}{\partial s}\,, \qquad c^2\mathcal{B} = -2\frac{\partial \Lambda}{\partial n^2} = \frac{-1}{n}\frac{\partial \Lambda}{\partial n}\,.$$
$$c^2\mathcal{A} = -\frac{\partial \Lambda}{\partial x^2} = \frac{-1}{2x}\frac{\partial \Lambda}{\partial x}\,, \tag{2.38}$$

It frequently occurs in practice that one needs to construct a conducting fluid model on the basis only approximate knowledge of the appropriate value for the resistivity scalar, Z, together with prior knowledge of the *thermal equilibrium states* in which there is no relative current flow, so that the 4-vectors, $\vec{s}$ and $\vec{n}$ are parallel. In terms of the independent scalar variables s, x, n, such (local) thermal states correspond to the zero limit of the non-negative combination $x^2 - sn$, i.e.

$$(\,0 \leq\,)\; x^2 - sn \to 0\,. \tag{2.39}$$

It can be seen from the form of (2.22) and (2.23) that in such a state the 3-variable function $\Lambda(n,x,s)$ will go over to minus the the ordinary equilibrium energy density which depends only on the equilibrium values of n and s, i.e. we shall have

$$\Lambda(s,x,n) \to -c^2\rho(s,n) \quad \text{as} \quad x^2 \to sn\,. \tag{2.40}$$

However the reverse process is not a priori well defined: if as we have supposed, our prior knowledge is limited to the form of the 2-variable function of state $\rho(n,s)$, we need additional input in the form of some independent physical information before the functional form of Λ can be obtained. If the required information is not available from experiment or from a microscopic theoretical model, or if it is available only in a form that is more comlicated than is worthwhile for the purose in hand, one will want to use some general purpose "off the peg" prescription in the form of a simplifying ansatz for reducing the number of independent variables from three to two.

Although they were originally derived by a quite different route, the conductivity theories of Eckart[1] and of Landau and Lifshitz[2] can be interpreted as corresponding (within the order of accuracy reqired for a Hiscock-Lindblom type equilibrium-perturbation analysis) to the result of applying different simplifying ansatzes of this kind. One of the ways in which their mode of derivation differed from the approach followed here was in the departure from manifest covarience by the adoption of (different) preferred rest-frames for reference purposes, which not only biassed these authors towards choosing what we shall see to have been quite inapproriate choices of simplifying ansatz, but also led to much debate in the subsequent litterature as to whether the respective theories were essentially equivalent modulo a change of reference system, or whether they were fundamentally different. By working in terms of the two covariantly defined 4-momenta given by (2.37), (and by leaving out viscosity, which is an irrelevant complication as far as these results of different choices of preferred rest-frame are concerned), we shall have no difficulty in seeing from the start that (as was made clear by the recent causality analysis of Hiscock and Lindblom[13]) the Eckart and Landau- Lifshitz models are in fact fundamentally distinct, as well as being able to see why they are both fundamentally wrong. The present approach will also enable us to see that there is another equally simple (and, from this covariant point of view, even more obvious) simplifying ansatz [10,11] which does everything one wants in a uniquely natural manner. After describing this third "regularity" ansatz, the consideration that its less satisfactory predecessors are so much better known makes it virtually obligatory that we should digress to describe them also.

In addition to pure mathematical simplicity, the physical principle that lead to the "regularity" ansatz was the consideration that any general purpose "off the peg" prescription should be adaptable to the weakly-coupled limit in which virtually all the entropy is contained in a "black-body radiation" gas (so that the "entropons" would be identifiable with photons, or at higher temperatures with electron-positron pairs et cetera) in very weak interaction with a background flux of comparatively heavy particles which, though numerically sparce might nevertheless provide the dominant contribution to the mass density. (The radiation pressure dominated gas in the interior of an upper main sequence star is a well known example). In such a weakly coupled gas, the action density Λ would split up into a thermal contribution depending only on s (it could be taken to be approximately proportional to s^4 in the quoted upper-main-sequence example) and a conserved particle contribution depending only on n (it could be taken approximately to be linearly proportional to n itself in the upper main-sequence example), and the third (cross product) variable x would not be involved at all. As a result the thermal momentum covector components Θ_μ would be proportional to the covariant components s_μ of the corresponding number flux, and hence (by (2.18)) similarly χ_μ would be proportional to n_μ i.e.

$$\Theta_{[\mu}s_{\rho]} = 0 \qquad \Leftrightarrow \qquad \chi_{[\mu}n_{\rho]} = 0 \,. \tag{2.41}$$

This suggests that in the absence of any indications to the contrary we should postulate the proportionality relation (2.41) between $\underline{\Theta}$ and $\vec{s}$ (and hence also the analogous proportionality relation between $\underline{\chi}$ and $\vec{n}$) even in the strongly interacting case. It is evident from (2.37) that this postulate, which is what we shall refer to as the *Regularity Ansatz* is equivalent to the restriction

$$\mathcal{A} = 0 \quad \Leftrightarrow \quad \frac{\partial \Lambda}{\partial x^2} = 0 \tag{2.42}$$

on the equation of state for the action density, which means that it is to be taken as being determined in terms of the (supposedly already known) equilibrium density function $\rho(s,n)$ by the relation

$$\Lambda(s,x,n) = -c^2\rho(s,n) \tag{2.43}$$

Although (in view of the favorable results of the causality analysis described below) I reccommend it as as "off the peg" prescription for wide range of general purposes, this regularity ansatz is not to be considered as a universal panacea for use in situations for which more elaborate tailoring might be absolutely necessary for physical realism. For example our general model, as set up in the preceeding section, should be capable of providing a good description of relativistic superfluid behaviour in the force-free limit in which the thermal resistance coefficient Z is set equal to zero, so that it would be consistent with (2.24) to impose the necessary irrotationality requirement on the particle 4-momentum form $\underline{c}hi$, which would be interpreted as specifying the so called "superfluid velocity" as given in terms of the gradient of an order parameter ϕ say by $\underline{\chi} = \partial\phi$. (In this example the "entropons" would be interpreted as being identifiable with phonons, or at higher temperatures with "rotons" et cetera). In such a cases non-alignment between the "superfluid velocity" (the inverted commas being motivated by the consideration that it is merely a *phase-velocity* and not a true (group) velocity) and the ordinary velocities (i.e the velocity $\vec{u}^1$ of the conserved particles and the "normal component" velocity $\vec{u}^0$ of the entropy) is a phenomenon that is already familiar from experience with the non-relativistic Landau "two-fluid" model that is obtained in the Newtonian limit. When such non-alignment is present, i.e. when the "regularity" condition does not apply, the 4-momenta may appropriately be described as being "anomalous", the "anomaly" being evidently (by (2.37)) proportional to the coefficient $\mathcal{A} = -c^{-2}\partial\Lambda/\partial x^2$, which we shall refer to as the "anomaly coefficient", the other coefficients (i.e the only ones present in the "regular" case) being referred to as the "bulk coefficient", $\mathcal{B} = -2c^{-2}\partial\Lambda/\partial n^2$, and the "caloric coefficient", $\mathcal{C} = -2c^{-2}\partial\Lambda/\partial s^2$.

In contrast with the covariantly derived and trouble-free simplification ansatz that has just been introduced, both the Eckart and the Landau Lifshitz ansatzes are motivated by a (misguided) desire to simplify things as much as possible with respect to a *single* preferred rest-frame as specified by some preferred timelike-unit vector $\vec{u}$ say. It is because this leads to ignoring what in fact are important second order contributions from deviations from the chosen rest-frame that both the Landau-Lifshitz and the Eckart models are "first order theories" in the sense of Hiscock and Lindblom[13]. From the point of view of the present approach the Eckart ansatz is the simpler one to analyse, though it is also the one that is more arbitrary and ultimately less satisfactory. It is based on taking $\vec{u}$ simply to coincinde with the unit vector allong the flow lines of the conserved particles, i.e. the vector $\vec{u}^1$ as defined above by (2.33). Such a choice may seem natural when only one species of material particle is involved, but its arbitrarness becomes evident as soon as one wonders how it should be generalised to situations of the type to be considered later on in the final section, where several species are involved. Arbitrary though it may be, the Eckart frame is certainly convenient for detailed analysis, and we shall use it in the next subsection for the purpose of translating the rather unconventional concepts with which we have been working into the more traditional terminology that will be familiar to most readers. Although it was of course originally phrased in such traditional terminology, the simplifying ansatz chosen by Eckart[1] can be expressed in terms of the more covariant terminology that we have been using here as the postulate that the thermal momentum covector that we have denoted by $\underline{\Theta}$ should be aligned with this preferred timelike direction, i.e. that Θ_μ should be proportional to n_μ,

$$\Theta_{[\mu}n_{\rho]} = 0 \,. \tag{2.44}$$

From the point of view of the primary equation of state, the Eckart ansatz is therefore expressible as absence of the "caloric" coefficient, $\mathcal{C}$,

$$\mathcal{C} = 0 \qquad \Leftrightarrow \qquad \frac{\partial \Lambda}{\partial s^2} = 0 \,. \tag{2.45}$$

Thus whereas the "regularity ansatz" (2.41) makes Λ independent of x while dependent on s, the Eckart ansatz (2.44) would make it independent of s while dependent on x. In integrated form, as the analogue of the "regular" form (2.43) of the equation of state, the Eckart ansatz thus gives

$$\Lambda(s, x, n) = -c^2 \rho(\frac{x^2}{n}, n) \,. \tag{2.46}$$

When (2.44) is compared with the "correct" (regularity) ansatz (2.41) it can be seen that Eckart effectively made the mistake of getting the thermal and particle parts mixed up, which is obviously incompatible with a well behaved weakly couple limit. What Eckart did not realise moreover was that (except in strict thermal equilibrium) as a consequence of the identity (2.18) his ansatz effectively prevented the particle-momentum components χ_μ could *not* from being proportional to n_μ, so that the apparent absence of "second order terms" is rather deceptive, and it is therefore not surprising that the Hiscock-Lindblom causality analysis[13] (of which a simplified version[11] will be described in the next subsection) reveals that Eckart type models will exhibit quasi-elliptic behaviour, as partial differential systems.

Unlike the Eckart ansatz, the choice proposed by Landau and Lifshiz[2] cannot be criticised for being arbitrary. It shares with the regularity ansatz described above the quality of being entirely "democratic" as between the particle and entropy currents involved (so that it could in principle be extended in an unambiguous way to systems involving an arbitrary number of currents). Although originally specified with respect to a preferred rest-frame defined in terms of a unit-vector $\vec{u}$ chosen as the timelike eigenvector of the material stress-momentum-energy tensor, the Landau-Lifshitz ansats can be given a direct covariant characterisation within the present framework as the condition that the thermal and particle 4-momenta $\underline{\Theta}$ and $\underline{\chi}$ should simply be proportional to each other

$$\Theta_{[\mu}\chi_{\rho]} = 0 \,. \tag{2.47}$$

(The unique direction thereby defined is the same as that of the timelike eigenvector of the stress-momentum-energy tensor on which the original formulation of this prescription was based.) Although unassailable on ground of mathematical elegance, this condition shares with that of Eckart the physical drawback of being obviously incompatible with the weakly coupled limit. Within the framework of the present discussion it also has the inconvenience that when expressed analogously to (2.42) and (2.45) as a condition on the partial derivatives of Λ it takes the non-linear form

$$\mathcal{C}\mathcal{B} - \mathcal{A}^2 = 0 \qquad \Leftrightarrow \qquad 4\frac{\partial \Lambda}{\partial s^2}\frac{\partial \Lambda}{\partial n^2} - \left(\frac{\partial \Lambda}{\partial x^2}\right)^2 = 0 \tag{2.48}$$

which cannot easily be integrated to give a fully explicit form analogous to (2.43) and (2.46) in terms of the variables with which we have been working. However the most serious objection to the Landau-Lifshitz ansatz is that, as one might have guessed, the imposition of the degeneracy condition (2.47) between what would normally be independent 4-momenta leads to a corresponding degeneracy of the causal behavior which, as had long been guessed (wrongly in the Eckart case but rightly in this case) is indeed[13] of parabolic type, which is nearly, albeit not quite, as seriously pathological as the quasi-elliptic behaviour of the Eckart case.

2.4 Causal Behaviour

The full set of equations of motion for the general conducting fluid model (unrestricted by the regularity postulate or any other simplifying ansatz) as set up in section 2.2 may be written out in explicit coordinate form as the set

$$2s^\rho(\partial_{[\rho}\Theta_{\mu]} + Z\sigma_{[\rho}\Theta_{\mu]}) = 0\,, \qquad \nabla_\mu s^\mu = Z\sigma_\mu\sigma^\mu\,,$$

$$2n^\rho(\partial_{[\rho}\chi_{\mu]} + \frac{Zs^\nu\Theta_\nu}{c^2n^2}s_{[\rho}n_{\mu]}) = 0\,, \qquad \nabla_\mu n^\mu = 0\,. \tag{2.49}$$

where the entropy and particle 4-momentum covectors Θ_μ and χ_μ are as given by (2.37), and the entropy transport covector components σ_μ are as given by (2.31).

A convenient method (introduced long ago by Hadamard) of investigating the characteristic hypersurfaces of possible discontinuity in such a partial differential system is to consider the first order case in which the algebraically related variables s^μ, n^μ, Θ_μ, χ_μ are themselves continuous but have space-time derivatives that are weakly discontinuous across some characteristic hypersurface with tangent direction specified by some normal covector $\underline{\lambda}$ say. For any component ϕ say the discontinuities $[\partial_\mu\phi]$ in its gradient components will have to be proportional to the normal compoments λ_μ, i.e. we shall have $[\partial\phi] = \hat{\phi}\underline{\lambda}$ for some scalar $\hat{\phi}$. Applying this to the relevant variables in the present case, we see that we shall have

$$[\partial_\mu\Theta_\rho]=\hat\Theta_\rho\lambda_\mu\,,\qquad [\partial_\mu s^\nu]=\hat s^\nu\lambda_\mu\,,$$

$$[\partial_\mu\chi_\rho]=\hat\chi_\rho\lambda_\mu\,,\qquad [\partial_\mu n^\nu]=\hat n^\nu\lambda_\mu\,,\tag{2.50}$$

for some set of vectors $\hat{\vec s}$, $\hat{\vec n}$ and covectors $\hat{\underline{\Theta}}$, $\hat{\underline{\chi}}$ on the hypersurface. The resulting discontinuities in the set of equations (2.49) will therefore be given by

$$2s^\rho\lambda_{[\rho}\Theta_{\mu]}=0\,,\qquad \lambda_\mu s^\mu=0\,,$$

$$2n^\rho\lambda_{[\rho}\chi_{\mu]}=0\,,\qquad \lambda_\mu n^\mu=0\,.\tag{2.51}$$

The discontinuity covectors $\hat{\underline{\Theta}}$, $\hat{\underline{\chi}}$ will not however be independent of the corresponding discontinuity vectors $\hat{\vec s}$, $\hat{\vec n}$ in view of the algebraic relationship (2.37), which shows that infinitesimal changes $d\vec s$ and $d\vec n$ will induce corresponding infinitesimal changes of the form

$$d\Theta_\mu=d\pi^0{}_\mu=\mathcal{P}^{00}{}_{\mu\nu}ds^\nu+\mathcal{P}^{01}{}_{\mu\nu}dn^\nu\,,$$

$$d\chi_\mu=d\pi^1{}_\mu=\mathcal{P}^{10}{}_{\mu\nu}ds^\nu+\mathcal{P}^{11}{}_{\mu\nu}dn^\nu\,,\tag{2.52}$$

which implies that the discontinuity amplitudes inn (2.51) will correspondingly be related by

$$\hat\Theta_\mu=\mathcal{P}^{00}{}_{\mu\nu}\hat s^\nu+\mathcal{P}^{01}{}_{\mu\nu}\hat n^\nu\,,$$

$$\hat\chi_\mu=\mathcal{P}^{11}{}_{\mu\nu}\hat n^\nu+\mathcal{P}^{10}{}_{\mu\nu}\hat s^\nu\,.\tag{2.52}$$

By differentiating (2.37) the coefficients can be worked out to be given by the explicit expressions

$$\mathcal{P}^{00}{}_{\mu\rho}=\mathcal{C}g_{\mu\rho}-\frac{2\partial\mathcal{C}}{c^2\partial s^2}s_\mu s_\rho-\frac{4\partial\mathcal{A}}{c^2\partial s^2}s_{(\mu}n_{\rho)}-\frac{\partial\mathcal{A}}{c^2\partial x^2}n_\mu n_\rho\,,$$

$$\mathcal{P}^{11}{}_{\mu\rho}=\mathcal{B}g_{\mu\rho}-\frac{2\partial\mathcal{B}}{c^2\partial n^2}n_\mu n_\rho-\frac{4\partial\mathcal{A}}{c^2\partial n^2}n_{(\mu}s_{\rho)}-\frac{\partial\mathcal{A}}{c^2\partial x^2}s_\mu s_\rho\,,$$

$$\mathcal{P}^{01}{}_{\mu\rho}=\mathcal{P}^{10}{}_{\rho\mu}=\mathcal{A}g_{\mu\rho}-\frac{2\partial\mathcal{B}}{c^2\partial s^2}s_\mu n_\rho-\frac{2\partial\mathcal{A}}{c^2\partial n^2}n_\mu n_\rho-\frac{2\partial\mathcal{A}}{c^2\partial s^2}s_\mu s_\rho-\frac{\partial\mathcal{A}}{c^2\partial x^2}n_\mu s_\rho\,,\tag{2.53}$$

(which can be seen to simplify considerably if one imposes the regularity postulate to the effect that $\mathcal{A}$ vanishes).

With respect to a rest frame determined by some preferred timelike unit vector $\vec u$ the velocity v say of propagation in the direction of some orthogonal unit spacelike vector $\vec\nu$ will be given for suitably normalised $\underline{\lambda}$ by

$$\lambda_\mu=\nu_\mu+vc^{-2}u_\mu\,,\quad \nu_\mu\nu^\mu=1\,,\ \nu_\mu u^\mu=0\,,\ u_\mu u^\mu=-c^2\,.\tag{2.54}$$

If, as usual[11,13], our main concern is with small perturbations from the thermal equilibrium state in which no relative transport, then we may most conveniently take $\vec u$ along the corresponding (common) unperturbed flow direction, so that the unperturbed background currents will be expressible in the form

$$s^\mu = s u^\mu \,, \qquad n^\mu = n u^\mu \tag{2.55}$$

It will at this point be convenient to express the characteristic equations (2.51) in terms of an orthonormal rest-frame aligned with the direction of propagation at the point under consideration, so that the coordinates of $\vec{u}$ and $\vec{\nu}$ are given by the Kronecker delta expressions

$$u^\mu = \delta^\mu_4 \,, \qquad \nu^\mu = \delta^\mu_3 \,. \tag{2.56}$$

In such a system the only non-zero metric coefficients are

$$g_{11} = g_{22} = g_{33} = 1 \,, \; g_{44} = -c^2 \tag{2.57}$$

and the only non-zero coefficients in (2.53) will be

$$\mathcal{P}^{00}{}_{11} = \mathcal{P}^{00}{}_{22} = \mathcal{P}^{00}{}_{33} = \mathcal{C} \,, \quad \mathcal{P}^{11}{}_{11} = \mathcal{P}^{11}{}_{22} = \mathcal{P}^{11}{}_{33} = \mathcal{B} \,,$$

$$\mathcal{P}^{01}{}_{11} = \mathcal{P}^{10}{}_{11} = \mathcal{P}^{01}{}_{22} = \mathcal{P}^{10}{}_{22} = \mathcal{P}^{01}{}_{33} = \mathcal{P}^{10}{}_{33} = \mathcal{A} \,, \tag{2.58}$$

and

$$\mathcal{P}^{00}{}_{44} = -2c^2\tilde{\mathcal{C}} \,, \quad \mathcal{P}^{11}{}_{44} = -2c^2\tilde{\mathcal{B}}$$

$$\mathcal{P}^{01}{}_{44} = \mathcal{P}^{10}{}_{44} = -c^2\tilde{\mathcal{A}} \tag{2.59}$$

with

$$\tilde{\mathcal{C}} = \mathcal{C} + 2s^2\frac{\partial \mathcal{C}}{\partial s^2} + 4sn\frac{\partial \mathcal{A}}{\partial s^2} + n^2\frac{\partial \mathcal{A}}{\partial x^2} \,,$$

$$\tilde{\mathcal{B}} = \mathcal{B} + 2n^2\frac{\partial \mathcal{B}}{\partial n^2} + 4ns\frac{\partial \mathcal{A}}{\partial n^2} + s^2\frac{\partial \mathcal{A}}{\partial x^2} \,,$$

$$\tilde{\mathcal{A}} = \mathcal{A} + 2sn\frac{\partial \mathcal{B}}{\partial s^2} + 2n^2\frac{\partial \mathcal{A}}{\partial n^2} + 2s^2\frac{\partial \mathcal{A}}{\partial s^2} + ns\frac{\partial \mathcal{A}}{\partial x^2} \,. \tag{2.60}$$

It can now easily be seen from (2.51) that there can be no transverse modes, i.e. we must have

$$\hat{\Theta}_1 = \hat{\Theta}_2 = 0 \,, \qquad \hat{s}^1 = \hat{s}^2 = 0 \,,$$

$$\hat{\chi}_1 = \hat{\chi}_2 = 0 \,, \qquad \hat{n}^1 = \hat{n}^2 = 0 \,, \tag{2.61}$$

and that the characteristic equation for the longitudinal modes will take the form

$$\hat{\Theta}_4 + v\hat{\Theta}_3 = 0 \,, \qquad \hat{s}^3 - v\hat{s}^4 = 0 \,,$$

$$\hat{\chi}_4 + v\hat{\chi}_3 = 0 \,, \qquad \hat{n}^3 - v\hat{n}^4 = 0 \,, \tag{2.62}$$

with

$$\hat{\Theta}_3 = \mathcal{C}\hat{s}^3 + \mathcal{A}\hat{n}^3 \,, \quad \hat{\Theta}_4 = -c^2(\tilde{\mathcal{C}}\hat{s}^4 + \tilde{\mathcal{A}}\hat{n}^4) \,,$$

$$\hat{\chi}_3 = \mathcal{B}\hat{s}^3 + \mathcal{A}\hat{n}^3 \,, \quad \hat{\chi}_4 = -c^2(\tilde{\mathcal{B}}\hat{s}^4 + \tilde{\mathcal{A}}\hat{n}^4) \,. \tag{2.63}$$

The resulting eigenvalue equation for the propagation velocity v is

$$(v^2\mathcal{C} - c^2\tilde{\mathcal{C}})(v^2\mathcal{B} - c^2\tilde{\mathcal{B}}) - (v^2\mathcal{A} - c^2\tilde{\mathcal{A}})^2 = 0 \,. \tag{2.64}$$

In the decoupled limit for which Λ splits up as a sum of two separate single variable functions, respectively of s only and of n only, we shall have not only $\mathcal{A} = 0$ but also $\tilde{\mathcal{A}} = 0$, which implies the existence of the two well behaved propagation modes that one would expect, which are interpretable as "heat" (or "second sound")modes (with $\hat{\tilde{s}}$ but not $\hat{\tilde{n}}$) with velocity given by $v^2/c^2 = \tilde{\mathcal{C}}/\mathcal{C}$ and "ordinary" (or "first") sound modes (with $\hat{\tilde{n}}$ but not $\hat{\tilde{s}}$) with velocity given by $v^2/c^2 = \tilde{\mathcal{B}}/\mathcal{B}$, provided of course that requirements $0 \leq \tilde{\mathcal{C}}/\mathcal{C} \leq 1$ and $0 \leq \tilde{\mathcal{B}}/\mathcal{B} \leq 1$ are satisfied. In this case, as in the more general coupled but still *regular* case for which we have non-zero $\tilde{\mathcal{A}}$ but for which the anomally coefficient itself is still zero, i.e.

$$\mathcal{A} = 0 \,, \tag{2.65}$$

we shall have

$$\hat{\mathcal{C}} = \frac{\partial(s\mathcal{C})}{\partial s} = \frac{-1}{c^2}\frac{\partial^2\Lambda}{(\partial s)^2} \,, \qquad \hat{\mathcal{B}} = \frac{\partial(n\mathcal{B})}{\partial n} = \frac{-1}{c^2}\frac{\partial^2\Lambda}{(\partial n)^2} \,,$$

$$\hat{\mathcal{A}} = \frac{\partial(n\mathcal{B})}{\partial s} = \frac{\partial(s\mathcal{C})}{\partial n} = \frac{-1}{c^2}\frac{\partial^2\Lambda}{(\partial n)(\partial s)} \,. \tag{2.66}$$

Rewriting the characteristic propagation velocity equation (2.64) in the standard form

$$A\left(\frac{v^2}{c^2}\right)^2 + 2B\left(\frac{v^2}{c^2}\right) + C = 0 \tag{2.67}$$

we can express the "hyperbolicity" condition to the effect that both of the roots for $(v/c)^2$ should be stricty positive as the requirement

$$B^2 \geq 4AC > 0 > AB \,, \tag{2.68}$$

and the further condition that neither root should exceed the speed of light can be expressed as the additional requirement

$$A + B + C \geq 0 \geq B^2 - 4A^2 \,, \tag{2.69}$$

where the coefficients are expressible (both in the regular case characterised by (2.66) and in the general case given by (2.60)) as

$$A = \mathcal{C}\mathcal{B} - \mathcal{A}^2 \,,$$

$$B = -(\mathcal{C}\tilde{\mathcal{B}} + \mathcal{B}\tilde{\mathcal{C}} - 2\mathcal{A}\tilde{\mathcal{A}}) \,,$$

$$C = \tilde{\mathcal{C}}\tilde{\mathcal{B}} - \tilde{\mathcal{A}}^2 \,. \tag{2.70}$$

These conditions will naturally be satisfied in ordinary physical situations with not more than moderate coupling, for which we should expect $\mathcal{C}$ and $\mathcal{B}$ to be positive with product large compared to the square of $\mathcal{A}$ (trivially so in the regular case for which $\mathcal{A}$ vanishes) so that the quadratic coefficient A in (2.70) will be strictly *positive*. (The associated requirement that C in (2.70) should be positive can be seen in the regular case to be interpretable (by (2.66) and (2.43)) to be equivalent just to the condition that in thermal equilibrium the mass-energy density ρ should be convex upwards as a function of s and n.) It would require an excessively strong coupling to attain the condition postulated by the Landau-lifshitz ansatz, (2.48) which can be seen to correspond to the "parabolic" limit $A \to 0$ for which one of the roots tends to infinity. The Eckart ansatz (2.45) would require an even more excessively strong coupling for which the quadratic coefficient A must actually be *negative*, which accounts for the unstable quasi elliptic behaviour brought to light by Hiscock and Lindblom[13,14].

[Preliminary results of work in progress at the time of writing these notes would indicate that positivity of the coefficient A in (2.70) is precisely the the condition for avoiding the existence of the anti-damping instability in the non-propagating transverse-current perturbation mode that was found[13] in the Eckart case.]

2.5 Eckart-frame interpretation of the model.

Having given a rather abstract introduction to the variational two-constituent model for a thermally conducting fluid using the fully covariant treatment that is most convenient for general mathematical purposes, we shall now adopt a more frame-dependent point of view for the purpose of interpreting the model in terms of previously familiar physical concepts (so as to be able to reply, for example, to questions about the relation between heat flux and temperature gradient after such quantities have been suitably defined).

Among the many possible choices of "rest-frame" as specified by a corresponding preferred unit vector $\vec{u}$, there are two that are particularly simple to use for the presentation of the two-constituent conductivity theory that has just been presented, which are the *thermal rest-frame* unit vector, $\vec{u}_0$ say, and the *particle rest-frame* unit vector, $\vec{u}_1$ say, as defined in terms of the basic entropy and particle current vectors $\vec{s}$ and $\vec{n}$ by

$$s^\mu = s u_0^{\ \mu} , \qquad n^\mu = n u_1^{\ \mu} , \qquad u_0^{\ \mu} u_{0\mu} = -c^2 = u_1^{\ \mu} u_{1\mu} , \tag{2.71}$$

Although (as we shall see in the next section) the special nature of the entropy current (as distinct from ordinary "material" particle currents) makes the thermal unit vector $\vec{u}_0$ the most convenient choice for the specification of a preferred rest frame for general theoretical purposes, the most traditional choice is to use a particle rest-frame, whose specification would have the disadvantage of arbitrariness if several species were involved, but which leads uniquely to taking $\vec{u} = \vec{u}_1$ in the present case. This choice, which will be used as the basis of the analysis in the present subsection, was the one used as the starting point for Eckart's historic approach[1] to the problem, and we shall therefore refer to $\vec{u}_1$ as the Eckart unit vector, and will refer to the values of frame dependent quantities specified in terms of $\vec{u}_1$ as their Eckart values.

(The Landau-Lifshitz[2] unit vector, defined as the timelike eigenvector of the stress-momentum-energy tensor, has the advantage over the Eckart unit vector of being free of arbitrariness even when several species are involved, but its use has traditionally been less popular, perhaps because it is not so easy to work with analytically. When a prefered reference frame is needed in a thermodynamic context, the the present author strongly reccommends the use of the *thermal* unit vector, which avoids both the arbitrariness of the Eckart unit vector and the analytic inconvenience of the Landau-Lifshitz unit vector as well as having other advantages that will begin to emerge later on in the work of the next section.)

Having decided to express the theory in terms of Eckart-frame quantities, we introduce the use of the symbol $\|$ whenever necessary to distinguish Eckart-frame components from natural frame components as previously defined. Thus we start by defining the Eckart-frame entropy and particle number densities, $s^{\|}$ and $n^{\|}$, and the entropy transfer velocity, $\vec{v}$ by

$$s^{\mu} = s^{\|}(u^{\mu} + v^{\mu}) , \qquad n^{\mu} = n^{\|}u^{\mu} , \qquad v^{\mu}u_{\mu} = 0 , \qquad u^{\mu}u_{\mu} = -c^2 \tag{2.72}$$

These Eckart densities are evidently related to their covariantly defined analogues (as previously introduced in (2.15)) by

$$s^{\|} = \frac{u_{\mu}s^{\mu}}{-c^2} = \frac{x^2}{n} = \gamma s , \qquad n^{\|} = \frac{u_{\mu}n^{\mu}}{-c^2} = n , \tag{2.73}$$

and the entropy transfer velocity vector $\vec{v}$ is related to the associated contravariant version, $\vec{\sigma}$ of the entropy transfer covector $\underline{\sigma}$ (as previously introduced in (2.31)) by

$$\sigma^{\mu} = s^{\|}v^{\mu} , \tag{2.74}$$

where the corresponding Lorentz factor, γ is defined by

$$\gamma = \left(1 - \frac{v^2}{c^2}\right)^{-\frac{1}{2}} , \qquad v = (v^{\mu}v_{\mu})^{\frac{1}{2}} , \tag{2.75}$$

Continuing in the same way, we introduce quantities interpretable as the Eckart-frame temperature $\Theta_{\|}$ and the effective mass per particle μ by the definitions

$$\Theta_{\|} = -u^{\mu}\Theta_{\mu} , \qquad \mu c^2 = -u^{\mu}\chi_{\mu} . \tag{2.76}$$

These quantities turn up naturally as partial derivatives of the action scalar Λ when (instead of using our previously chosen set of variables s, x, n) it is expressed in terms of the set of Eckart-frame variables n, $s^{\|}$, σ, where σ is the magnitude of the entropy transfer vector (2.74) as defined by

$$\sigma = (\sigma_{\mu}\sigma^{\mu})^{\frac{1}{2}} = s^{\|}v , \tag{2.77}$$

since the general differential of Λ is expressible as

$$d\Lambda = -c^{-2}\mu dn - \Theta_{\|}ds^{\|} + pd\sigma , \tag{2.78}$$

where the new Eckart frame coefficients are expressible in terms of the previously introduced coefficients $\mathcal{C}$, $\mathcal{B}$, $\mathcal{A}$ by

$$\Theta_{\parallel} c^{-2} = \mathcal{C} s^{\parallel} + \mathcal{A} n \,,$$

$$\mu = \mathcal{B} n + \mathcal{A} s^{\parallel} \,,$$

$$p = \mathcal{C} s^{\parallel} v \,. \tag{2.79}$$

The interpretations of $\Theta_{\parallel}$ as a temperature in the usual sense may be justified by translating into terms of the Eckart-frame mass-energy density ρ, as defined by

$$\rho c^4 = T_{\mu\rho} u^{\mu} u^{\rho} \,, \tag{2.80}$$

which can be seen to be obtainable from Λ by a Legendre-type transformation of the form

$$\rho\, c^2 = p\, \sigma - \Lambda \,. \tag{2.81}$$

The resulting expression for ρ as a function of n, $s^{\parallel}$, p will evidently have a general differential expressible as

$$c^2 d\rho = c^2 \mu \, dn + \Theta_{\parallel} \, ds^{\parallel} + \sigma \, dp \tag{2.82}$$

which leads naturally to the interpretation of the effective mass μ as the relativistic chemical potential and of $\Theta_{\parallel}$ as the temperature with respect to the Eckart frame. The third coefficient p appearing here is itself interpretable as the magnitude of the "effective 3-momentum per entropon" in the Eckart frame, whose covectorial form $\underline{p}$ is defined to be

$$p_{\mu} = \Theta_{\mu} - \Theta_{\parallel} c^{-2} u_{\mu} = \Theta_{\nparallel} c^{-2} v_{\mu} \,, \qquad \Theta_{\nparallel} c^{-2} = \frac{p}{v} = \mathcal{C} s^{\parallel} \tag{2.83}$$

so that the full thermal 4-momentum $\underline{\Theta}$ is expressible as

$$\Theta_{\mu} = c^{-2} (\Theta_{\parallel} u_{\mu} + \Theta_{\nparallel} v_{\mu}) \,, \tag{2.84}$$

the corresponding decomposition for the particle 4-momentum $\underline{\chi}$ being

$$\chi_{\mu} = \mu u_{\mu} - \frac{\alpha}{n} v_{\mu} \,, \tag{2.85}$$

where α is the anomally parameter that was introduced in our original presentation[10,11], whose definition and relationship to the anomally coefficient $\mathcal{A}$ of the variational presentation are given by

$$\alpha = s^{\parallel} (\Theta_{\nparallel} - \Theta_{\parallel}) c^{-2} = - s^{\parallel} n \mathcal{A} \,. \tag{2.86}$$

It can be seen that the regularity ansatz (2.42) is equivalent to the postulate that this quantity α should vanish, or in other words that the quantity denoted by $\Theta_{\nparallel}$ should be *the same* as the ordinary Eckart frame temperature $\Theta_{\parallel}$ as read out from (2.82). This contrasts with the situation in Eckart's own theory, which was based simply on the unfortunate supposition that ρ could be treated as a function only of n and $s^{\parallel}$, which tacitly amounted to the outright neglect of the $\sigma\, dp$ term in (2.78), and was thus equivalent to the postulate that the quantity denoted by $\Theta_{\nparallel}$ should be *zero.*

In so far as most of the situations to which the continuum models under consideration will be realistically applicable can involve at most *small* deviations from thermal equilibrium, so that it will be appropriate to use equations of motion that are linearised with respect to such deviations, one might naively suppose that since deviations from thermal equilibrium contribute to the mass-energy density ρ only at quadratic and higher order, their contribution in ρ might therefore safely be neglected. This commonly (albeit usually subconsciously) followed line of reasonning[33] is erroneous because it overlooks the fact that the quadratic order deviation terms give rise to derivative terms that are of the same, linear, order as other first order contributions to the equations of motion. In so much as their inclusion can affect the very sign of the coefficient a in the characteristic equation (2.67), the contribution to ρ of the terms of quadratic order in the deviation amplitudes σ and p is in fact of vital importance. By leaving them out altogether Eckart obtained a model that actually had the wrong (i.e. negative) sign for this coefficient a. Among all models that are of first order in the Hiscock-Lindblom sense, meaning that they are based on the neglect of such contributions in some (Eckart or other) frame, the least pathological would appear[13] to be that of Landau and Lifshitz[2], using the stress -momentum-energy eigenvector frame (whose 3-velocity $v_{\rm LL}$ say relative to the Echart frame works out as $v_{\rm LL} = (1 + n\mu/s^{\parallel}\Theta_{\parallel})v)$, which corresponds in the present Eckart-frame terminology to postulating that $\Theta_{\nparallel}$ should have the intermediate value $\Theta_{\nparallel} = \Theta_{\parallel}/(1 + n\mu/s^{\parallel}\Theta_{\parallel})$ (giving $a = 0$) between the Eckart value, $\Theta_{\nparallel} = 0$ (giving $a < 0$), and the regular value $\Theta_{\nparallel} = \Theta_{\parallel}$ that is compatible with the good behavior requirement $a > 0$.

The complete specification of the general model in Eckart-frame terminology starts with the prescription of the basic equation of state for ρ, as a function of n, $s^{\parallel}$, and p, thereby fixing secondary equations of state for μ, $\Theta_{\parallel}$, and σ by (2.82), and then goes on to specify the full stress-momentum-energy tensor in the standard form

$$T^{\mu\rho} = \rho u^{\mu} u^{\rho} + 2Q^{(\mu}u^{\rho)}c^{-2} + P^{\mu\rho} , \tag{2.87}$$

with

$$u_{\mu}Q^{\mu} = 0 , \qquad u_{\mu}P^{\mu\nu} = 0, \quad P^{[\mu\rho]} = 0 , \tag{2.88}$$

by prescribing the heat transfer vector $\vec{Q}$ to be

$$Q^{\mu} = \Theta_{\parallel}\sigma^{\mu} = \Theta_{\parallel}s^{\parallel}v^{\mu} , \tag{2.89}$$

and the pressure tensor $\mathbf{P}$ to be

$$P^{\mu\rho} = P\gamma^{\mu\rho} + p^{\mu}\sigma^{\rho} , \qquad \gamma^{\mu\rho} = g^{\mu\rho} + c^{-2}u^{\mu}u^{\rho} \tag{2.90}$$

where the Eckart-frame pressure scalar P is given by

$$P = s^{\parallel}\Theta_{\parallel} + (n\mu - \rho)c^2 . \tag{2.91}$$

The specification of the system is completed by giving the equations of motion, which are just the usual conservation laws

$$\nabla_{\mu}(nu^{\mu}) = 0 , \qquad \nabla_{\mu}T^{\mu\nu} = 0 , \tag{2.92}$$

together with a thermal conductivity equation whose Eckart-frame version may be written as

$$\gamma_\mu^\rho \nabla_\rho \Theta_{\|} + c^{-2}\Theta_{\|} \dot{u}_\mu + \dot{p}_\mu = -Y_{\mu\nu} Q^\nu \tag{2.93}$$

where $\mathbf{Y}$ is a positive-indefinite thermal resistivity tensor whose explicit form will be given below, and where we have used the notation

$$\dot{u}_\mu = \vec{u}\mathcal{L}u_\mu = u^\rho \nabla_\rho u_\mu \tag{2.94}$$

for the Eckart-frame acceleration, and

$$\dot{p}_\mu = \vec{u}\mathcal{L}p_\mu = u^\rho \nabla_\rho p_\mu + p_\rho \nabla_\mu u^\rho \tag{2.95}$$

for the rate of change of the thermal 3-momentum covector as analogously defined in terms of Lie differentiation with respect to the Eckart-frame unit vector $\vec{u}$. The first term in (2.93) is just a traditional Fourier type temperature-gradient driving force contribution. The second term is the relativistic frame-acceleration correction term that was discovered by Eckart himself[1], which clearly exhibits the effect of a thermal inertia $\mu_{\|}^{0}$ given by a formula of the same form, $\mu_{\|}^{0} = c^{-2}\Theta_{\|}$ as was shown in Section 1.3 to apply even in the non-conducting case: it is ironic that dispite his own revelation of the relativistic inertia associated with the transport of heat, Eckart's original theory was vitiated by his failure to take it into account in a systematic manner. It is the presence of the third term in (2.93) as specified by (2.95) that distinguishes the more general theory[10,11] given here from Eckart's original pathological special case.

Appart from an antisymmetric frame-adjustment contribution that has no effect on the rate of entropy generation as expressed by

$$\nabla_\mu s^\mu = Y_{\mu\rho}\sigma^\mu \sigma^\rho \ , \tag{2.96}$$

the Eckart-frame thermal resistance tensor is just proportional to the resistivity scalar Z as introduced in (2.32): its precise detailed expression is

$$Y_{\mu\rho} = Z\gamma_{\mu\rho} + \frac{2}{s\|c^2}\omega_{\mu\nu} + \frac{2}{s\|\Theta_{\|}}\gamma_\mu^{[\nu}\gamma_\rho^{\sigma]}\nabla_\nu p_\sigma \tag{2.97}$$

in which the first antisymmetric (non-disipative) frame adjusment contribution (interpretable as a thermal analogue of the well known Hall effect in electomagnetic theory) just takes account of the Eckart-frame Coriolis force proportional to the relevant kinematic rotation tensor $\underline{\omega}$ as defined by

$$\omega_{\mu\rho} = \gamma_\mu^{[\nu}\gamma_\rho^{\sigma]}\nabla_\nu u_\sigma \ . \tag{2.98}$$

The final term in (2.98) is of genuinely second order in deviations from thermal equilibrium and therefore (unlike the last term in (2.82) whose neglect by Eckart was disastrous) could safely and consistently be dropped in an approximate linearised treatment, being necessary only if we require an exactly consistent non-linear theory.

The need for the inclusion in (2.93) of some term that, like the term (2.95) obtained here, would be roughly proportional to an appropriately defined time derivative of the heat flux (thereby introducing a new timescale interpretable microscopically in terms of relaxation towards thermal equilibrium) was pointed out many years ago by Cattaneo, and various suggestions for the specification of such a term were put foreward on an ad hoc basis[3,4]. What the present approach does is to superceed such ad hoc methods by drawing attention to the existence of the uniquely natural regularity ansatz which gives a simple and unambiguous prescription for a term of the required type as given by substitution of

$$\Theta_{\nparallel} = \Theta_{\parallel} \tag{2.99}$$

in (2.83). A similar but smaller Cattaneo type term is also specified in a simple and unambigous way by the Landau-Lifshitz ansatz, which, instead of (2.99) corresponds to setting $\Theta_{\nparallel} = \Theta_{\parallel}/(1+n\mu/s^{\parallel}\Theta_{\parallel})$ in (2.83) but as we saw in the previous section, (as well as the physical disadvantage of being incompatible with the weakly coupled limit) this much smaller inertial contribution is marginally insufficient to restore hyperbolicity, and falls a long way short of what is needed for ordinary subluminal causality.

3 CONDUCTIVITY IN MULTICONSTITUENT FLUID OR SOLID MEDIA.

3.1 Mathematical requisites: convective differentiation.

The uniquely simple 2-constituent fluid model described in detail in the preceeding section can be generalised in a natural way for application to a wide range of more complicated situations. Because the viscous effects that are discussed in the accompanying course by Israel were not of dominant importance in the particular (low Prandtl number) astrophysical contexts that originally motivated my interest in conductivity (and because inclusion of a wider range of disipative effects inevitably entails more physical ambiguity than does allowance for other phenomena of a conservative nature) I have have not yet incorporated that aspect within the present approach, but have given priority to the physically more straightforeward problem of extending the treatment to multiconstituent (e.g. electrically as well as thermally) conducting fluids, and to the case when the entropy and other relevant fluxes are moving through a background that may be of elastic solid type (as is the case for the ionic lattice that is believed to occur in the crust of a neutron star).

The allowance for anisotropic (as opposed to isotropic fluid pressure type) elastic restoring forces makes it more difficult to work in a covariant manner (even in the restricted Galilean sense in Newtonian theory) and most treatments of solid media are heavily dependent on the use of comoving ("Lagrangian") coordinates X^I say ($I = 1, 2, 3$) (which are traditionally denoted by capitals to distinguish them from the general space-time coordinates x^μ say, $\mu = 1, 2, 3, 4$). For specific computational purposes the explicit use of such comoving coordinates is virtually inevitable, but it is nevertheless possible to retain a considerable degree of covariance in general theoretical analysis by using an approach pioneered by Oldroyd[34]. Although originally developped in the specialised context of a simple perfectly elastic solid[35], this methods of this "rheometric" approach are now available[36] in a form suitable for general purposes, including the treatment of conductivity with which we are concerned here.

The basic tool in the "rheometric" approach[36] is the concept of *convective variation* as defined with respect to the natural projection $\mathcal{P}$ say of the 4-dimensional space-time manifold, $\mathcal{M}$ on to the (not necessarily Hausdorff) 3-dimensional quotient manifold, $\mathcal{X}$ say, whose elements are the flow world-lines of the medium (as labelled by the comoving coordinates X^I) that are to be identified in the abstract with the corresponding set of idealised particles or lattice positions. (The possibility of $\mathcal{X}$ having a *non-Hausdorff* topological structure arises for example when one wishes to deal with a crystal that partially melts and then resolidifies, so that a marginal lattice point that has just survived intact may find that its neighbourhood at a later time is structurally quite different from that with which it started out.)

Using the symbol $\vec{u}$ (without any of the distinguishing marks we shall reserve for other rest frames to be introduced later on) for the unit tangent to the lattice flow lines, and introducing the corresponding orthogonal projection tensor

$$\gamma_\mu^\rho = g_\mu^\rho + c^{-2} u_\mu u^\rho \,, \qquad u_\mu u^\mu = -c^2 \,, \tag{3.1}$$

we may represent *any* (mixed) tensor $\mathbf{S}$ say, with components $S^{\mu\rho\ldots}_{\lambda\ldots}$ in terms of a set of *orthogonally projected* parts (whose contractions with u^μ or u_μ all vanish) to be denoted by $^\perp\mathbf{S}$, $\mathbf{S}^{\|}$, $\mathbf{S}_{.\|}$, $\mathbf{S}^{\|}_{.\|}$, et cetera, in the form

$$S^{\mu\rho\ldots}_{\lambda\ldots} = {}^\perp S^{\mu\rho\ldots}_{\lambda\ldots} + u^\mu S^{\|\rho\ldots}_{\lambda\ldots} + c^{-2} u_\lambda S^{\mu\rho\ldots}_{\|\ldots} + c^{-2} u^\mu u_{\lambda\ldots} S^{\|\rho\ldots}_{\|\ldots} + \cdots \,. \tag{3.2}$$

with

$$\begin{aligned} {}^\perp S^{\mu\rho\ldots}_{\lambda\ldots} &= S^{\nu\sigma\ldots}_{\kappa\ldots}\gamma_\nu^\mu \gamma_\lambda^\kappa \gamma_\sigma^\rho \ldots \,, \qquad S^{\|\rho\ldots}_{\lambda\ldots} = -c^{-2} u_\mu S^{\mu\sigma\ldots}_{\kappa\ldots}\gamma_\lambda^\kappa \gamma_\sigma^\rho \ldots \,, \\ S^{\mu\rho\ldots}_{\|\ldots} &= -u^\lambda S^{\nu\sigma\ldots}_{\lambda\ldots}\gamma_\nu^\mu \gamma_\sigma^\rho \ldots \,, \quad \cdots \,. \end{aligned} \tag{3.3}$$

Since any such orthogonally projected part can be unambiguously represented, via the projection $\mathcal{P}$, by a unique corresponding 3-dimensional tensor in $\mathcal{X}$, ${}^\perp S^{\mu\nu\ldots}_{\lambda\ldots} \leftrightarrow {}^\perp S^{JK\ldots}_{I\ldots}$, et cetera, the arbitrary non-orthogonal tensor with which we started can correspondingly be unambiguously represented by a *finite set* of such 3-dimensionally projected image tensors:

$$\mathcal{P}: \quad S^{\mu\rho\ldots}_{\lambda\ldots} \leftrightarrow \{{}^\perp S^{JK\ldots}_{I\ldots}, S^{\|K\ldots}_{I\ldots}, S^{JK\ldots}_{\|\ldots}, S^{\|K\ldots}_{\|\ldots}, \cdots\} \tag{3.4}$$

This way of representating $\mathcal{X}$ enables us to make unambiguous comparisons between values of $\mathbf{S}$ at different points on the same world line or for different values of the gravitational metric $\mathbf{g}$ at the same point (independently of, and in general quite distinctly from any notion of parallel or Fermi transport). In particular, for infinitesimal variations we define the *convected differential*, which we shall denote here by d_C, in terms of corresponding base-space differentials (for which no ambiguity arises since they are evaluated directly at a fixed position in X) which will be denoted simply by d, in the obviously corresponding manner by

$$\mathcal{P}: \quad d_C S^{\mu\rho\ldots}_{\lambda\ldots} \leftrightarrow \{d^{\perp} S_I^{JK\cdots}, dS^{\|K\cdots}_{I\ldots}, dS^{JK\cdots}_{\|\ldots}, dS^{\|K\cdots}_{\|\ldots}, \cdots\} \tag{3.5}$$

This concept is specially adapted for studying the working of *equations of state*, which must be formulated in an "objective" manner, i.e. independently of the space-time configuration of particular solutions of the dynamic equations of motion, which (except in the purely fluid limit case for which only scalars are involved) means that they need to be expressed ultimately in terms of tensor (or for more sophisticated purposes even more general) structures defined intrinsically on the 3-dimensional particle or lattice space $\mathcal{X}$ which represents the medium in the abstract sense.

For the purpose of studying *equations of motion* (as opposed to equations of state) howeveer, it is usually most convenient to think in terms of *Eulerian* variations, which we shall denote by d_E, as evaluated at a point that is held "fixed" in space-time (according to some – inevitably not generally covariant – prescription). As a convenient intermediate stepping stone between convective and general Eulerian variations, it is useful also to work with *Lagrangian* variations, which we shall denote by d_L, and which may be considered as a special subcategory of Eulerian variations that are characterised by the requirement that the meaning of the word "fixed" is restricted to include the condition "on the world-line of the same idealised particle or lattice position", which leaves only the time-variation to be fixed by a freely chosen prescription. In the strictly infinitesimal case the relative displacement between a (freely) prescribed Eulerian comparison point an associated (comoving) Lagrangian comparison point will be represented by a corresponding vector field, $\vec{\xi}$ say, and the difference between the two variations will be given in terms of this vector field by

$$(d_L - d_E) = \vec{\xi}\mathcal{L} \tag{3.6}$$

(which may be memorised as "Lagrange less Euler is Lie"[30]) so that in the case of an ordinary tensorial field of the form (3.2) we have

$$(d_L - d_E)S^{\mu\rho\ldots}_{\lambda\ldots} = (\vec{\xi}\cdot\nabla)S^{\mu\rho\ldots}_{\lambda\ldots} + S^{\mu\rho\ldots}_{\alpha\ldots}\nabla_\lambda\xi^\alpha + \cdots$$
$$- S^{\beta\rho\ldots}_{\lambda\ldots}\nabla_\beta\xi^\mu - \cdots . \tag{3.7}$$

Although the Lagrangian variations denoted by d_L are closer than the general Eulerian ones denoted by d_E to the convected variations denoted by d_C that are obtained from the equations of state, a certain amount of work is required, at least in the relativistic case, to make their relationship explicit. What one obtains[36] is expressible as

$$(d_C - d_L) S^{\mu\rho\ldots}_{\lambda\ldots} = S^{\mu\rho\ldots}_{\alpha\ldots} c^{-2} u^\alpha d_L u_\lambda + \cdots$$
$$- S^{\beta\rho\ldots}_{\lambda\ldots} c^{-2} u^\mu d_L u_\beta - \cdots \tag{3.8}$$

where, as in (3.7) there is a positively signed term for each covariant index and a negatively signed term for each contravariant index. The Lagrangian variation of the contravariant unit flow velocity 4-vector $\vec{u}$ is very simple, consisting of at most a metric scale adjustment,

$$d_L u^\mu = \frac{1}{2c^2} u^\mu u^\nu u^\sigma d_L g_{\nu\sigma} \,, \tag{3.9}$$

but the formula (3.8) involves the Lagrangian variation of the associated covector $\underline{u}$, which is given[30] by the slightly more complicated expression

$$d_L u_\mu = (\gamma^\nu_\mu - \frac{1}{2c^2} u_\mu u^\nu) u^\sigma d_L g_{\nu\sigma} \,. \tag{3.10}$$

As an elementary application of the foregoing general formula one has

$$d_C g_{\mu\rho} = d_L \gamma_{\mu\rho} \,, \qquad d_L \gamma_{\mu\rho} = \gamma^\nu_\mu \gamma^\sigma_\rho d_L g_{\nu\sigma} \,. \tag{3.11}$$

An important class of applications of the foregoing formalism consists of those for which the Lagrangian displacement is simply generated by a proper time displacement, $d\tau$ say, along the world lines, i.e. by taking

$$d_L = (\vec{u} d\tau) \mathcal{L} \,. \tag{3.12}$$

In this case, even if $d\tau$ is a variable function of position, one obtains a simple proportionality relation[35] of the form

$$d_C S^{\mu\rho\ldots}_{\lambda\ldots} = (d\tau) D_C S^{\mu\rho\ldots}_{\lambda\ldots} \,, \tag{3.13}$$

in which the gradient of $d\tau$ is not involved, where the *convective time derivative* denoted here by D_C is given explicitly in terms of ordinary Riemannian covariant differentiation by

$$(D_C - \vec{u}\cdot\nabla) S^{\mu\rho\ldots}_{\lambda\ldots} = S^{\mu\rho\ldots}_{\alpha\ldots} (c^{-2} \dot{u}_\lambda + \nabla_\lambda) u^\alpha + \cdots$$
$$- S^{\beta\rho\ldots}_{\lambda\ldots} (c^{-2} \dot{u}_\beta + \nabla_\beta) u^\mu - \cdots \tag{3.14}$$

where $\underline{\dot{u}}$ is the acceleration covector of the flow, as defined in the usual way by

$$\dot{u}_\mu = (\vec{u}\mathcal{L}) u_\mu = (\vec{u}\cdot\nabla) u_\mu \,. \tag{3.15}$$

For our present purpose the application with which we shall be principally concerned is the study of variations of a Lagrangian density scalar, Λ say. In so far as it is not given merely in terms of scalar functions of the relevant fields (as in the purely fluid case of the previous section) the objective specification of such a Λ must necessarily be expressed exclusively in terms of the projections of the relevant fields on the lattice 3-manifold $\mathcal{X}$, which is the only receptacle of the necessary "objective memory" of the medium. Without (yet) restricting ourselves to any particular model theory, let us now consider any variational scalar Λ that is given as a function of the metric g and of at least one othe tensorial field **S** say. To be well defined in a manner that is "objective" in the sense of being strictly *intrinsic* to the medium, the dependence must be expressible in terms of the corresponding projections onto $\mathcal{X}$ under $\mathcal{P}$, i.e. in terms of the components γ_{IJ} (the metric has no other variable projected components) and the finite sequence ${}^{\perp}S^{JK\cdots}_{\ I\ldots}$, $S^{\|K\cdots}_{I\ldots}$, et cetera. This means that a general infinitesimal variation of Λ will be expressible in the form

$$d\Lambda = \frac{\partial\Lambda}{\partial\gamma_{IJ}}\, d\gamma_{IJ} + \frac{\partial\Lambda}{\partial\, {}^{\perp}S^{JK\cdots}_{\ I\ldots}}\, d\, {}^{\perp}S^{JK\cdots}_{\ I\ldots} + \frac{\partial\Lambda}{\partial S^{\|K\cdots}_{I\ldots}}\, d\, S^{\|K\cdots}_{I\ldots} + \cdots . \tag{3.16}$$

The formalism of convected derivation not only allows us to translate this directly into 4-dimensional space-time tensorial notation as

$$d_{C}\Lambda = \frac{\partial\Lambda}{\partial\gamma_{\nu\varphi}}\, d_{C}\gamma_{\nu\varphi} + \frac{\partial\Lambda}{\partial\, {}^{\perp}S^{\mu\rho\ldots}_{\ \lambda\ldots}}\, d_{C}\, {}^{\perp}S^{\mu\rho\ldots}_{\ \lambda\ldots} + \frac{\partial\Lambda}{\partial S^{\|\rho\ldots}_{\lambda\ldots}}\, d_{C} S^{\|\rho\ldots}_{\lambda\ldots} + \cdots . \tag{3.17}$$

under $\mathcal{P}^{-1}$, but also allows us to compactify this expression in the form

$$d_{C}\Lambda = \frac{\partial\Lambda}{\partial\gamma_{\nu\varphi}}\, d_{C}\gamma_{\nu\varphi} + \frac{\partial\Lambda}{\partial S^{\mu\rho\ldots}_{\ \lambda\ldots}}\, d_{C} S^{\mu\rho\ldots}_{\ \lambda\ldots} \tag{3.18}$$

where

$$\frac{\partial\Lambda}{\partial S^{\mu\nu\ldots}_{\ \lambda\ldots}} = \frac{\partial\Lambda}{\partial^{\perp} S^{\mu\nu\ldots}_{\ \lambda\ldots}} - \left(u_{\mu}\frac{c^{-2}\partial\Lambda}{\partial S^{\|\rho\ldots}_{\lambda\ldots}} + u^{\lambda}\frac{\partial\Lambda}{\partial S^{\mu\rho\ldots}_{\ \|\ldots}} + \cdots \right)$$
$$+ \left(c^{-2} u_{\mu} u^{\lambda} \frac{\partial\Lambda}{\partial S^{\|\rho\ldots}_{\|\ldots}} + \cdots \right) - \cdots . \tag{3.19}$$

Our formula (3.8) can now be used to convert this into terms of ordinary Lagrangian variations in the form

$$d_{L}\Lambda = \frac{\partial\Lambda}{\partial g_{\nu\varphi}}\, d_{L} g_{\nu\varphi} + \frac{\partial\Lambda}{\partial S^{\mu\rho\ldots}_{\ \lambda\ldots}}\, d_{L} S^{\mu\rho\ldots}_{\ \lambda\ldots} \tag{3.20}$$

where the second coefficient is given by (3.19) and the first is given by

$$\frac{\partial\Lambda}{\partial g_{\nu\varphi}} = \frac{\partial\Lambda}{\partial\gamma_{\nu\varphi}}$$
$$-\frac{u^{\alpha}}{c^{2}}\left(S^{\kappa\rho\ldots}_{\ \lambda\ldots}\frac{\partial\Lambda}{\partial S^{\alpha\rho\ldots}_{\ \lambda\ldots}} + \cdots - S^{\mu\rho\ldots}_{\ \alpha\ldots}\frac{\partial\Lambda}{\partial S^{\mu\rho\ldots}_{\ \kappa\ldots}} + \cdots \right)\left(\gamma_{\kappa}^{(\nu} u^{\varphi)} - u_{\kappa}\frac{u^{\nu}u^{\varphi}}{2c^{2}}\right) . \tag{3.21}$$

3.2 Multi-worldline displacement variation formulation for conducting medium.

In order to set up a theory from scratch, the experience of the work in the preceeding sections would suggest that a very generally useful recipe for a mathematically consistent and elegant result (and one that is unlikely to let us down even from the more objective point of view of physical applicability) is to start on the basis of a variational formulation, since even if dissipative effects must then be added in later "by hand", the result will still be better than building the entire model "by hand" from the outset. This means that until one comes to add in dissipation coefficients, which require subsidiary equations of state of their own, one need only to postulate a single equation of state function, which by itself will govern the characteristic (i.e. causal) structure, though not the stability, which is affected by dissipation. The dependent function appearing in this primary equation of state may be chosen in various ways (e.g. as the energy density with respect to some preferred reference system) but the most generally covariant choice (whose earlier use would, as we have seen, have avoided the confusion that arose as to whether the Landau-Lifshitz theory is different from the Eckart theory) is to work with the Lagrangian action density scalar Λ itself.

The general purpose convective variation formalism described in the preceeding section was originally developed for handling comparatively complicated electromagnetic tensors in a polarisable medium, but for our present investigation of the simplest kinds of conducting model we need only consider the cases in which, appart from the metric itself, Λ depends just on a set of current vectors $\vec{n}_X$ where (as in our earlier discussion of the non-conducting fluid case in the first part of this course) X is an index running over several values of which one, let us say $X = 0$ so as to conform to the convention adopted earlier, corresponds to the *entropy*, while others (given by strictly positive integer values of X) might represent correspond to whatever independent particle species (such as conduction electrons for example) may be relevant. At the level of the 3-dimensional base space $\mathcal{X}$ that carries the intrinsic structure of the medium, we shall therefore have an action variation expression of the form

$$d\Lambda = \frac{\partial \Lambda}{\partial \gamma_{IJ}}\, d\gamma_{IJ} + \frac{\partial \Lambda}{\partial\, {}^{\perp}n_X{}^I}\, d\,{}^{\perp}n_X{}^I + \frac{\partial \Lambda}{\partial n_X^{\parallel}}\, d\, n_X^{\parallel} \tag{3.22}$$

where, as usual, summation over all relevant values of X is understood. Translated into the notation of convected differentials in ordinary 4-dimensional space-time, this becomes

$$\begin{aligned} d_C\Lambda &= \frac{\partial \Lambda}{\partial \gamma_{\mu\rho}}\, d_C\gamma_{\mu\rho} + \frac{\partial \Lambda}{\partial\, {}^{\perp}n_X^{\mu}}\, d_C^{\perp} n_X^{\mu} + \frac{\partial \Lambda}{\partial n_X^{\parallel}}\, d_C n_X^{\parallel} \\ &= \frac{\partial \Lambda}{\partial \gamma_{\mu\rho}}\, d_C\gamma_{\mu\rho} + \frac{\partial \Lambda}{\partial n_X^{\mu}}\, d_C n_X^{\mu} \end{aligned} \tag{3.23}$$

with

$$\frac{\partial \Lambda}{\partial n_X^{\mu}} = \frac{\partial \Lambda}{\partial\, {}^{\perp}n_X^{\mu}} - \frac{u_\mu}{c^2}\frac{\partial \Lambda}{\partial n_X^{\parallel}} \tag{3.24}$$

where

$${}^{\perp}n_X^{\mu} = n_X^{\nu}\gamma_\nu^{\mu}\,, \qquad n_X^{\parallel} = -c^{-2}u_\mu n_X^{\mu}\,. \tag{3.25}$$

It therefore follows from the general formula (3.19) that the corresponding Lagrangian variation formula will be expressible as

$$d_L\Lambda = \chi^{\rm x}_{\mu}\, d_L n^{\mu}_{\rm x} + \frac{\partial\Lambda}{\partial g_{\mu\nu}}\, d_L g_{\mu\nu} \tag{3.26}$$

where the coefficients are given in terms of those appearing in (3.23) by

$$\chi^{\rm x}_{\mu} = \frac{\partial\Lambda}{\partial n^{\mu}_{\rm x}} \tag{3.27}$$

and

$$\frac{\partial\Lambda}{\partial g_{\mu\nu}} = \frac{\partial\Lambda}{\partial\gamma_{\mu\nu}} - \frac{1}{c^2} u^{\rho}\chi^{\rm x}_{\rho} n^{\sigma}_{\rm x}(\gamma_{\sigma}^{\ (\mu} u^{\nu)} - \frac{1}{2c^2} u_{\sigma} u^{\mu} u^{\nu}) \,. \tag{3.28}$$

The physical interpretation of the 4-momentum covectors defined by (3.27) can be clarified by writing them in the decomposed form

$$\chi^{\rm x}_{\mu} = \mu^{\rm x} u_{\mu} + p^{\rm x}_{\mu} \tag{3.29}$$

in which the *effective inertia* $\mu^{\rm x}$ and the *3-momentum* $\underline{p}^{\rm x}$ associated with each species are defined by setting

$$\frac{\partial\Lambda}{\partial n^{\|}_{\rm x}} = -\mu^{\rm x} c^2 \,, \qquad \frac{\partial\Lambda}{\partial\, {}^{\perp}n^{\mu}_{\rm x}} = p^{\rm x}_{\mu} \,. \tag{3.30}$$

In so much as we are restricting our attention to the simplest kind of situation in which effects of polarisation[36] are excluded, we we may suppose that any external electromagnetic forces acting on the medium can be allowed for simply by supplementing the internal Lagrangian action density scalar Λ by an electromagnetic coupling term of the usual form

$$\Lambda_{\rm elec} = j^{\mu} A_{\mu} \tag{3.31}$$

where $\underline{A}$ is the electromagnetic 4-potential (as introduced by 1.61)) and $\vec{j}$ is the total electric current 4-vector, which will have the form

$$j^{\mu} = e n^{\mu} + e^{\rm x} n^{\mu}_{\rm x} \,, \tag{3.32}$$

where for each conduction current the constant $e^{\rm x}$ is the corresponding charge per particle, while e is the analogous fixed charge per lattice point, in terms of a normalisation convention with respect to which

$$n^{\mu} = n u^{\mu} \tag{3.33}$$

is the corresponding current density of lattice points, as defined in terms of some 3-volume measure on the base space $\mathcal{X}$ that will determine the number density as a function of the base projected metric with components γ_{IJ} by an equation of state of the form $n = \nu\|\gamma\|^{\frac{1}{2}}$ for some scalar ν defined as a constant field on the material base space $\mathcal{X}$, thus inducing a spacetime field ν that is constant along the world-lines, $D_C\nu = 0$ in the notation of (3.13), as required for consistency with the lattice point number conservation law

$$\nabla_{\mu}(n u^{\mu}) = 0 \tag{3.34}$$

Under these circumstances the analogue of (3.26) for the total Lagrangian will be expressible as

$$d_L(\Lambda + \Lambda_{\rm elec}) = \pi^{X}_{\mu}\, d_L n^{\mu}_{X} + (\frac{\partial \Lambda}{\partial g_{\mu\nu}} - \frac{1}{2} e n u^{\rho} A_{\rho} g^{\mu\nu})\, d_L g_{\mu\nu} + j^{\mu} d_L A_{\mu} \tag{3.35}$$

where the total 4-momenta as thus defined are given by

$$\pi^{X}_{\mu} = \chi^{X}_{\mu} + e^{X} A_{\mu}\ . \tag{3.36}$$

Let us now adopt the same principles that were used for our earlier perfect fluid model (so that it will be included in the present model as a special case) by supposing that appart from Eulerian variations $d_E \mathbf{g}$ of the gravitational background, the only independent variations to be considered are those produced by displacements $\vec{\xi}$ of the world-lines of the flow (with tangent vector $\vec{u}$ of the basic material medium itself, and by independent displacements $\vec{\xi}_X$ of the world-lines of the flows (with tangent vectors $\vec{n}_X$) of the relevant currents, so that (by (3.6)) the Lagrangian variations required for the evaluation of (3.26) will be given by

$$d_L g_{\mu\nu} = d_E g_{\mu\nu} + 2\nabla_{(\mu}\xi_{\nu)} \tag{3.37}$$

and

$$d_L n^{\mu}_{X} = -n^{\mu}_{X}(\tfrac{1}{2} g^{\mu\nu} d_L g_{\mu\nu} + \nabla_{\nu}\eta^{\nu}_{\{X\}}) + n^{\nu}_{X}\nabla_{\nu}\eta^{\mu}_{\{X\}} - \eta^{\nu}_{\{X\}}\nabla_{\nu} n^{\mu}_{X} \tag{3.38}$$

where we have introduced the use of curly brackets $\{\}$ to distinguish indices that are to be left out of account as far as the activation of the summation convention is concerned, and where we have used the abbreviation

$$\eta^{\mu}_{X} = \xi^{\mu}_{X} - \xi^{\mu}\ . \tag{3.39}$$

The expression (3.32) can be seen (again using (3.6)) to correspond to a purely Eulerian current variation expression of the same form

$$d_E n^{\mu}_{X} = -\frac{1}{2} n^{\mu}_{X} g^{\mu\nu} d_E g_{\mu\nu} + n^{\nu}_{X}\nabla_{\nu}\xi^{\mu}_{\{X\}} - \nabla_{\nu}(n^{\mu}_{X}\xi^{\nu}_{\{X\}}) \tag{3.40}$$

as has already been used in (2.20).

We have now assembled all the pieces needed for the evaluation (using (3.6)) of the Eulerian variation of the Lagrangian integrand given by the scalar-density $\|\mathbf{g}\|\Lambda$, which works out to have the form

$$\|\mathbf{g}\|^{-\frac{1}{2}} d_E(\Lambda\|\mathbf{g}\|^{\frac{1}{2}}) = \frac{1}{2} T^{\mu\nu} d_E g_{\mu\nu} + f_{\mu}\xi^{\mu} + f^{X}_{\mu}\xi^{\mu}_{X}$$

$$+\nabla_{\mu}\{(T^{\mu\nu} - \Lambda g^{\mu\nu})\xi_{\nu} + 2n^{[\mu}_{\{X\}}\eta^{\nu]}_{X}\pi^{X}_{\nu}\} \tag{3.41}$$

where the variational stress-momentum-energy tensor is read out as

$$T^{\mu\nu} = 2\frac{\partial \Lambda}{\partial g_{\mu\nu}} + (\Lambda - \chi^{X}_{\sigma} n^{\sigma}_{X}) g^{\mu\nu} \tag{3.42}$$

and the force-densities of the conduction currents are read out in the form (which by our previous work should by now be familiar)

$$f^{X}_{\mu} = \chi^{X}_{\mu}\nabla_{\nu}n^{\ \nu}_{\{X\}} + 2n^{\ \nu}_{\{X\}}\nabla_{[\nu}\chi^{X}_{\ \mu]} \tag{3.43}$$

while finally the force acting back on the medium itself will be given by

$$f_{\mu} = \nabla_{\nu}T^{\nu}_{\ \mu} - \sum_{X} f^{X}_{\mu}\,, \tag{3.44}$$

this force balance relation being interpretable as the Noether identity resulting from general covariance. The consideration that the system is unaffected by displacements of the flow-lines along themselves (by setting $\vec{\xi} = (d\tau)\vec{u}$) leads to the additional Noether identity

$$f_{\mu}u^{\mu} = 0\,. \tag{3.45}$$

For purposes of interpretation, it is interesting to convert the expression (3.36) defining the variational stress-momentum-energy tensor to the less covariant but more familiar orthogonally decomposed form

$$T^{\mu\nu} = \rho u^{\mu}u^{\nu} + \frac{2}{c^2}Q^{[\mu}u^{\nu]} + P^{\mu\nu} \tag{3.46}$$

subject to the usual stipulations

$$P^{[\mu\nu]} = 0\,, \qquad P^{\mu\nu}u_{\nu} = 0\,, \qquad Q^{\nu}u_{\nu} = 0\,. \tag{3.47}$$

One finds that the pressure tensor defined in this way is given by

$$P^{\mu\nu} = 2\frac{\partial\Lambda}{\partial\gamma_{\mu\nu}} + (\Lambda - \chi^{X}_{\sigma}n^{\ \sigma}_{X})\gamma^{\mu\nu} \tag{3.48}$$

and that the energy transport vector is given by

$$Q^{\mu} = \mu^{X}c^{2}\,{}^{\perp}n^{\ \mu}_{X} \tag{3.49}$$

while finally the mass density in the rest frame of the medium will be given by

$$\rho = {}^{\perp}n_{X}p^{X}_{\ \mu} - \Lambda \tag{3.50}$$

The last of these formulae shows that all the quantities involved could have been obtained in an alternative (less covariant) approach based on the equation of state for ρ, starting from its fundamental variation formula, as evaluated in the 3-dimensional material base space $\mathcal{X}$, which can be seen to have the form

$$d\rho = \mu^{X}dn^{\parallel}_{X} + {}^{\perp}n^{\ I}_{X}\,dp^{X}_{\ I} - \frac{\partial\Lambda}{\partial\gamma_{IJ}}\,d\gamma_{IJ} \tag{3.51}$$

To determine the dynamics of the model, all that remains to be specified is a law prescribing the form of the relevant force densities as defined by (3.43) and (3.44). The strictly conservative limit case of the model may be obtained by postulating the rigorous application of the variational principle, which amounts simply to the requirement that the back force, $\underline{f}$ acting on the medium, as well as the forces $\underline{f}^{X}$ acting on the relevant conduction currents, should all vanish, which by (3.44) will automatically ensure "covariant conservation" of the variational stress-momentum-energy tensor as given by (3.42). To allow for dissipative effects resulting from non-zero resistive forces and chemical reactions we need a more general ansatz, which must of course be compatible with the identity (3.45) and which must also satisfy the force balance condition

$$f_\mu + \sum_X f^X_{\ \mu} = f^{\rm ext}_\mu \tag{3.52}$$

if we are to satisfy the condition

$$\nabla_\nu T^{\nu\mu} = f^{\rm ext}_\mu \tag{3.53}$$

where $\underline{f}^{\rm ext}$ is the external force acting on the medium, which, assuming it is just due to the effect of the electromagnetic coupling (3.31), will be given simply by

$$f^{\rm ext}_\mu = j^\nu F_{\nu\mu}\,. \tag{3.54}$$

A natural ansatz for doing this in a manner that is also consistent with the "second law of thermodynamics" will be described in the final subsection that follows.

3.3 Chemical and Resistive dissipation.

This final subsection will describe the way to incorporate the treatment of dissipative effects of two physically different kinds within the general variational framework that has just been set up: as well as allowing for resistance between the currents we shall also allow for the possibility of "chemical" (which, in the astrophysical applications such as supernovae that I have in mind, is in practice likely to mean *nuclear*) interactions between the constituent currents.

It is at this stage that we must cease to treat the entropy, with current vector $\vec{n}_0 = \vec{s}$ say, and 4-momentum covector $\underline{\chi}^0 = \underline{\Theta}$ say, on the same footing as the other "material" currents involved, for which (in accordance with the convention adopted in the first part of this course) we shall use early alphabetic index letters A, B which, unlike the index letter X used in the previous subsection, are restricted to run only over strictly positive chemical index values, so as to exclude the zero value that has been chosen to label the entropy. The total force equilibrium (stress-energy-momentum "conservation") requirement (3.46) is thus expressible in this rather more explicit notation system as

$$-f^0_{\ \mu} = f_{\ \mu} + \sum_A f^A_{\ \mu} - f^{\rm ext}_\mu \tag{3.55}$$

where the term on the left represents the thermal force density, the first term on the right represents the force density effectively acting back on the material lattice ofthe medium, the terms in the summation over the allowed (strictly positive) values of A are the force densities acting on the other (non-thermal) conduction currents that may be relevant, and the final term on the right is the external force contribution.

Since we want to make sure that our model is consistent with the entropy inequality (1.77) expressing the "second law", we see that it will be advantageous to work in terms of a *thermal reference frame*, characterised by the timelike unit vector $\vec{u}_0$ given by

$$s^\mu = s^\emptyset u_0^{\ \mu} \, , \qquad s^\emptyset = s \, , \qquad u_0^\mu u_{0\mu} = -c^2 \tag{3.56}$$

where we introduce the convention that components defined with respect to the thermal reference frame are to be indicated by a the symbol $\emptyset$ (as distinct from the symbol $||$ symbol used analogously in the preceeding subsections for the material medium rest frame). This kind of reference frame derives its importance from the fact that it is the most obviously natural all-purpose generalisation to non-equilibrium situations of the unambiguously preferred common rest frame of a thermal equilibrium state . This leads to the definition of the natural *thermal temperature*, $\Theta_\emptyset$, the most natural out-of-equilibrium generalisation of the unambiguously defined thermal equilibrium temperature, as the thermal-frame component

$$\Theta_\emptyset = -u_0^\mu \Theta_\mu \, . \tag{3.57}$$

Thus introducing the corresponding inverse temperature vector

$$\beta_0^\mu = \frac{1}{\Theta_\emptyset} u_0^\mu = \frac{s^\mu}{(-s^\nu \Theta_\nu)} \tag{3.58}$$

we see (by (3.43) that the second law inequality is expressible by

$$\nabla_\mu s^\mu = -f^0{}_\mu \beta_0^\mu \geq 0 \, . \tag{3.59}$$

Hence decomposing each of the relevant forces as the sum of the part that is of purely internal origin, which (generalising the notation scheme of section 2.1) will be indicated by a tilde , and the external electromagnetic part, in the form

$$f_\mu = \tilde{f}_\mu + enF_{\mu\nu}u^\nu \, , \tag{3.60}$$

and

$$f^A{}_\mu = \tilde{f}^A{}_\mu + e^A F_{\mu\nu} n_A^\nu \tag{3.61}$$

(and noting that there is no external contribution to the entropy force, $\underline{f}^0$ since the entropy current is presumed to have no associated charge, $e^0 = 0$) we see from (3.44) that (3.55) can be rewritten as

$$-f^0{}_\mu = -\tilde{f}^0{}_\mu = \tilde{f}_\mu + \sum_A \tilde{f}^A{}_\mu \tag{3.62}$$

and hence that the "second law" requirement (3.59) is equivalent to the condition that internal contributions to the non-thermal forces should be such that

$$(\tilde{f}_\mu + \sum_A \tilde{f}^A{}_\mu)\beta_0^\mu \geq 0 \, . \tag{3.63}$$

To convert this basic thermodynamic requirement into a more tractable form, the thermal reference system can be used for defining a convenient decomposition of each of the relevant currents, starting with the conserved lattice-point current, whose thermally decomposed form is

$$n^\mu = nu^\mu = n(u_0^\mu + v_0^\mu) \,, \qquad n = \gamma_0 n \,, \qquad v_{0\mu} u_0^\mu = 0 \,, \tag{3.64}$$

while similarly for the conduction currents we shall have

$$n_A^{\ \mu} = n_A u_{\{A\}}^\mu = n_A (u_0^\mu + v_{0\{A\}}^{\ \mu}) \,, \qquad n_A = \gamma_{0\{A\}} n_A \,, \qquad v_{0A\mu} u_0^\mu = 0 \,, \tag{3.65}$$

where the Lorentz factors are given explicitly by

$$\gamma_0 = (1 - v_{0\mu} v_0^\mu c^{-2})^{-\frac{1}{2}} \,, \qquad \gamma_{0A} = (1 - v_{0A\mu} v_{0\{A\}}^{\ \mu} c^{-2})^{-\frac{1}{2}} \,, \tag{3.66}$$

We may use this terminology to express $\vec{\beta}_0$ diversely as

$$\beta_0^\mu = \frac{1}{\Theta_\emptyset}\Big(\frac{1}{\gamma_0} u^\mu - v_0^\mu\Big) = \frac{1}{\Theta_\emptyset}\Big(\frac{1}{\gamma_{0\{A\}}} u_A^{\ \mu} - v_{0A}^{\ \mu}\Big) \tag{3.67}$$

thereby using (3.62) and (3.45) to obtain

$$f_\mu^0 \beta_0^\mu = \frac{1}{\Theta_0}\Big(\tilde{f}_\mu v_0^\mu + \tilde{f}_\mu^A v_{0A}^{\ \mu} - \frac{1}{\gamma_{0\{A\}}} f_\mu^A u_A^{\ \mu}\Big) \tag{3.68}$$

The "second law" requirement (3.63) to the effect that the quantity in (3.67) be non-positive is therefore expressible, using (3.43), as the requirement that the non-thermal forces should be such as to satisfy

$$-(\tilde{f}_\mu v_0^\mu + \tilde{f}^A v_{0A}^{\ \mu}) \geq \frac{\chi_\natural^A}{\gamma_{0\{A\}}} \nabla_\nu n_A^{\ \nu} \tag{3.69}$$

where we have introduced the covariantly defined "natural" chemical potential of each species as the component of the corresponding 1-form $\underline{\chi}^X$ along the direction of the corresponding current $\vec{n}_X$, so that

$$\chi_\natural^A = -u_{\{A\}}^{\ \mu} \chi^A_{\ \mu} \,, \qquad u_A^{\ \mu} = \frac{n_A^{\ \mu}}{n_{\{A\}}} \tag{3.70}$$

which one would in general expect to be positive, the sign of the inequality in (3.69) being actually dependent on the presumption of positivity of the analogously defined "natural" temperature $\Theta_\natural$, which is the same as the thermal rest frame temperature Θ already introduced, i.e. on the supposition

$$\Theta_\natural = \Theta_\emptyset > 0 \tag{3.71}$$

which may safely be assumed to hold in any realistic application.

Now if *all* the currents $\vec{n}_A$ are conserved, the right hand side of (3.69) will be zero, which suggests in the usual way that we should postulate a linear law relating the negatives of the relative velocity vectors via a positive resistivity matrix to the orthogonally projected components

$$^{\dashv}\tilde{f}_\mu = \gamma_{0\mu}^{\ \nu} \tilde{f}_\nu \,, \qquad ^{\dashv}\tilde{f}_\mu^A = \gamma_{0\mu}^{\ \nu} \tilde{f}_\nu^A \tag{3.72}$$

of the forces, using the notation

$$\gamma_{0\mu}^{\ \ \nu} = g_\mu^\nu + c^{-2} u_\mu u^\nu \tag{3.73}$$

and using the symbol $\dashv$ to indicate orthogonal projection with respect to the thermal rest-frame (as distinct from orthogonal projection with respect to the material rest-frame, which we have indicated by the symbol $\perp$). Such a relation will thus be expressible as

$$\begin{pmatrix} {}^{\dashv}\tilde{f}_\mu \\ {}^{\dashv}\tilde{f}^A_\mu \end{pmatrix} = -\begin{pmatrix} K_{\mu\nu} & K_\mu{}^B{}_\nu \\ K_\mu{}^A{}_\nu & K_\mu{}^{AB}{}_\nu \end{pmatrix} \begin{pmatrix} v_0^\nu \\ v_{0B}^{\ \nu} \end{pmatrix} \tag{3.74}$$

in which, for the simplest applications, we may plausibly assume that the matrix satisfies an Onsager type summetry condition with the coefficients $K_{\mu\nu}$, $K_\mu{}^A{}_\nu$, $K_\mu{}^{AB}{}_\nu$ given as algebraic equations of state, i.e. as functions of the (material projections of the) space-time metric and the conduction currents, being isotropic (proportional to $\gamma_{\mu\nu}$) in the thermal equilibrium (zero v^μ and $v_A^{\ \mu}$) limit of special case of a strictly fluid medium. It follows from (3.62) that the ensuing drag on the heat current will be given by

$${}^{\dashv}f^0_\mu = (K_{\mu\nu} + \sum_A K_\mu{}^A{}_\nu)v_0^\nu + (K_\mu{}^B{}_\nu + \sum_A K_\mu{}^{AB}{}_\nu)v_{0B}^{\ \nu} \tag{3.75}$$

The positivity requirement on the entire matrix as it appears in (3.74) (which evidently needs $K_{\mu\nu}$ and $K_\mu{}^{AB}{}_\nu$ to have separate positivity properties of their own) ensures that we shall have

$$-(\tilde{f}_\mu v_0^\mu + \tilde{f}^A_\mu v_{0A}^\mu) \geq 0\,. \tag{3.76}$$

If we wish to envisage situations in which there are non-zero particle creation rates

$$\nabla_\mu n^\mu = r_A = -r_{[C]} N_A^{[C]} \tag{3.77}$$

where the $N_A^{[C]}$ are the chemical reaction coefficients introduced in (1.76), the requirement that (3.67) should hold for arbitrary values of the relative currents implies in particular that we must have

$$r_A \chi_\natural^A \frac{1}{\gamma_{0\{A\}}} \leq 0 \tag{3.78}$$

and in the simplest situations, for which cross coupling between chemical and resistive effects need not be taken into account, (3.76) and (3.78) together will be sufficient to ensure (3.69). To obtain (3.78) separately, it is evident that we should simply generalise the non-conducting chemical reaction law (1.78) to

$$r_{[C]} = -\kappa_{[CD]} N_B^{[D]} \chi_\natural^B \frac{1}{\gamma_{0\{B\}}} \tag{3.79}$$

for a positive, presumably symmetric, reaction-rate matrix $\kappa_{[CD]}$ given, like the resisitivity matrix as a function of state, which should reduce in the thermal equilibrium limit to the non-conducting reaction rate matrix already introduced in (1.78). The situation could however be much more complicated than this, since there is no reason in principle why there should not exist cross couplings between reactions and currents, so that for situations involving i reactions between j currents (as well as the entropy) we might need a fully $(i+3j+3)\times(1+3j+3)$ dissipation matrix. If nevertheless we still restrict ourselves to cases without cross-coupling between resistivity and reactivity, and if we are not interested in the detailed chemical pathways but only in the outcome, we can simplify the formalism by using a composite chemical reaction matrix with components Ξ_{AB} as defined by (1.79), which enables us to replace (3.78) by

$$r_A = -\Xi_{AB}\mu^B_\natural \frac{1}{\gamma_{0\{B\}}}, \qquad \mu^A_\natural = c^{-2}\chi^A_\natural \tag{3.80}$$

where the property of being positive indefinite is inherited by Ξ_{AB}. The combined effect of (3.74) and (3.80) can be seen by (3.68) to imply that the entropy creation rate will be given by

$$f^0_\emptyset = \Theta\nabla_\mu s^\mu = \frac{\mu^A_\natural \Xi_{AB}\mu^B_\natural c^2}{\gamma_{0\{A\}}\gamma_{0\{B\}}} + v^\mu_0 K_{\mu\nu}v^\nu_0 + 2v_{0A}{}^\mu K^A_\mu{}_\nu v^\nu_0 + v_{0A}{}^\mu K^{AB}_\mu{}_\nu v_{0B}{}^\nu, \tag{3.81}$$

extending our notation scheme in an obvious way by the use of the abbreviation

$$f^0_\emptyset = -f^0_\mu u_0{}^\mu = f^0_\natural \tag{3.82}$$

so that the total thermal force density will be determined by (3.75) and (3.81) in the form

$$f^0_\mu = {}^\dashv f^0_\mu + f^0_\emptyset c^{-2} u_{0\mu}. \tag{3.83}$$

In an analogous way the total force density on eack particle current will be determined by (3.74) and (3.80) in terms of the components

$$\tilde f^A_\emptyset = -f^A_\mu u_0{}^\mu = \frac{1}{\gamma_{0\{A\}}}\tilde f^A_\natural + {}^\dashv \tilde f^A{}_\mu v_{0\{A\}}{}^\mu, \qquad \tilde f^A_\natural = -\tilde f^A_\mu u_{\{A\}}{}^\mu \tag{3.84}$$

in the form

$$\tilde f^A_\mu = {}^\dashv f^A_\mu + f^A_\emptyset c^{-2} u_{0\mu} \tag{3.85}$$

with

$$\tilde f^A_\natural = f^A_\natural = -f^A_\mu u_{\{A\}}{}^\mu = \chi^A_\natural r_{\{A\}} \tag{3.86}$$

which works out by (3.80) as

$$\tilde f^A_\natural = -\mu^A_\natural \Xi_{\{A\}B}\mu^B_\natural \frac{c^2}{\gamma_{0\{B\}}}. \tag{3.87}$$

There have been several studies of the causal properties of a pure elastic solid, which has good (subluminal) hyperbolic behaviour for reasonable equations of state[37,38]. There has not yet been time to investigate the more general theory of a conducting and chemically active medium whose specification has just been completed here, but in view of the good behaviour of the purely conducting fluid limit described in the preceeding section there is every reason for confidence.

Acknowledgements

This work was was supported by N.S.F grant No PHY82-17854 supplemented by N.A.S.A., and by the C.N.R.S., Equipe de Recherche 176.

The author would like to thank many participants of the C.I.M.E. session on Relativistic Hydrodynamics, and most particularly Marcelo Anile, Darryl Holm, Werner Israel and Lee Lindblom, for a number of stimulating discussions.

References

1. C. Eckart, Phys. Rev. **58**, 919 (1940).

2. L. Landau, E.M. Lifshitz, *Fluid Mechanics*, section **127** (Addison-Wesley, Reading, Massachussetts, 1958).

3. C. Cattaneo, C.R. Acad. Sci. Paris, **247**, 431 (1958).

4. M. Kranys, Nuovo Cimento, **B50**, 48 (1967).

5. I. Muller, Z. Physik, **198**, 329 (1967).

6. W. Israel, Ann. Phys. **100**, 310 (1976).

7. J.M. Stewart, Proc. Roy. Soc. Lond., **A357**, 59 (1977).

8. W. Israel, J.M. Stewart, Proc. Roy. Soc. **A365**, 43 (1979).

9. W. Israel, J.M. Stewart, Ann. Phys. (N.Y.), **118**, 341 (1979).

10. B. Carter, in *Journées Relativistes 1976*, ed. M. Cahen, R. Debever, J. Geheniau, pp 12-27 (Université Libre de Bruxelles, 1976).

11. B. Carter, in *A Random Walk in Relativity and Cosmology*, ed N. Dadhich, J. Krishna Rao, J.V. Narlikar, C.V. Vishveshwara (Wiley Eastern, Bombay, 1983).

12. W.A. Hiscock, L. Lindblom, Ann. Phys. (N.Y.), **151**, 466 (1983)

13 W.A. Hiscock, L. Lindblom, Phys. Rev. **D31**, 725 (1985)

14 W.A. Hiscock, L. Lindblom, preprint, (1987).

15 B.F. Schutz, *Geometrical Methods of Mathematical Physics*, (Cambridge University Press, 1980)

16 B. Carter, in *Active Galactic Nuclei*, ed C. Hazard, S. Mitton, pp 273-299 (Cambridge University Press, 1979)

17 H. Ertel, Met. Z. **59**, 277 (1942)

17 J. Katz, D. Lynden-Bell, Proc. Roy. Soc. Lond., **A381**, 263 (1982).

19 J. L. Friedman, Commun. Math. Phys., **62**, 247 (1978).

20 J. Katz, Proc. Roy. Soc. Lond., **A391**, 415 (1984).

21 L. Woltjer, Proc. Nat. Acad. Sci. U.S.A., **44**, 489; 833 (1958).

22 H.K. Moffat, J. Fluid Mech. **35**, 117 (1969)

24 A. Lichnerowicz, *Relativistic Hydrodynamics and Magnetohydrodynamics* (Benjamin, New York, 1967).

25 B. Carter, B. Gaffet, J. Fluid Mech. (1987).

26 A.H. Taub, Phys. Rev., **94**, 1468 (1954).

27 J.L. Friedman, B.F. Schutz., Astroph. J., **200**, 204 (1975).

28 B.F. Schutz, Phys. Rev., **D2**, 2762.

29 B.F. Schutz, R. Sorkin, Ann. Phys. (N.Y.), **107**, 1.

30 B. Carter, Commun. Math. Phys. **30**, 261 (1973).

31 B. Carter, in *Journées Relativistes 1979*, ed. I. Moret-Bailly, C. Latremolière, pp 166-182 (Faculté des Sciences, Anger, 1979).

32 A. Trautman, *Lectures in Theoretical Physics (Brandeis Summer School Notes)* ed H. Bondi, F. Pirani, A. Trautman, (Prentice Hall, New Jersey, 1965).

33 S. Weinberg, Gravitation and Cosmology, pp 53-57 (Wiley, New York, 1972).

34 J.G. Oldroyd, Proc. Roy. Soc. Lond., **A272**, 44 (1970).

35 B. Carter, H. Quintana, Proc. Roy. Soc. Lond., **A 331**, 57 (1972).

36 B. Carter, Proc. Roy. Soc. Lond., **A 372**, 169 (1980).

37 B. Carter and H. Quintana, Phys. Rev., **D16**, 2928 (1977).

38 B. Carter, in *Gravitational Radiation (Les Houches 1982)*, ed. N. Deruelle and T. Piran, pp 455-464 (North Holland, Amsterdam, 1983).

HAMILTONIAN TECHNIQUES FOR RELATIVISTIC FLUID DYNAMICS AND STABILITY THEORY

Darryl D. Holm

Center for Nonlinear Studies
and
Theoretical Division, MS B284
Los Alamos National Laboratory
Los Alamos, New Mexico 87545 USA

Lectures given at the
Centro Internazionale Matematico Estivo
Session on Relativistic Fluid Dynamics
Noto, Sicily

May 25 - June 3, 1987

PART I
HAMILTONIAN FORMALISM FOR SPECIAL - RELATIVISTIC ADIABATIC FLUIDS

ABSTRACT

This part of the lectures derives the Hamiltonian structures of nonrelativistic and special - relativistic adiabatic fluids in the Eulerian representation in Riemannian space by using standard variational principles. The evolution in each case is generated by a Hamiltonian that is equivalent to that obtained from a canonical analysis. The nonrelativistic and relativistic fluid theories share the same Lie - Poisson bracket, when expressed in the appropriate spaces of physical variables constructed here. (A Lie - Poisson bracket is a Poisson bracket defined on the dual space to a Lie algebra). The Lie - Poisson bracket for fluids is associated to the dual space of the semidirect - product Lie algebra of vector fields acting on differential forms. An immediate consequence of this shared structure is that both the fluid theories treated here possess an infinite family of conservation laws: the so - called "Casimirs" that belong to the kernel of the Lie-Poisson bracket. The role of these Casimirs in the study of Lyapunov stability (or dynamic stability) for fluid equilibria is discussed. The relationship of this approach to other approaches in the literature is also discussed.

INTRODUCTION

These lectures derive the noncanonical Hamiltonian structure for special relativistic adiabatic fluids, by starting from a physically-motived action principle and using standard variational techiques.

In relativistic fluid theories, Hamiltonian methods and initialvalue procedures can be applied once the theory has been translated into 3+1 language. The variational action principle approach then leads to a Hamiltonian formalism,

$$\partial_t F = \{H,F\}_c ,$$

for the dynamical evolution, where $\{,\}_c$ is the canonical (symplectic) Poisson bracket, H is the Hamiltonian, and F is any fuctional on the space of (canonically conjugate) dynamical variables.

The equations for both nonrelativistic and special relativistic adiabatic fluids turn out to be Hamiltonian with a symplectic Poisson bracket when the fluid variables are represented using Lagrangian coordinates, regarded as maps of spatial ponts in the current fluid configuration to reference points in the reference, or initial configuration.

These lectures construct the map for adiabatic fluids from the symplectic Poisson bracket to the Lie - Poisson bracket (A Lie - Poisson bracket is a Poisson bracket defined on the dual space to a Lie algebra). Holm, Marsden, and Ratiu (1986a) describe how this map fits into the mathematical framework of Marsden, Ratiu, and Weinstein (1984a) and compare this map to the one discussed for nonrelativistic fluids in Holm, Kupershmidt, and Levermore (1983), Marsden et. al. (1983), and Marsden, Ratiu, and Weinstein (1984a, b).

The aim of Part I of the lectures is to derive the nonsymplectic Lie - Poisson structures of nonrelativistic adiabatic fluid (NRAF) and special relativistic adiabatic fluids (SRAF) in the Eurian representation, by using standard variational techniques. One objective of such a unified treatment is to keep the similarities, differences, and limiting processes betwenn the two levels of description apparent at every stage. Another objective is to provide an esplicit basis for: (1) extensions, such as seeking Lie - Poisson structures for general relativistic systems as in Holm (1985) and including additional physics such as magnetohydrodynamics, electromagnetic interactions, and Yang - Mills interactions; and (2) "Thechnology transfer", such as the use in relativistic systems of Hamiltonian methods for studying Lyapunov stability as developed in Holm, Marsden, Ratiu and Weistein (1985), or approximation techniques such as Whitham - averaged action principles as used in Similon, Kaufman, and Holm (1984).

The main structural similarity between the two theories is that the resulting Hamiltonian structures for NRAF and SRAF share a **common** Lie - Poisson bracket when expressed in the appropriate spaces of fluid variables constructed here. The NRAF Lie - Poisson bracket is due to Iwinski and Turski (1976), altrough it was rediscovered by Dzyaloshinskii and Volovick (1980) and by Morrison and Greene (1980) (who also treated NR magnetohydrodynamics). The SRAF Lie - Poisson bracket is due to Bialynicki - Birula and Iwinski (1972) for free-streaming pressureless fluids, and to Iwinski and Turski (1976) for SRAF including pressure forces and electromagnetic interactions. The Poisson structure shared by these two descriptions of adiabatic fluids is the Lie - Poisson bracket shown in Holm and Kupershmidt (1983) to be associated to the dual of the semidirect - product Lie algebra of vector fields acting on differential forms. (This type of Lie - Poisson bracket is given a group theoretical interpretation in Marsden, Ratiu, and Weinstein (1984a,b)). An immediate consequence of this shared structure is that both of these theories possess an infinite family of conservation laws: the so-called "Casimirs" that belong to the kernel of the Lie - Poisson bracket. The role of these Casimirs in the study of Lyapunov stability for fluid equilibria is discussed at the end of Part I of the lectures and applied to determine stability criteria for relativistic fluid plasmas in Part II. The exstension of this Lie - Poisson structure to describe special - relativistic ideal fluids carrying Yang - Mills charges and interacting self - consistently with a Yang - Mills field is discussed in Part III of the lectures.

Remark: Although there has been considerable progress in the development of noncanonical Hamiltonian structures in fluid dynamics since 1980, the idea of noncanonical Poisson bracket in continuum physics is not new. A catalog of noncanonical Poisson brackets of Lie - Poisson type used in continuum physics before 1980 should include: Landau (1941) in the macroscopic theory of superfluids; Arnold (1966) in incompressible flow; Dashen and Sharp (1968) and Goldin and Sharp (1972) in the classical theory of current algebras; Bialynicki-Birula and Iwinski (1973) in special relativistic theories of both charged and neutral, but pressureless, fluids; and Iwinski and Turski (1976) in special - relativistic, charged - fluid plasmas interacting both electromagnetically and thermodynamically. The last citation, of course, includes NRAF and SRAF.

In fact, noncanonical Poisson bracket were introduced into classical mechanics long ago by Lie (1880) in his study of general composition laws satisfying the Jacobi identity. A modern perspective on Poisson structures is given in Weinstein (1984). Application of noncanonical Poisson brackets to the classical heavy top appear in Sudarshan and Mukunda (1974) and Ratiu (1981).

Part I of the lectures is organized into three sections: section 1 treats NRAF; section 2 treats SRAF; and section 3 presents conclusions and comments. Each section is organized in a parallel fashion into subsections that treat: 1) the starting equations of motion and notation; 2) introduction of a configuration space of position and velocity fields (these are the Lagrangian positions and velocities defined in **Eulerian** space); 3) an action principle in the Lagrangian configuration space; 4) the Legendre trasformation to find the Hamiltonian density in the space od symplectic (canonically conjugate) fields; 5) definition of the map from the space of symplectic fields to the space of physical variables, such as the fluid mass density, specific entropy, and momentum density (this map determines the Lie - Poisson bracket and expresses the Hamiltonian density from step 4) as a function in the space of physical variables); 6) calculation of the variational derivatives of the Hamiltonian with respect to the physical varibles; 7) demonstration that using the Lie - Poisson bracket and Hamiltonian in the space of physical variables recovers the original equations of motion.

Form - invariance of the map in step 5) from the symplectic fields to the physical fields in Eulerian space in both of the fluid theories results in form - invariance of the Lie - Poisson bracket obtained from the map. Consequently, the two fluid theories share the same Lie - Poisson bracket when expressed in the appropriate physical variables. Some of the implication of this shared structure for the dynamic stability of equilibrium solutions are discussed in section 3.

Acknowledgments. Incisive criticism from Boris Kupershmidt and instructive encouragement from Jerry Marsden have been invaluable in the development of the material covered in these lectures.

1. NONRELATIVISTIC ADIABATIC FLUIDS

1.1 Equations of Motion and Notation.

In adiabatic fluid dynamics, the fundamental variables are: the mass density ρ; the specific entropy η; and momentum density $M_i, i=1,2,\ldots n$. The fluid moves through an n-dimensional Riemannian space, with positions $x^i, i=1,2,\ldots,n$, and metric tensor $g_{ij}, ij=1,2,\ldots,n$.

In the nonrelativistic case, the fluid velocity v_i is related to the momentum density by

$$M_i = \sqrt{g}\,\rho v_i \,, \tag{1.1}$$

where $\sqrt{g} := \sqrt{\det g_{ij}}$. The Eulerian hydrodynamics equations are expressed as

$$\partial_t \rho = -(\rho v^i)_{;i} = -\frac{1}{\sqrt{g}}\left[\rho\sqrt{g}\,v^i\right]_{,i} \,, \tag{1.2a}$$

$$\partial_t \eta = -v^i \eta_{,i}, \tag{1.2b}$$

$$\partial_t v_i = -v^j v_{i;j} - \frac{1}{\rho} p_{,i} - \phi_{,i}, \tag{1.2c}$$

$$\Delta\phi = (\phi_{,j})^{;j} = 4\pi G\rho, \tag{1.2d}$$

where partial time derivative is denoted by $\partial/\partial t$, partial space derivative is denoted by subscript comma (e.g., $\partial\eta/\partial x^i=\eta_{,i}$), covariant derivative compatible with the time - independent Riemannian metric

g_{ij} (i.e., $\partial_t g_{ij}=0$) is denoted by subscript semicolon (;), and we sum on repeated indices over their indicated ranges. Indices are raised as for v^i in (1.2a) by the inverse metric tensor g^{ij}, which satisfies $g^{ij}g_{jk}=\delta^i_k$ and gives $v^i=g^{ij}v_j$. Equation (1.2a) is the continuity equation expressing conservation of mass, and Eq. (1.2b) is the adiabatic contition, so that each fluid element exchanges no heat with its surroundings. Equation (1.2c) is the hydrodynamic motion equation expressed in covariant form. The fluid pressure p is determined as a function of ρ and η from a prescribed relation for the specific internal energy $e(\rho,\eta)$ (i.e. from an equation of state) via the first law of thermodynamics,

$$de = e_\rho d\rho + e_\eta d\eta = \rho^{-2} p d\rho + T d\eta, \tag{1.3}$$

where T is temperature. The motion equation (1.2c) also includes a Newtonian gravitational potential, ϕ, determined by (1.2d) as a functional of ρ and a function of $\underline{\mathbf{x}}$.

1.2 Action Principle and Legendre Trasformation to Canonical Variables

We show next that Eqs. (1.2a-d) follow from a stationary variational principle, $\delta S=0$, expressed in a space of different variables from those in (1.2a-d). Let us define Lagrange coordinates $q^A(\underline{\mathbf{x}},t)$, $A=1,2,\ldots,n$, as time - dependent maps of spatial points x^i to reference points q^A. Then, density, specific entropy, and velocity are defined in terms of these Lagrange coordinates by the following expressions,

$$\rho(\underline{x},t)\sqrt{g(\underline{x})} = \bar{\rho}([q^A])\det(\partial q^A/\partial x^i) =: \bar{\rho}(q)\det(q^A_i), \tag{1.4a}$$

$$q(\underline{x},t) = \bar{\eta}([q^A])\bar{\eta}(q), \tag{1.4b}$$

where $\bar{\rho}$ and $\bar{\eta}$ are prescribed functions of the argument $q:=(q^A)$ determined by the values of the Lagrange coordinates q^A at some initial time, and q^A satisfies

$$\partial_t q^A + q^A_j v^j = 0, \tag{1.4c}$$

with $q^A_j := \partial q^A/\partial x^j$. Regarding (q^A_j) as an invertible matrix allows Eq. (1.4c) to be solved for v^j as

$$v^j = -(q^{-1})^j_A \dot{q}^A, \qquad (1.4d)$$

where $\dot{q}^A := \partial_t q^A$. We introduce definition (1.4a-d) into the following action density,

$$\mathcal{L} = \sqrt{g}\left[\frac{1}{2}\rho|\underline{v}|^2 - \rho e(\rho,\eta) - \frac{1}{2}\rho\phi\right]. \qquad (1.5)$$

Variation of the corresponding action $S=\int dt\ d^n x\mathcal{L}$ with respect to $\dot{q}^A$ determines the canonical momentum varible p_A conjugate to q^A as

$$p_A := \frac{\delta L}{\delta \dot{q}^A} = -\sqrt{g}\ \rho v_j (q^{-1})^j_A. \qquad (1.6)$$

Therefore, by taking the matrix product of (1.6) with q^A_j, the physical momentum density (1.1) is expressible as

$$M_j = \sqrt{g}\ \rho v_j = -p_A \partial_j q^A \qquad (1.7)$$

We shall see that this type of momentum density relation is a generic feature of our approach.

One passes to the Hamiltonian formulation by Legendre transforming the action density (1.5), in which the velocities $\dot{q}^A$ need to be expressed in terms of (p_A, q^A). For this, we first rewrite (1.6) using (1.4d) as

$$p_A = \left[-\sqrt{g}\right]\left[-g_{jk}(q^{-1})^k_B \dot{q}^B (q^{-1})^j_A\right] =$$

$$= \sqrt{g}\ \rho\left[(q^{-1})^j_A\ g_{jk}(q^{-1})^k_B\right]\dot{q}^B =: \sqrt{g}\,\rho\ (\Delta^{-1})_{AB}\dot{q}^B, \qquad (1.8)$$

where, in the notation of Künzle and Nester (1984), the quantity

$$(\Delta^{-1})_{AB} = (q^{-1})^j_A\ g_{jk}(q^{-1})^k_B \qquad (1.9)$$

is the pull back metric under the map q_A. Hence,

$$\sqrt{g}\,\rho\,\dot{q}^A = \Delta^{AB} p_B, \tag{1.10}$$

where $\Delta^{AB}(\Delta^{-1})_{BC} = \delta^A_C$ and $g^{ab}g_{bc} = \delta^a_c$.

Whe have, therefore, the following relations,

$$\frac{1}{\rho\sqrt{g}}\,|\underline{p}|^2 := \frac{1}{\rho\sqrt{g}}\, p_A\Delta^{AB}p_B(\text{by }(1.10)) = p_A\dot{q}^A(\text{by }(1.4c)) =$$

$$= -p_A q^A_j\, v^j(\text{by }(1.7)) = v^j M_j(\text{by }(1.1)) =$$

$$= \rho\sqrt{g}\,|\underline{v}|^2\ (\text{by }(1.1)) = \frac{1}{\rho\sqrt{g}}\,|\underline{M}|^2. \tag{1.11}$$

Equating the first and last of these expressions implies $|\underline{p}|^2 = |\underline{M}|^2$; so the momentum magnitudes are equal. Thus, the Hamiltonian that arises from Legendre trasforming (1.5) and using (1.11) is

$$H = \int d^n x\,\mathcal{H} = \int d^n x\left\{\frac{|\underline{p}|^2}{2\rho\sqrt{g}} + \sqrt{g}\left[\rho e(\rho,\eta) + \frac{1}{2}\rho\phi\right]\right\}. \tag{1.12}$$

Note that this Hamiltonian is invariat under replacing p_A by $-p_A$, which we do now for later convenience. This replacement removes the minus signs in (1.6) and (1.7).

1.3. Map from Canonical to Physical Fluid Variables

At this stage in the standard Hamiltonian formalism for fluids in terms of canonically conjugate variables, the starting system should be shown to be expressible as $\partial_t F(p,\rho) = \{H,F\}_C$ with canonical (symplectic) Poisson bracket

$$\{H,F\}_C = \int d^n x \begin{vmatrix} \delta F/\delta P_A \\ \delta F/\delta q^A \end{vmatrix}^t \begin{vmatrix} 0 & -\delta^B_A \\ \delta^A_B & 0 \end{vmatrix} \begin{vmatrix} \delta H/\delta P_B \\ \delta H/\delta q^B \end{vmatrix} \tag{1.13}$$

In other words, the canonical equations

$$\partial_t q^A = \frac{\delta H}{\delta P_A}\,,\quad \partial_t P_A = -\frac{\delta H}{\delta q^A}\,, \tag{1.14}$$

together with H given in terms of (P_A, q^A) by (1.12) should be shown to lead to the starting equations (after cumbersome algebraic manipulations, because of the complicated q^A dependence of H).

There is, however, an alternative and less cumbersome route for checking that the canonical equations (1.14) imply the physical motion equations (1.2a-d) via the relations (1.4a-d). This route, which we now explain, uses the noncanonical Hamiltonian formalism in the space of physical variables, and leads to considerable insight into the mathematical structure of the physical equations.

We seek a Hamiltonian description of the fluid motion equations (in the present case (1.2a-d)) directly in terms of the physical variables (in this case $\underline{M}, \rho\eta$). Such a description will result provided the following three requirements are satisfied:

1) The canonical Poisson bracket (1.13) induces a noncanonical Poisson bracket {,} in the physical space, in this case by the map:

$$M_i = P_A \partial_i q^A, \quad \hat{\rho} := \rho\sqrt{g} = \bar{\rho}(q)\, \det(q^A_i), \quad \eta = \bar{\eta}(q). \tag{1.15}$$

2) The Hamiltonian function in the canonical space can be expressed in terms of physical variables only, in the present case through (1.15) using (1.11).

3) This Hamiltonian function, H, expressed in terms of physical variables, generates in the physical space correct equations of motion according to the rule $F_t = \{H, F\}$, using the noncanonical Poisson bracket {,} of requirement 1).

The second and third of these requirements are self explanatory. The first requirement amounts to checking the following formula (see, e.g., Holm and Kupershmidt (1983))

$$\phi(\underline{\underline{B}}) = \frac{D\underline{Z}}{D\underline{Y}}\, \underline{\underline{b}} \left[\frac{D\underline{Z}}{D\underline{Y}} \right]^{\dagger} \tag{1.16}$$

where: ϕ is the map from the canonical space with coordinates $\underline{Y}$, into the physical space with coordinates $\underline{Z}$; $\underline{\underline{b}}$ is the canonical matrix $\begin{bmatrix} 0 & -1 \\ 1 & 0 \end{bmatrix}$; $\underline{\underline{B}}$ is the Hamiltonian matrix in the physical space; $\phi(\underline{\underline{B}})$ is computed by applying the map ϕ to each matrix element of $\underline{\underline{B}}$; $\frac{D\underline{Z}}{D\underline{Y}}$ is the Fréchet derivative of the variables $\underline{Z}$ with respect to the variables $\underline{Y}$; and the symbol $\dagger$ stands for the adjoint with respect to the measure $d^n x = dx^1 \wedge \ldots \wedge dx^n$.

In the present case, the Hamiltonian matrix in the space of

physical variables which results from this procedure is (cf. Holm and Kupershmidt (1983))

$$-\underline{\underline{B}} = \begin{bmatrix} M_j\partial_i+\partial_j M_i & \hat{\rho}\partial_i & -\eta_{,i} \\ \partial_j\hat{\rho} & 0 & 0 \\ \eta_{,j} & 0 & 0 \end{bmatrix}, \tag{1.17}$$

where $\eta_{,j}:=(\partial\eta(\partial x^j)$ and ∂_j is regarded now as a differential operator. The Poisson bracket corresponding to the Hamiltonian matrix 81.17) is given by

$$\{H,F\}=-\int d^nx\{\delta F\delta M_i[(M_j\delta_i+\partial_j M_i)\delta H/\delta M_j+\hat{\rho}\partial_i\delta H/\delta\hat{\rho} -$$

$$- \eta_{,j}\delta H/\delta\eta]+[(\delta F/\delta\hat{\rho})\partial_j\hat{\rho}+(\delta F/\delta\eta)\eta_{,j}]\delta H/\delta M_j\} \tag{1.18}$$

To see how (1.17) arises, notice that applyng formula (1.16) to the case when the matrix $\underline{\underline{b}}$ is canonical as in (1.13), we obtain, for the Z_i-Z_j entry of the Hamiltonian matrix $\phi(\underline{\underline{B}})$

$$\phi(\underline{\underline{B}})_{Z_i-Z_j} = \sum_A \frac{DZ_i}{Dq^A}\frac{DZ_j^{\dagger}}{DP_A} - \frac{DZ_i}{DP_A}\frac{DZ_j^{\dagger}}{Dq^A} \tag{1.19}$$

In the present case, the variables $(\underline{Z})$ are given in (1.15)

$$M_i = P_A q^A_{,i}, \bar{\rho} = \bar{\rho}(q)x, \quad \eta=\bar{\eta}(q), \tag{1.20}$$

where $q=(q^A, A=1,2,...,n)$.

$$x = \det\,(q^A_{,i}), \tag{1.21}$$

and $\bar{\rho}(q)$ and $\bar{\eta}(q)$ are given functions. Then, we have

$$\frac{DM_i}{Dq^A} = P_A\partial_i,\ \frac{DM_i}{DP_A} = q^A_{,i},\ \frac{D\hat{\rho}}{Dq^A} = \frac{\partial\bar{\rho}}{Dq^A}\chi + \bar{\rho}\,\frac{\partial x}{\partial q^A_{,j}}\,q^A_{,j},$$

$$\frac{\partial\hat{\rho}}{DP_A} = \frac{\partial\eta}{DP_A} = 0,\ \frac{\partial q}{Dq^A} = \frac{\partial\bar{\eta}}{Dq^A}, \tag{1.22}$$

so that substituting Eqs. (1.22) into (1.19) and using the standard identity

$$x^{-1}\sum_A \frac{\partial x}{Dq^A_{,i}}\, q^A_{,j} = \delta^i_j, \tag{1.23}$$

we obtain (1.17).

Remark. The Poisson bracket (1.18) is the natural Poisson bracket on the dual to the semidirect product Lie algebra $D\ s|\Lambda^0 \cdot \Lambda^n|$ (see, e.g., Holm and Kupershmidt (1983)), where $D=D(R^n)$ represents vector fields on R^n (X_j denotes elements of D) and $\Lambda^k=\Lambda^k(R^n)$ denotes k-forms on R^n. D acts on itself by commutation of vector fields denoted by (,) and acts upon Λ^k by Lie derivation, denotes, e.g., $X(\xi)$ for $\xi\varepsilon\Lambda^k$. The symbol s denotes semidirect product. The Lie algebraic commutator corresponding to the Poisson bracket (1.18) is, thus,

$$|(\underline{X};\xi^{(0)}\cdot\xi^{(n)}),\ (\underline{\overline{X}};\xi^{(0)}\cdot\xi^{(n)})| = (|\underline{X},\underline{\overline{X}}|;(\underline{X}(\xi^{(0)})-\underline{\overline{X}}(\xi^{(0)}))$$

$$\cdot(\underline{X}(\xi^{(n)})-\underline{\overline{X}}(\xi^{(n)}))). \tag{1.18'}$$

Dual coordinates are: M_i dual to $X_i\varepsilon D$, $\hat{\rho}$ dual to Λ^0, and η dual to Λ^n.

Poisson brackets such as (1.18) associated to the dual of a Lie algebra are called "Lie - Poisson brackets," see Marsden et al. (1983) for an exposition and Weinstein (1984) for more detail.

1.4 Hamiltonian Formulation in Physical Fluid Variables.

Using relations (1.11), the Hamiltonian (1.12) can be expressed in terms of physical variables (1.15) as

$$H = \int d^n x \left[\frac{1}{2\hat{\rho}}|\underline{M}|^2 + \hat{\rho}e(\rho/\sqrt{g}\ ,\eta) + \frac{1}{2}\hat{\rho}\phi\right] \tag{1.24}$$

The variational derivatives of this Hamiltonian are

$$\frac{\delta H}{\delta M_i} = v^i,$$

$$\frac{\delta H}{\delta\hat{\rho}} = -\frac{1}{2}|\underline{v}|^2 + e + p/\rho + \phi, \tag{1.25}$$

$$\frac{\delta H}{\delta \eta} = \hat{\rho} T,$$

where T is temperature, defined in (1.3). Substituting the variational derivatives (1.25) into the Lie - Poisson bracket (1.18) leads immediately to

$$\partial_t \hat{\rho} = \{H, \hat{\rho}\} = -(\hat{\rho} v^j)_{,j}, \tag{1.26a}$$

$$\partial_t \eta = \{H, \eta\} = -\eta_{,j} v^j , \tag{1.26b}$$

$$\partial_t M_i = \{H, M_i\} = -M_j v^j_{,i} - (M_i v^j)_{,j} - \hat{\rho} \Big[-\frac{1}{2} |\underline{v}|^2 + + e + p/\rho + \phi \Big]_{,i} + \hat{\rho} T \eta_{,i}. \tag{1.26c}$$

Equations (1.26a,b) reproduce (1.2a,b), the continuity equation and adiabatic condition, respectively. Equation (1.26c) can be re-expressed in terms of velocity to recover (1.2c). Using (1.26a) and (1.3) we find from (1.26c) that

$$\partial_t v_i = -\frac{1}{\rho} p_{,i} - \phi_{,i} - v^j v_{i,j} + \frac{1}{2} g_{jk,i} v^j v^k. \tag{1.27}$$

We rearrange the last two terms in (1.27) as follows. Upon setting

$$y_i := v^j v_{i,j} - \frac{1}{2} g_{jk,i} v^j v^k = v^j (g_{ik} v^k)_{,j} - \frac{1}{2} g_{jk,i} v^j v^k$$

$$= g_{ik} v^j v^k_{,j} + g_{ik,j} v^j v^k - \frac{1}{2} g_{jk,i} v^j v^k, \tag{1.28}$$

symmetrizing the middle term in (1.28) gives

$$y_i = g_{ik} v^j v^k_{,j} + \frac{1}{2} (g_{ij,k} + g_{ik,j} - g_{jk,i}) v^j v^k, \tag{1.29}$$

which we recognize upon raising indices as

$$y^i = g^{im} y_m = v^j v^i_{,j} + \{^{\,i}_{jk}\} v^j v^k =: v^j \nabla_j v^i, \tag{1.30}$$

i.e., the covariant derivative of v^i in the direction v^j. Hence, the motion Eq. (1.26c) becomes

$$\partial_t v^i = \frac{1}{\rho} g^{ij} p_{,j} - g^{ij}\phi_{,j} - v^j \nabla_j v^i, \tag{1.31}$$

which reproduces (1.2c).

Thus, the Lie - Poission bracket (1.18) and Hamiltonian (1.24) yield the nonrelativistic adiabatic fluid (NRAF) Eqs. (1.2a-c) directly in terms of the Eulerian physical variables. Since we have already seen that the map (1.15) is canonical (in the sense of preserving Poisson brackets), the Hamilton's Eqs. (1-14) will also imply the fluid equations (1.2a-c).

1.5 Remark on Casimir Functionals of the Lie-Poisson Bracket (1.18)

First, observe that using (1.26a) allows (1.26c) to be re-expressed as

$$\partial_t v_i = -(v^j v_{i,j} + v_j v^j_{,i}) - \left[-\frac{1}{2} |\underline{v}|^2 + e + p/\rho + \phi \right]_{,i} + T\eta_{,i} \tag{1.32}$$

Notice that the first term on the right hand side of (1.32) is the Lie - derivative with respect to velocity of the circulation one - form $v_i dx^i$, namely

$$\mathcal{L}_v (v_i dx^i) = (v^j v_{i,j} + v_j v^j_{,i}) dx^i. \tag{1.33}$$

Consequently, the motion Eq. (1.32) can be written as

$$\partial_t (v_i dx^i) = \mathcal{L}_v (v_i dx^i) - d \left[-\frac{1}{2} |\underline{v}|^2 + e + p/\rho + \phi \right] + T d\eta \tag{1.34}$$

in terms of the Lie derivative, $\mathcal{L}$, and the spatial exterior derivative, d. Likewise, equations (1.26a,b) can be expressed as

$$\partial_t (\hat{\rho} d^n x) = - \mathcal{L}_v (\hat{\rho} d^n x), \tag{1.35a}$$

$$\partial_t \eta = - \mathcal{L}_v \eta \tag{1.35b}$$

Now, taking the exterior product of d times (1.34) with d times (1.35b) (using the properties $d^2 = 0 = [d, \mathcal{L}_{\underline{v}}]$) implies taht the three-form $d(v_i dx^i) - d\eta =: \hat{\rho}\Omega d^3x$, with $\Omega := \hat{\rho}^{-1}$ curl $\underline{v} \cdot \underline{\nabla}\eta$, is conserved along flow lines, i.e.,

$$\partial_t \, (\hat{\rho}\Omega d^3x) = - \mathcal{L}_v(\hat{\rho} d^3x). \tag{1.36}$$

Consequently, in three dimensions (n=3) we may use (1.35a) and (1.36) to show that

$$\partial_t \Omega = - \mathcal{L}_{\underline{v}} \Omega \tag{1.37}$$

for the scalar function Ω defined above to be

$$\Omega = \hat{\rho}^{-1} \, \mathrm{curl}\underline{v}\cdot\underline{\nabla}\eta. \tag{1.38}$$

Thus, we have the following conservation law for adiabatic fluid flow in a three dimensional Riemannian space,

$$\partial_t C=0, C= \int d^3x \, \hat{\rho}\phi(\eta,\Omega), \tag{1.39}$$

for an arbitrary function ϕ of the two indicated arguments, η and Ω.

The presence of the arbitrary function ϕ in (1.39) is a clue that this conservation law is kinematical; depending only on having expressed the Hamiltonian in the space of physical variables, rather than depending on the dynamics generated by the particular choice of Hamiltonian (1.24). Indeed, the conserved functional C in (1.39) is a "Casimir" in the sense that

$$\{C,F\} = 0 \ \forall \ F(\{\hat{\rho},\eta,M_i\}) \tag{1.40}$$

That is, C is in the kernel of the Lie - Poisson bracket (1.18) and, thus, is conserved **independently of the choice of Hamiltonian** in the space of physical fluid variables $\{\eta,\hat{\rho},M_i\}$. The advected quantity Ω is the generalization for three dimensional Riemannian space
and compressible fluids of Ertel's invariant, the so-called potential vorticity (see, e.g., kochin, Kibel, and Roze (1964)).

The conservation law (1.39) can also be understood as resulting via Noether's theorem from the symmetry of the action density (1.5) under transformations in the Lagrangian configuration space $(q^A,\dot{q}^A)$ that leave invariant the Eulerian variables $\{\hat{\rho},\eta,M_i\}$. Such so-called "trivial" tranformations (in the nomenclature of Friedman (1978) and Friedman and Schutz (1978)) can be considered as gauge transformations under the group of diffeomorphisms of the Lagrangian fields q^A preserving the value of the density $\hat{\rho}$ and specific entropy η at each Eulerian point (cf. Marsden, Ratiu, and Weinstein (1984a)). The corresponding momentum map is then (1.39). Explicitly, the allowed infinitesimal transformations are those satisfying (see Holm (1984b))

$$\delta\rho = \frac{\rho}{\bar{\rho}(q)} \frac{\partial}{\partial q^A} (\bar{\rho}(q)\delta q^A)=0 \tag{1.41a}$$

$$\delta\eta = \frac{\partial\bar{\eta}}{\partial q^A} \delta q^A=0, \tag{1.41b}$$

$$\delta v^i = -(q^{-1})^i_A \frac{\partial\delta q^A}{\partial t} = 0, \tag{1.41c}$$

That is, the action

$$S = \int dtd^n x\mathcal{L}, \tag{1.42}$$

with $\mathcal{L}$ given in (1.5) is invariant under the infinitesimal transformation

$$q^A(\underline{x},t)+\bar{q}^A(\underline{x},t)=q^A(\underline{x},t)+\delta q^A(\underline{x},t), \tag{1.43}$$

with

$$\delta q^A = \frac{1}{\bar{\rho}(q)} \text{curl}_q[f(\bar{\eta})\underline{\nabla}_q a(q)] \tag{1.44}$$

satisfying (1.41a-c), where $f(\bar{\eta})$ and $a(q)$ are arbitrary functions, and subscript q on $\underline{\nabla}_q$ denotes gradient in Lagrangian coordinates. The corresponding conserved density is then $\hat{\rho}\phi(\Omega)$ with Ω given in (1.38) and ϕ an arbitrary function.

Recently, Casimirs such as (1.39) have been used to study the Lyapunov stability of fluid and plasma equilibria in a variety of situations, see, e.g., Abarbanel et al. (1984a,b), Arnold (1965, 1969), Hazeltine et al. (1984), Holm, Marsden, Ratiu, and Weinstein (1983, 1951), and Holm and Kupershmidt (1986). Further comments concerning Casimirs and Lyapunov stability are given at the conclusion of this part of the lectures.

2. SPECIAL RELTIVISTIC ADIABATIC FLUIDS.

2.1 Equations of Motion and Notation.

The special generalization of the adiabatic fluid equations (1.2a-c) in Riemannian space is, in Lorentz - covariant form,

$$(\rho_f u^{\mu})_{;\mu} = 0, \tag{2.1a}$$

$$u^{\mu}\eta_{f,\mu} = 0, \tag{2.1b}$$

$$T^{\mu\nu}_{\ \ ;\nu} = 0, \tag{2.1c}$$

where u^{μ}, with $\mu=0,1,\ldots,n$, (Greek indices run from 0 to n) denotes the timelike Lorentz vector for the fluid velocity, which becomes $u^0=1$, $u^i=0$, $i=1,2,\ldots,n$, (Latin indices run from 1 to n) in the reference frame of the fluid. The vector u^{μ} satisties

$$g_{\mu\nu} u^{\mu}u^{\nu} = -1. \tag{2.2}$$

The spacetime metric tensor $g_{\mu\nu}$ is given by the expression $-d\tau^2=g_{\mu\nu}dx^{\mu}dx^{\nu}$ for the proper time interval, $x^0=ct$ being the real timelike coordinate. In (2.1), covariant derivatives with respect to the spacetime metric $g_{\mu\nu}$ are denoted by a semicolon subscript (;). Ordinary partial derivatives are denoted by a comma subscript (,). Subscript f denotes variables as measured in the reference frame moving with the fluid. For example, ρ_f is proper mass density and η_f is proper specific entropy, each in the fluid frame. The quantity $T^{\mu\nu}$ in Eq. (2.1) is the energy - momentum tensor, given by

$$T^{\mu\nu} = \rho_f c^2 w u^{\mu}u^{\nu} + p_f g^{\mu\nu} , \tag{2.3}$$

where c is the speed of light, p_f is the pressure in the fluid frame, and w is the relativistic specific enthalpy, defined as

$$w = 1+(e_f+p_f/\rho_f)c^{-2} , \tag{2.4}$$

with specific internal energy in the fluid frame e_f prescribed by an equation of state $e_f(\rho_f, \eta_f)$ satisfying

$$de_f = T_f d\eta_f + (\rho_f)^{-2} p_f d\rho_f . \tag{2.4'}$$

Equation (2.1a) expresses relativistic mass conservation, Eq. (2.1b) is the adiabatic condition for the relativistic fluid, and (2.1c) expresses the covariant conservation laws for energy and momentum. There is a well - known redundency among equations (2.1a-c): a linear combination of Eqs. (2.1a,b) can be obtained from the projection of (2.1c) along u_μ (see, e.g., Taub (1967)). Contracting u_μ with (2.1c) using $T^{\mu\nu}$ given by (2.3) and the condition $u_\mu u^\mu{}_{;\nu} = 0$ (implied by (2.2) contributes the relation

$$\begin{aligned} 0 = u_\mu T^{\mu\nu}{}_{;\nu} &= u_\mu c^2 w u^\mu (\rho_f u^\nu)_{;\nu} + \rho_f u^\nu (c^2 w u^\mu)_{;\nu} u_\mu + u^\nu p_{f,\nu} = \\ &= -c^2 w (\rho_f u^\nu)_{;\nu} + \rho_f u^\nu (c^2 w)_{,\nu} + u^\nu p_{f,\nu} = \\ &= c^2 w (\rho_f u^\nu)_{;\nu} - \rho_f T_f (u^\nu \eta_{f,\nu}), \end{aligned} \tag{2.5}$$

where we have used the definition of w in (2.4) and the thermodynamic relation (2.4′) in the fluid frame for the last step. According to (2.5), the projection of (2.1c) parallel to u_μ and one of either (2.1a), or (2.1b), implies the remaining equation in (2.1a,b).

Expanding (2.1c) using the chain rule gives

$$0 = T^{\mu\nu}{}_{;\nu} = c^2 w u^\mu (\rho_f u^\nu)_{;\nu} + \rho_f u^\nu (c^2 w u^\mu)_{;\nu} + g^{\mu\nu} p_{f,\nu} . \tag{2.6}$$

Upon lowering the free index μ in (2.6) and using (2.1a), we have (2.1c) in the form we shall use,

$$0 = T^\nu_{\mu;\nu} = \rho_f u^\nu (c^2 w u_\mu)_{;\nu} + p_{f,\mu} . \tag{2.7}$$

The n spatial components of (2.7) comprise the relativistic counterpart of the motion equation (1.2c) (excliding the Newtonian potential); the time component of (2.7) is a consequence of the space components.

For the purposes of the Hamiltonian formulation, we express Eqs. (2.1a,b) and the n space components of Eq. (2.7) as a dynamical system in a fixed frame. This we choose to be the laboratory frame, in which

u^μ becomes $u^0=\gamma$, $u^i=\gamma v^i/c$, with

$$\gamma:=(1-v_i v^i/c^2)^{-1/2}, \tag{2.8}$$

where $v_i=g_{ij}v^j$ and g_{ij} denotes the (fixed, time-indipendent) spatial Riemannian metric. In the laboratory frame, fluid state variables will be unadorned and related to their fluid - frame counterparts by $\eta=\eta_f$ and $\rho=\gamma\rho_f$. Upon using $u^\mu=\gamma(1,v^i/c)$, $\partial_t g_{ij}=0$, and the identity $(\rho_f u^\mu)_{;\mu}=(\sqrt{g}\ \rho_f u^\mu)_{,\mu}/\sqrt{g}$, the dynamical system resulting from (2.1) in the laboratory frame becomes, with $\hat{\rho}=\gamma\rho_f\sqrt{g}$. $\eta=\eta_f$, $\sqrt{g}$ = $=\sqrt{\det\ (g_{ij})}$,

$$\partial_t\hat{\rho}=-(\hat{\rho}v^i)_{,i}, \tag{2.9a}$$

$$\partial_t\eta=-\eta_{,i}v^i, \tag{2.9b}$$

$$\partial_t(\gamma w v_i)=-v^j(\gamma w v_i)_{;j}-\frac{\sqrt{g}}{\hat{\rho}}\,p_{f,i} \tag{2.9c}$$

where subscript semicolon (;) adjoined to Latin indices denotes covariant derivative compatible with the Riemannian metric tensor g_{ij}, as in section 1. In the nonrelativistic limit $(c^{-1}\to 0), \gamma=1+0(c^{-2})$, $w=1+0(c^{-2})$, and as c^{-2} tends to zero each equation in (2.9a-c) tends to its nonrelativistic counterpart in (1.2a-c).

2.2 Action Principle and Ledendre Trasformation to Canonical Variables.

As an auxiliary step in constructing the Hamiltonian formalism for Eqs. (2.9a-c), we introduce a stationary variational principle, $\delta L=0$ expressed in a space of Lagrangian fields $q^A(\underline{x},t)$, A=1,2,...,n, just as in section 1. First, we define the variables $\bar{\rho}$, η, v^i in terms of these Lagrangian fields, via expressions analogous to (1.4a-d), namely,

$$\hat{\rho}(\underline{x},t) = \bar{\rho}(q)\ \det(q^A_i), \tag{2.10a}$$

$$\eta(\underline{\mathbf{x}}, t) = \bar{\eta}(q), \tag{2.10b}$$

$$\dot{q}^A + q^A_j v^j = 0, \tag{2.10c}$$

$$v^j = -(q^{-1})^j_A \dot{q}^A, \tag{2.10d}$$

where $q^A_i = \partial q^A/\partial x^i$, $\dot{q}^A = \partial q^A/\partial t$, $(q^{-1})^j_A$ is the matrix inverse of q^A_i so that $\left[(q^{-1})^j_A q^A_i = \delta^j_i \text{ and } (q^{-1})^j_A q^B_i = \delta^B_A\right]$ and $\bar{\rho}(q)$ and $\bar{\eta}(q)$ are prescribed functions of of q: $=\{q^A\}$. We introduce definitions (2.10a-d) into the following Lagrangian density,

$$\mathcal{Y} = \sqrt{g}\ \varepsilon(\rho_f, \eta_f). \tag{2.11}$$

Here, ε is the total energy density for special for relativisty

$$\varepsilon = \rho_f(c^2 + e(\rho_f, \eta_f)), \tag{2.12}$$

so that, by (2.4′),

$$d\varepsilon = (c^2 + e_f + p_f/\rho_f)d\rho_f + \rho_f T_f d\eta_f$$

$$= c^2 w d\rho_f + \rho_f T_f d\eta_f. \tag{2.13}$$

Variation of the Lagrangian $L = \int dt d^n x \mathcal{L}$ from (2.11) with respect to $\dot{q}^A$ yields the canonical momentum variable P_A, conjugate to q^A, as

$$P_A := \frac{\delta L}{\delta \dot{q}^A} = -\sqrt{g}\ \frac{\partial \varepsilon}{\partial \rho_f}\bigg|_{\eta_f} \frac{\partial \rho_f}{\partial \gamma^{-1}}\bigg|_{\hat{\rho}} \frac{\partial \gamma^{-1}}{\partial v^j} \frac{\partial v^j}{\partial \dot{q}^A}$$

$$=- \sqrt{g}\ c^2 w(\hat{\rho}/\sqrt{g}\ ) \left[-\frac{\gamma}{c^2} v_j\right] (-(q^{-1})^j_A)$$

$$= -\hat{\rho}\gamma w v_j (q^{-1})^j_A =: -\hat{\vartheta} v_j (q^{-1})^j_A, \tag{2.14}$$

where we have used (2.13), the definitions $\rho_f = \hat{\rho}(\sqrt{g}\ \gamma)^{-1}$ and $\gamma^{-1} = =(1 - v_i v^i/c^2)^{1/2}$, and introduced the notation

$$\hat{\vartheta} := \hat{\rho}\gamma w. \tag{2.15}$$

By taking the matrix product of Eq. (2.14) with q^A_i, we find the physical momentum density in the laboratory frame, M_i, to be (cf. (1.7))

$$M_i := \frac{\sqrt{g}}{c} T^0_i (\text{by (2.3)}) = \hat{\vartheta} v_i \ (\text{by (2.14)}) = -P_A q^A_i. \tag{2.16}$$

We now seek the canonical Hamiltonian formalism in terms of (P_A, q^A). To solve for $\dot{q}^A$ in terms of (P_A, q^A), we first rewrite P_A in (2.14) using (2.10d) as

$$P_A = (-\hat{\vartheta})(-g_{jk}(q^{-1})^k_B \dot{q}^B)\ ((q^{-1})^j_A)=$$

$$= \hat{\vartheta}(\Delta^{-1})_{AB}\dot{q}^B. \tag{2.17}$$

where Δ^{-1} is defined as in (1.9). Hence, with Δ^{AB} defined by $\Delta^{AB}(\Delta^{-1})_{BC}=\delta^A_C$, we have

$$\Delta^{AB}P_B = \hat{\vartheta}\dot{q}^A \tag{2.18}$$

and we find the following relations:

$$\hat{\vartheta}^{-1}|\underline{P}|^2 := \hat{\vartheta}^{-1}P_A\Delta^{AB}P_B \ (\text{by (2.18)}) = P_A\dot{q}^A (\text{by (2.10c)} =$$

$$= -P_A q^A_i v^i \tag{2.19}$$

$$(\text{by (2.16)}) = M_i v^i \ (\text{by (2.16)}) = \hat{\vartheta}|\underline{v}|^2 \ (\text{by (2.16)}) = \hat{\vartheta}^{-1}|\underline{M}|^2,$$

where we have introduced the momentum magnitudes $|\underline{P}|^2 = P_A\Delta^{AB}P_B$ and $|\underline{M}|^2 = M_i g^{ij} M_j$ related by $|\underline{P}|^2 = |\underline{M}|^2$, as shown in (2.19).

Using (2.18), we see that $P_A\dot{q}^A = \hat{\vartheta}^{-1}|\underline{P}|^2$, so that the Hamiltonian obtained by Legendre transforming the action density (2.11) takes the form

$$H = \int d^n x \mathcal{H} = \int d^n x\ (\hat{\vartheta}^{-1}(\underline{P})^2 + \sqrt{g}\ \varepsilon(\rho_f, \eta_f))\ . \tag{2.20}$$

Here $\hat{\vartheta}$ is given in (2.15) and depends on the relativistic factor, γ, and the other dynamical variables. By the definition of γ in (2.8) we have

$$1-\gamma^{-2}=c^{-2}v_iv^i \text{ (by (2.10d) and (1.9))}=c^{-2}(\Delta^{-1})_{AB}\dot{q}^A\dot{q}^B \text{ (by (2.18))}$$

$$= (c\hat{\vartheta})^{-2}|\underline{P}|^2 = (c\hat{\vartheta})^{-2}|\underline{M}|^2. \qquad (2.21)$$

Thus, γ ia expressible in terms of $\hat{\vartheta}$ and the momentum magnitudes. Consequently, either for the canonical variables (P_A, q^A), or for the noncanonical physical variables, $\{\hat{\rho}, \eta, M_i\}$, the only implicit dependence in the Hamiltonian (2.20) is in $\hat{\vartheta}$.

2.3 Map from Canonical to Physical Fluid Variables.

We could now consider the relation to the fluid Eqs. (2.9a-c) of the canonical equations that follow from the Hamiltonian (2.20) in the space of canonical variables (P_A, q^A). However, the additional complexity due to implicit dependence in $\hat{\vartheta}$ makes this task even less perspicuous than in the earlier, nonrelativistic case in section 1. Instead, just as in that earlier case, we shall seek a Hamiltonian description of the fluid motion equations (in the present case, (2.9a-c)) directly in terms of the physical variables (in the present case, the relativistic laboratory - frame quantities $\{\hat{\rho}, \eta, M_i\}$). For this, we first replace P_A by $(-P_A)$ in the Hamiltonian (2.20) (which leaves the Hamiltonian invariant) and in Eq. (2.16) (which conveniently changes the last minus sign in (2.16) to plus). Then, collecting equations (2.10a,b) and the revised Eq. (2.16) results in the following analog of the Lagrange - to - Euler map

$$M_i = P_A \partial_i q^A, \; \hat{\rho} = \rho(q) \det(q^A_i), \; \eta = \bar{\eta}(q), \qquad (2.22)$$

which is identical in form to the map (1.15), but now the earlier nonrelativistic variables on the left - hand sides of the map are replace by relativistic, laboratory - frame variables. The Hamiltonian functional (2.20) in the canonical space can be expressed in terms of physical variables only, through the last equality in (2.21)), resulting in

$$H = \int d^nx[(|\underline{M}|^2/\hat{\vartheta} + \sqrt{g}\; \varepsilon(\rho_F, \eta_F)], \qquad (2.23)$$

where $\rho_F = \hat{\rho}(\gamma\sqrt{g})^{-1}$ and $\eta_F=\eta$. The Hamiltonian density in (2.23) is related to T^{00}, the time - time component of $T^{\mu\nu}$ in (2.3), by

$$\mathcal{H} := |M|^2/\hat{\vartheta} + \sqrt{g}\ \varepsilon(\rho_F, \eta_F) = \sqrt{g}\ T^{00}. \tag{2.24}$$

This relation is shown by a direct computation, which produces a convenient expression for (2.23) as a bonus. In the laboratory frame, we have $g^{00} = -1$, so that

$$\sqrt{g}\ T^{00}\ (\text{by } (2.3)) = \sqrt{g}\ (\rho_f c^2 w\gamma^2 - p_f)$$

$$= \sqrt{g}\ [(\gamma^2-1)\rho_f c^2 w + \rho_f c^2 w - p_f]\ (\text{by } (2.4) \text{ and } (2.12))$$

$$= \sqrt{g}\ [(\gamma^2-1)\rho_f c^2 w + \varepsilon(\rho_f, \eta_f)]\ (\text{by } (2.15) \text{ and } (2.21))$$

$$= \sqrt{g}\left[\frac{|\underline{M}|^2/c^2}{(\hat{\rho}w)^2}\rho_f c^2 w + \varepsilon(\rho_f, \eta_f)\right]\ (\text{by } \hat{\rho} = \gamma\sqrt{g}\ \rho_f)$$

$$= \sqrt{g}\left[\frac{|\underline{M}|^2/c^2}{\sqrt{g}\ \hat{\rho}\gamma w} + \varepsilon(\rho_f, \eta_f)\right]\ (\text{by } (2.15))$$

$$= |\underline{M}|^2/\hat{\vartheta} + \sqrt{g}\ \varepsilon(\rho_f, \eta_f).$$

This computation proves relation (2.24, demostrates the physical significance of the Hamiltonian density, , and results in the following convenient expression for the Hamiltonian (2.23),

$$H = \int d^n x\ \sqrt{g}\ T^{00} = \int d^n x\ (c^2\hat{\vartheta} - \sqrt{g}\ p_f). \tag{2.25}$$

2.4 Hamiltonian Formulation in Physical Fluid Variable.

Just as in Section 1, the "Lagrange - to - Euler" map (2.22) takes the canonical Hamiltonian matrix in (1.13) to that in (1.17), resulting in the Lie - Poisson bracket (1.18). The variational derivatives of the Hamiltonian (2.23) with respect to the physical variables are shown below to be

$$\frac{\delta M}{\delta M_i} = v^i, \tag{2.26a}$$

$$\frac{\delta H}{\delta \hat{\rho}} = c^2 w/\gamma, \tag{2.26b}$$

$$\frac{\delta H}{\delta \eta} = \sqrt{g}\ \rho_f T_f. \tag{2.26c}$$

We shall soon see that substituting the variational identities (2.26a-c) into $\partial_t F=\{H,F\}$ with Lie - Poisson bracket (1.18) and Hamiltonian (2.23) will yield the relaticistic adiabatic system in the laboratory frame (2.9a-c). However, first we show how these identities arise. By (2.4) and (2.4′), we find, as a preliminary step, that

$$c^2 dw = \rho_f^{-1} \frac{\partial p_f}{\partial \rho_f} d\rho_f + \left[\rho_f^{-1} \frac{\partial p_f}{\partial \eta_f} + T_f \right] d\eta_f. \tag{2.27}$$

Also, from (2.15) and (2.21) we hawe

$$1 - (\hat{\rho}/\hat{\vartheta})^2 = 1-\gamma^{-2} = |\underline{v}|^2/c^2 = |\underline{M}|^2/(c\hat{\vartheta})^2, \tag{2.28}$$

so that

$$\hat{\vartheta} = \sqrt{(\hat{\rho} w)^2 + |\underline{M}|^2/c^2}. \tag{2.29}$$

Using formulae (2.27), (2.29), and $\hat{\rho}=\rho_f\gamma\sqrt{g}$, $\eta=\eta_f$ we obtain from (2.25)

$$\frac{\delta H}{\delta M_i} = c^2 \frac{\partial \hat{\vartheta}}{\partial M_i} - \sqrt{g}\ \frac{\partial p_f}{\partial M_i} = \frac{c^2}{\hat{\vartheta}} \left[M^i/c^2 + \hat{\rho}^2 w \frac{\partial w}{\partial \rho_f} \frac{\partial \rho_f}{\partial M_i} \right] - \sqrt{g}\ \frac{\partial p_f}{\partial \rho_f} \frac{\partial \rho_f}{\partial M_i}$$

$$= M^i/\hat{\vartheta} + \frac{\partial \rho_f}{\partial M_i} \left[\frac{c^2 \hat{\rho}}{\gamma} \frac{\partial w}{\partial \rho_f} - \sqrt{g}\ \frac{\partial p_f}{\partial \rho_f} \right] = M^i/\hat{\vartheta} = v^i \tag{2.30}$$

which proves (2.26a). Next, by (2.27) and $\hat{\rho}=\rho_f\gamma\sqrt{g}$.

$$\frac{\delta H}{\delta \hat{\rho}} = \frac{c^2}{\hat{\vartheta}} \hat{\rho} w \frac{\partial (\hat{\rho} w)}{\partial \hat{\rho}} - \sqrt{g}\ \frac{\partial p_f}{\partial \hat{\rho}} = \frac{c^2}{\gamma} \left[w + \hat{\rho} \frac{\partial w}{\partial \hat{\rho}} \right] -$$

$$+ \sqrt{g}\ \frac{\partial p_f}{\partial \hat{\rho}} = c^2 w/\gamma + \sqrt{g}\ \left[c^2 \rho_f \frac{\partial w}{\partial \rho_f} - \frac{\partial p_f}{\partial \rho_f} \right] \frac{\partial \rho_f}{\partial \hat{\rho}} = c^2 w/\gamma, \tag{2.31}$$

which is (2.26b). Finally

$$\frac{\delta H}{\delta \eta} = \frac{c^2}{\hat{\vartheta}} \hat{\rho} w \frac{\partial \hat{\rho} w}{\partial \eta} - \sqrt{g} \frac{\partial p_f}{\partial \eta} = \frac{c^2}{\gamma} \hat{\rho} \frac{\partial w}{\partial \eta} - \sqrt{g} \frac{\partial p_f}{\partial \eta} =$$

$$= \sqrt{g} \left[\rho_f c^2 \frac{\partial w}{\partial \eta} - \frac{\partial p_f}{\partial \eta} \right] = \sqrt{g}\, \rho_f T_f, \tag{2.32}$$

which is (2.26c).

Substituting the variational derivatives (2.26a-c) into the Lie - Poisson bracket (1.18) leads immediately to

$$\partial_t \hat{\rho} = \{H, \hat{\rho}\} = -(\hat{\rho} v^j)_{,j}, \tag{2.33}$$

$$\partial_t \eta = \{H, \eta\} = - \eta_{,j} v^j, \tag{2.34}$$

$$\partial_t M_i = \{H, M_i\} = -M_j v^j_{,i} - (M_i v^j)_{,j} - \gamma^{-1} \hat{\rho} c^2 w_{,i}$$

$$-\hat{\rho} c^2 w (\gamma^{-1})_{,i} + \gamma^{-1} \hat{\rho} T_f \eta_{,i}. \tag{2.35a}$$

Substituting (2.27) into (2.35) leads to

$$\partial_t M_i = -M_j v^j_{,i} - (M_i v^j)_{,j} - \sqrt{g}\, p_{f,i} - \hat{\rho} c^2 w (\gamma^{-1})_{,i} \tag{2.35b}$$

Using (2.33) and the definition $M_i = \hat{\rho} \gamma w v_i$ in Eq. (2.35b) yields

$$\partial_t (\gamma w v_i) = -\gamma w v_j v^j_{,i} - v^j (\gamma w v_i)_{,j} - \frac{\sqrt{g}}{\hat{\rho}} p_{f,i} - c^2 w (\gamma^{-1})_{,i} \tag{2.36}$$

Now, $(\gamma^{-1})_{,i}$ in (2.36) is given by

$$(\gamma^{-1})_{,i} = \frac{-\gamma}{c^2} v_j v^j_{,i} - \frac{\gamma}{2c^2} v^j v^k g_{jk.i}, \tag{2.37}$$

so that (2.36) becoms

$$\partial_t (\gamma w v_i) = -v^j (\gamma w v_i)_{,j} - \frac{\sqrt{g}}{\hat{\rho}} p_{f,i} + \frac{1}{2} \gamma w v^j v^k g_{jk,i} \tag{2.38}$$

Then, rearranging terms in (2.38) as in (1.28-30) gives

$$\partial_t (\gamma w v_i) = -v^j (\gamma w v_i)_{;j} - \frac{\sqrt{g}}{\hat{\rho}} p_{f,i}, \tag{2.39}$$

which is the equation of motion (2.9c).

Thus, the Lie - Poisson bracket (1.18) and Hamiltonian (2.25) yield the special relativistic adiabatic fluid Eqs. (2.9a-c), directly in terms of the Eulerian physical variables, $\{\hat{\rho}, \eta, M_i\}$. This implies, because the map (2.22) is canonical (i.e., preserves Poisson brackets), that the Lagrangian equations in the canonically conjugate variables (P_A, q^A) are also equivalent to Eqs. (2.9a-c).

Remarks A. The Lie - Poisson bracket (1.18) for SRAF appears in Iwinski and Turski (1976), where electromagnetic interactions via Maxwell's equations are included, as well. We will return to this Lie - Poisson braket in Part II. (Iwinski and Turski (1976) also presents a Lie - Poisson bracket for the SR Maxwell - Vlasov system, which is not discussed here). The present derivation of this Lie - Poisson bracket for SRAF illustrates its relation to the corresponding symplectic bracket in Lagrangian fields, $q^A(\underline{\mathbf{x}}, t)$ and $P_A(\underline{\mathbf{x}}, t)$.

B. Using (2.27), the motion Eq. (2.36) can be expressed in Lie - derivative form as

$$\partial_t(\gamma w v_i dx^i) = \mathcal{L}_{\underline{v}}(\gamma w v_i dx^i) - d(c^2 w/\gamma) + \gamma^{-1} T_f d\eta, \tag{2.40}$$

where d denotes exterior derivative and $\mathcal{L}_{\underline{v}}$ is the Lie derivative with respect to the vector field $v^i\partial_i$. Similarly, Eqs. (2.9a,b) can be expressed as

$$\partial_t(\hat{\rho} d^n x) = - \mathcal{L}_{\underline{v}}(\hat{\rho} d^n x), \tag{2.41a}$$

$$\partial_t \eta = \mathcal{L}_{\underline{v}} \eta. \tag{2.41b}$$

As a consequence of the properties $d^2=0=[d, \mathcal{L}_{\underline{v}}]$ of the exterior and Lie derivatives, and the antisymmetry or the exterior product, we find from (2.40) and (2.41a) that the three - form $d(\gamma w v_i dx^i)^\wedge d\eta$ is preserved along flow lines, i.e.,

$$\partial_t[d(\gamma w v_i dx^i)^\wedge d\eta] = - \mathcal{L}_{\underline{v}}[d(\gamma w v_i dx^i)^\wedge d\eta]. \tag{2.42}$$

Consequently, in three dimensions (n=3) we find as in section 1.5 that

$$\partial_t \Omega = -\mathcal{L}_{\underline{v}} \Omega \,, \tag{2.43}$$

where the scalar Ω is defined to be

$$\Omega = (\hat{\rho})^{-1} \mathrm{curl}(\gamma w \underline{v}) \cdot \underline{\nabla} \eta \tag{2.44}$$

Thus, for n=3 we have the following conservation law for special relativistic adiabatic fluids, $\partial C, /\partial t=0$ for

$$C = \int d^3x \hat{\rho} \phi(\eta, \Omega), \tag{2.45}$$

for an arbitrary function ϕ of the two indicated variables in (2.45). The conserved functional C in (2.45) is the special relativistic versin of the "Casimir" functional (1.39) for the Lie - Poisson bracket (1.18). Its use for determining stability criteria for special relativistic fluids is discussed in Holm and Kupershmidt (1984b, 1986) and in Part II of the lectures.

C. Having understood the special relativistic, case, the Hamiltonian structure of general relativistic adiabatic fluids can be investigated by using essentially the same method again, modulo changes to include the metric tensor g_{ij} and its canonically conjugate momentum as additional dynamical variables, see Holm (1985).

3. COMMENTS ON CASIMIRS AND LYAPUNOV STABILITY.

This part of the lectures has focused on the common features of the Hamiltonian structure of NRAF and SRAF. The main result of this unified treatment is that in both theories the fluid variables share the same Lie - Poisson bracket, when expressed in terms of the appropriate spaces of physical variables constructed here. As discussed in the Introduction, one of our motivations for presenting this work as explicitly as possible is to facilitate "technology transfer", i.e., so that Hamiltonian techniques can be more widely applied in fluid dynamics, particularly in the study of dynamic Lyapunov stability.

A first step in this direction is to notice that certain (nondegenerate) equilibrium adiabatic fluid flows can be associated to critical points of the sum of the Hamiltonian H in (2.25) and the Casimirs C in (2.45) with an appropriate choice of the fuction ϕ. Consider the following Proposition: **with H and C defined in (2.25) and (2.45), respectively, critical points of the sum H+C are equilibrium states of SRAF dynamics.**

To understand this Proposition, we first digress to discuss Casimirs further. Recall that the Casimirs have vanishing Poisson bracket (i.e., they "Poisson - commute") with any functional of the dynamical variables in the set $\{\hat{\rho}, \eta, M_i\}$. Thus, Casimirs are conservation laws, since they Poisson-commute withe the Hamiltonian; but they are only kinematic, since their conservation is independent of the choice of the Hamiltonian, H, which generates the dynamics in this space of variables, under the rule for the Hamiltonian formalism,

$$\partial_t F = \{H, F\}. \tag{3.1}$$

Here, { , } is the Lie - Poisson bracket for the fluid variables defined on the dual space of the semidirect product Lie algebra of vector fields acting on differential forms, and F is any functional of the dynamical variables. The Casimirs comprise an infinite family of these kinematic conservation laws, since they contain arbitrary (integrable) functions in their definition (2.45).

There are several explanations of how the Casimirs arise. From the wiewpoint of the present work, they appear via Noether's theorem. The action (2.11) in the Lagrangian configuration space admits the following symmetry trasformation: relabel the Lagrangian variables so as to preserve the values of the physical fluid variables. A standard computation using Noether's theorem yields the conservation of C by virtue of the Lagrangian relabelling symmetry of the action (2.11). This symmetry trasformation also plays a role in Lagrangian stability theory in astrophysics: Lagrangian variations that preserve the value of Ω in (2.44) are "nontrivial perturbations" in the sense of Friedman and Shutz (1978) and Friedman (1978), in Lagrangian perturbation theory.

Now, as for the Proposition. By definition, a Casimir C satisfies

$$\{C, F\} = 0, \forall F. \tag{3.2}$$

So C generates no dynamics. In particular, the sum $H_C := H+C$ generates the same dynamics as the Hamiltonian H does alone. The critical states of H_C are equilibrium states of the dynamics, since the Poisson bracket $\{H_C, F\} = \{H+C, F\}$ vanishes when the first variation of H_C vanishes, i.e., when

$$0 = \delta H_C := DH_C(\hat{\rho}, \eta, M_i) \cdot (\delta\hat{\rho}, \delta\eta, \delta M_i), \tag{3.3}$$

for arbitrary variations $(\delta\hat{\rho}, \delta\eta, \delta M_i)$. This observation proves the Proposition above.

Thus, the existence of Casimirs (2.45) with their freedom in the choice of the function associates classes of equilibrium states with the critical states of certain functionals, H_C. The Lyapunov stability of these equilibria can then be studied by establishing whether there esist sufficient conditions, imposable on an equilibrium states (i.e., the "stability conditions", under which we may obtain one (or both) of the following two situations. First, suppose $\delta^2 H_C$ is definite, i.e., the second variation of H_C evaluated at the equilibrium state is definite in sign, under certain conditions on the equilibrium state. This situation implies linearized Lyapunov stability in terms of the norm given by $\delta^2 H_C$, which is conserved by the linearized dynamics (since, for Lie - Poisson systems, $\delta^2 H_C$ is the Hamiltonian for the linearized dynamics, see Abarbanel et al. (1984b)). Second, suppose the conserved functional H_C is convex, i.e., the variation of H_C from its equilibrium value is bounded above and below by positive - definite quadratic forms; so that bounding norms exist for H_C. This situation implies Lyapunov stability for **finite** amplitude perturbations, in terms of the bounding norms. For a general description of Lyapunov analysis by this method as applied to nonrelativistic ideal fluid and plasma equilibria, see Holm, Marsden. Ratiu, and Weistein (1985). An application of this method to relativistic plasmas is given in Part II of these lectures.

REFERENCE (Part I)

H.D. I.Abarbanel, D.D. Holm, J.E.Marsden, and T. Ratiu (1984). Richardson number criterion for the nonlinear stability of three dimensional stratified flow. **Phys. Rev. Lett. 52,** 2352-2355.

H.D. I.Abarbanel, D.D. Holm, J.E.Marsden, and T. Ratiu (1986). Nonlinear stability analysis of stratified ideal fluid equilibria. Phil Trans. Roy. Soc. (London) A **318**, 349-409.

V.I. Arnold (1965), Conditions for nonlinear stability of stationary plane curvilinear flows in an ideal fluid. **Sov. Math. Dokl. 162** (5), 773-777.

V.I. Arnold (1966). Sur la géometrie differentielle des groupes de Lie de dimension infinie et ses applications a l'hydrodynamique des fluids parfaits. **Ann. Inst. Fourier** (Grenoble) 16, 319-361.

V.I. Arnold (1969). On an a priori estimate in the theory of hydrodynamical stability. **Am. Math. Soc. Transl. 79,** 267-269.

I. Bialynicki - Birula and Z. Iwinski (1973). Canonical formulation od relativistic hydrodynamics. **Rep. Math. Phys. 4,** 139-151.

R.F. Dashen and D.H. Sharp (1968). Currents as coordinates for hadrons. **Phys. Rev. 165,** 1857-1878.

T.E. Dzyloshinskii and G.E. Volovick (1980). Poisson bracket in condensed matter physics, **Ann. of Phys.** (New York) **125**, 67-97.

J.L. Friedman (1978). Generic instability of rotating relativistic stars. **Comm. Math. Phys. 62,** 247-278.

J.L. Friedman and B.F.Schutz (1978). Lagrangian perturbation theory of nonrelativistic fluids. **Astrophys. J. 221,** 937-957.

G.A. Goldin and D.H.Sharp (1972). Functional differential equations determining representations of local current algebras. In **Magic without Magic: J.A.Wheeler Festchrift,** ed. J.R.Klauder, pp. 171-185. W.H.Freeman: San Francisco.

R.D. Hazeltine, D.D. Holm, J.E. Marsden, and P.J. Morrison (1984) Generalized Poisson brackets and nonlinear Lyapunov stability - application to reduced MHD, ICCP Proceedings (Lausanne) **2,** 204-214.

D.D. Holm (1984a). Stability of planar multifluid plasma equilibria by Arnold's method. **Contemp. Math. AMS 28,** 25-50.

D.D. Holm (1984b). Notes on Noether's theorem (prepint).

D.D. Holm (1985). Hamiltonian formalism for general - relativistic adiabatic fluids. Physica D17, 1-36.

D.D. Holm and B.A.Kupershnidt (1983). Poisson brackets and Clebsch representations for magnetohydrodynamics, multifluid plasmas, and elasticity. **Physica** D **6,** 347-363.

D.D. Holm and B.A. Kupershmidt (1984). Relativistic fluid dynamics as a Hamiltonian system, **Phys. Lett.** A **101,** 23-26

D.D. Holm and B.A. Kupershmidt (1984b). Lyapunov stability conditions for a relativistic multifluid plasma, ICCP Proceedings (Lausanne), p.214.

D.D. Holm and B.A. Kupershmidt (1985). Relativistic magnetohydrodynamics as a Hamiltonian system, Comptes Rendus Acad. Sci. Paris, **300** (Ser. I) 153-156.

D.D. Holm and B.A. Kupershmidt (1986). Lyapunov stability of relativistic fluids and plasmas, Phys. Fluids **29**, 49-68.

D.D. Holm, B.A. Kupershmidt, and C.D. Levermore (1983). Canonical maps between Poisson brackets in Eulerian and Lagrangian descriptions of continuum mechanics. **Phys. Lett.** A 98, 389-395.

D.D. Holm, J.E. Marsden, and T. Ratiu (1986a). **Hamiltonian Structure and Lyapunov Stability for Ideal Continuum Dynamics** (Univ. Montreal Press).

D.D Holm, J.E.Marsden, and R. Ratiu (1986b). Nonlinear stability of kelvin - Stuart cat's eyes flows, Lect. in Appl. AMS **23**, 171-186.

D.D. Holm, J.E.Marsden, T. Ratiu, and A. Weistein (1983). Nonlinear stability conditions and a priori estimates for barotropic hydrodynamics. **Phys. Lett.** A **98,** 15-21.

D.D. Holm, J.E. Marsden, T. Ratiu, and A. Weistein (1985). Nonlinear stability of fluid and plasma equilibria. **Phys. Rep. 123,** 1-116.

Z. Iwinski and K. Turski (1976). Canonical theories of systems interacting electromagnetically. **Lett. in Appl. Sci and Eng. 4**, 179-191.

N.E. Kochin, I.A. Kibel, and N.V. Roze (1964) **Theoretical Hydromechanics.** Wiley: New York, pp. 176-178.

H.P. Künzle and J.M. Nester (1984). Hamiltonian formulation of gravitating perfect fluids and the Newtonian limit. **J.Math. Phys. 25,** 1009-1018.

L. Laudau (1941). The theory of superfluidity of Helium II. J. Phys. **Moscow, 5,** 71-90. Collected Works, pp. 185-204. Plenum: New York.

S. Lie (1980). Theorie der Transformationsgruppen, (Zweiter Abschnitt unter mitwirkung von Prof. Dr. Friendrich Engel), Teubner, Leipzig, Ch. 19 and Ch. 20.

J.E. Marsden, T. Ratiu, and A. Weistein (1984a). Semidirect products and reduction in mechanics, **Trans. Am. Math. Soc. 281.** 147-177.

J.E. Marsden, T. Ratiu, and A. Weinstein (1984b), Reduction and Hamiltonian structures on duals of semidirect product Lie algebras. **Cont. Math. AMS** 28, 55-100.

J.E. Marsden, A. Weistein, T. Ratiu, R. Schmid, and R.G.Spencer (1983). Hamiltonian systems with symmetry, coadjoint orbits, and plasma physics, Proc. IUTAM-ISIMM Symp. "Modern develpments in Analytical Mechanics", **Atti della Acad. Scienze di Torino, 117,** 289-340.

P.J. Morrison and J.M. Greene (1980). Noncanonical Hamiltonian density formulation of hydrodynamics and ideal magnetohydrodynamics.

Phys. Rev Lett. 45, 790-794.

T. Ratiu (1981). Euler - Poisson equations on Lie algebras and the N - dimension heavy rigid body, **Proc. Nat. Acad. Sci USA 78,** 1327-1328.

P. Similon, A.N. Kaufman, and D.D. Holm (1984). Ponderomotive Hamiltonian and Lyapunov stability for magnetically confined plasma in the presence of RF field. **Phys. Lett. 106A,** 29-33.

E.C.G. Sudarshan and N.Mukunda (1974). **Classical Dynamics: A Modern Perspective,** Wiley: New York.

A.H. Taub (1967). Relativistic hydrodinamics. In **Relativity Theory and Astrophysics 1. Relativity and Cosmology,** ed. J. Ehlers pp. 170-193. Lect Appl. Math. 8, AMS: New York.

A.Weistein (1984). The structure of Poisson manifolds, J. Diff. Geom. 18, 523-557.

PART II

LYAPUNOV STABILITY OF RELATIVISTIC FLUIDS AND PLASMAS

ABSTRACT

Lyapunov stability of relativistic ideal fluid and plasma equilibria is studied analitycally using the energy-Casimir method. Two dimensional relativistic equilibria in a fixed bounded domain are investigated whithin the framework of the macroscopic multifluid plasma model. Linearized Lyapunov stability conditions and stability norms are given, accounting for warm-plasma effects as well as relativistic and electromagnetic effects. The resulting Lyapunov stability conditions are compared to spectral stability analyses for relativistic cold plasmas in various examples and special cases, including 1) non-neutral electron flow in a planar diode, and 2) circulary-symmetric plasma flow enclosed in a coaxial waveguide. These linearized stability results can be extended readily to give nonlinear Lyapunov stability conditions for finite-amplitude perturbations by employng standard convexity arguments for the Lyapunov functions given here. The relativistic stability conditions are shown to reduce to their nonrelativistic counterparts.

INTRODUCTION AND DISCUSSION OF HAMILTONIAN STABILITY CONCEPTS

There are a number of fluid dynamical situations in which relativistic effects are important. Such situations occur when either the velocity of macroscopic motion is comparable to the speed of light, or when there is a sufficiently rapid microscopic motion. In astrophysics, for example, stars are commonly modeled as self-gravitating fluid bodies with relativistically high energy density and temperature. Relativistic fluid dynamics also applies in models of high-density charged-particle beams and free-electron lasers.

Here, in the context of charged-particle beams, we shall be interested in sufficiently high beam particle density that a hydrodynamic description of the beam is valid in a fixed, bounded domain. In particular, we shall be interested in small-amplitude disturbances in the neighborhood of an equilibrium state which is laminar. The case of relativistic streams of charged particles at low density, for which cooperative interactions dominate particle-particle scattering within the beam, has been studied extensively in the context of the Vlasov-Maxwell equations, beginning with Watson, Bludmam, and Rosenbluth[1]. Low density relativistic charged-particle beams are still being actively studied, but they are not our topic here.

This part of the lectures repeats some of the material in Holm and Kupershmidt[2]. The aim here is to examine stability of special relativistic adiabatic fluids, particularly how Lyapunov stability conditions for such fluids depend on the relativistic parameter $\gamma=(1-|\underline{v}|^2/c^2|^{-1/2}$ where $\underline{v}$ is fluid velocity and c is the speed of light. More specifically, we use the Hamiltonian formalism discussed in Part I to draw out the general properties of ideal fluids and show how these properties determine Lyapunov stability conditions for equilibrium states of charged and neutral relativistic fluids in two dimensions.

Before explaining the method we use and the types of results obtained from it, we recall some terminology regarding stability. An evolution equation

$$\partial_t \underline{u} = \underline{F}(\underline{u}) \tag{1}$$

for elements, $\underline{u}$ of a normed linear space may have equilibrium, or stationary solutions $\underline{u}_e$ satisfying $\underline{F}(\underline{u}_e)=0$. Such a solution is called <u>Lyapunov stable</u>, when any solution $\underline{u}(t)$ beginning near $\underline{u}_e$ at time t=0 remains near $\underline{u}_e$ for all time. Formally, this means that for every $\varepsilon>0$ there is a $\delta>0$, such that when the norm $\|\underline{u}(0)-\underline{u}_e\|<\delta$, then the norm $\|\underline{u}(t)-\underline{u}_e\|<\varepsilon$ for $-\infty<t<\infty$.

Assuming $\underline{F}$ is differentiable, the <u>linearized equation</u> at $\underline{u}_e$ is given by

$$\partial_t \delta\underline{u} = D\underline{F}(\underline{u}_e)\cdot\delta\underline{u} \ , \tag{1}$$

which describes the evolution of the infinitesimal disturbance, $\delta\underline{u}$, of the original equilibrium solution $\underline{u}_e$, where $D\underline{F}(\underline{u}_e)$ is the linearization of $\underline{F}$ evaluated at $\underline{u}_e$. If the solution of this linearized equations is Lyapunov stable, then the equilibrium solution $\underline{u}_e$ for the original problem is said to be linearized stable, or linearly stable. A necessary condition for such linearized Lyapunov stability is that the spectrum of the linearized operator $\underline{DF}(\underline{u}_e)$ lie on the imaginary axis. This situation is called neutral stability, or spectral stability. This spectral property suggests, but does not prove linearized Lyapunov stability; since algebraically unstable neutral modes can preclude Lyapunov stability. Spectral stability of relativistic cold plasma has been investigated recently in Chernin and Lau[3] and Davidson, Tsang, and Swegle[4].

In these lectures, Lyapunov stability is studied rather than spectral stability: both the stability conditions and the Lyapunov stability norms are derived here for two dimensional plasma equilibria that are nondegenerate (in a certain sense defined for each case in the text, but meaning essentially that the flow is not static and the gradient of Bernoull's law is not trivially zero). Also, warm-plasma effects are included, and the stability results are compared to those for relativistic cold plasmas in various examples and special cases, in section 3 for the two dimensional barotropic case.

The energy-Casimir stability method we use is a development of the traditional Lyapunov method of stability analysis. This development incorporates recent advances in the Hamiltonian theory of continuum systems for the study of fluid stability. In particular, our stability analysis takes advantage of the Hamiltonian structure of the relativistic problem as discussed in Iwinski and Turski[5] and Holm and Kupershmidt[6]. The main ideas of the energy - Casimir method were first proposed by Arnold[7,8] in the context of the incompressible Euler equations. The method was developed into an algorithm and used by Holm et al.[9,10] to study nonlinear stability of various models of nonrelativistic fluids and plasmas. A related method, used for investigating stability criteria of nonrelativistic low-density plasmas, is due to Newcomb[11] and Kruskal and Oberman[12]. The energy-Casimir method is based on a few general facts about equilibrium points of Hamiltonian systems which we now recall.

A hamiltonian system is a dynamical system which can be written as $\delta_t F=\{H,F\}$ where ∂_t is the time derivative, H is the Hamiltonian, and $\{,\}$ is the Poisson bracket, which is bilinear, skew-symmetric and satisfies the Jacobi identity. In particular, if $\underline{u}=\{u_i\}$ is the set of dynamical variables, their Hamiltonian motion is given by

$$\delta_t u_i = B_{ij} \frac{\delta H}{\delta u_j} , \tag{3}$$

where B_{ij} is a matrix differential operator, $\delta H/\delta u_j$ denotes functional derivative of H with respect to u_j, and we sum over repeated indices unless stated otherwise.

A functional C is called a Casimir for the Poisson bracket defined by the Hamiltonian matrix B_{ij}, if it produces null equations of motion, that is, if $B_{ij}(\delta C/\delta u_j)=0$. It follows that if $\underline{u}$ is a critical point of the functional $H_C=H+C$, i.e. if $\delta H_C/\delta \underline{u}=0$, then $\underline{u}$ is an equilibrium state of the motio equations (3). The converse statement, namely, than an equilibrium flow of (3) can be represented as a critical point of the energy plus Casimir, H_C, for appropriate C, requires separate analysis, which we give for several fluid systems in the course of the paper.

The next general fact of interest in the context of the energy-Casimir method concerns the linearization of hamiltonian systems. To linearize equations (3) abaut a solution $\underline{u}$, one substitutes into (3) (with H replaced by H_C) the quantity $\underline{u}+\varepsilon\underline{v}$ instead of $\underline{u}$ and retains terms up to first order in ε. The resulting equations are again Hamiltonian, as shown in Kupershmidt[13]. Indeed, they can be written in the form

$$\partial_t \begin{pmatrix} \underline{u} \\ \underline{v}_t \end{pmatrix} = \left(\begin{array}{c|c} \vartheta & B \\ \hline B & \lambda(B) \end{array}\right) \begin{pmatrix} d\tilde{H}/\delta\underline{u} \\ \delta\tilde{H}/\delta\underline{v} \end{pmatrix} , \tag{4}$$

where the linearized Hamiltonian $\tilde{H}$ is given by

$$\tilde{H}=\underline{v}\cdot \frac{\delta H}{\delta \underline{u}} C= \lim_{\varepsilon\to 0} \frac{H_C(\underline{u}+\varepsilon\underline{v})-H_C(\underline{u})}{\varepsilon} \tag{5}$$

and the quantity

$$\lambda(B)= \lim_{\varepsilon\to 0} \frac{B(\underline{u}+\varepsilon\underline{v})-B(\underline{u})}{\varepsilon}$$

is the linearization of the matrix B (so that λ(B) is linear in $\underline{v}$). In particular, when $\underline{u}_c$ is a critical point of H_C, i.e., $\frac{\delta H}{\delta u} C(\underline{u}_c)=0$ in

(5), then $\underline{u}_c$ is an equilibrium solution, $\delta\tilde{H}/\delta\underline{v}=0$, and (4) reduces to

$$\partial_t \underline{v} = B_c \left[\frac{\delta\tilde{H}}{\delta\underline{u}} \right] . \tag{6}$$

wher B_c here means B evaluated at u_c.

Since $\dfrac{\delta\tilde{H}}{\delta\underline{u}} = D\left(\dfrac{\delta H}{\delta\underline{u}}\ C\right)\cdot\underline{v}$, with D the Frechét differential, (6) becomes

$$\partial_t \underline{v} = B_c \left[D\left(\frac{\delta H}{\delta\underline{u}}\ C\right)\cdot\underline{v}\right] . \tag{7}$$

Now, since $D\left(\dfrac{\delta H}{\delta\underline{u}}\ C\right)$ is a symmetric operator, we have

$$D\left(\frac{\delta H}{\delta\underline{u}}\ C\right)\cdot\underline{v} = \frac{\delta}{\delta\underline{v}}\ \frac{1}{2}\ \underline{v}\cdot D\left(\frac{\delta H}{\delta\underline{u}}\ C\right)\cdot\underline{v} , \tag{8}$$

so that we finally obtain

$$\partial_t \underline{v} = B_c \left[\frac{\delta}{\delta\underline{v}} \left(\frac{1}{2}\ \underline{v}\cdot D\left(\frac{\delta H}{\delta\underline{u}}\ C\right)\cdot\underline{v}\right)\right] . \tag{9}$$

Thus, the linearization of the Hamiltonian system around an equilibrium state which is a critical point of its Hamiltonian-plus-Casimir, H_C, is again a Hamiltonian system (9). For the case when the Hamiltonian matrix B is of Lie-Poisson type, this linearization property is proven in Holm et al.[10] and discussed further in Abarbanel et al.[14]

Since, by definition,

$$\underline{v}\cdot D\left(\frac{\delta H}{\delta\underline{u}}\ \underline{C}\right)\cdot\underline{v} = \delta^2 H_C(\underline{v},\underline{v}) , \tag{10}$$

evaluated at equilibrium, we see that one-half the second variation $\frac{1}{2}\ \delta^2 H_C$ is the hamiltonian of the linearized system (9), and so, is preserved in time by the dynamics of the linearized flow. Consequently, if δ^2H_C is definite in sign as a quadratic form, then δ^2H_C provides a conserved norm that measures deviations from equilibrium of an initial disturbance under the linearized dynamics. Therefore, the conditions on the equilibrium flow for δ^2H_C to be definite are sufficient conditions for linear Lyapunov stability. That is, a flow that starts near an equilibrium solution satisfying these conditions and evolves under the linearized dynamics, will remain in a neighborhood of this solution, as measured by the norm derived from δ^2H_C. This is the essence of the energy-Casimir method.

In summary, the energy-Casimir stability method we use is based on two main ideas: 1) characterization of equilibrium flows as critical points of certain functionals; and 2) preservation in time of the second variations of these functionals, considered as norms for Lyapunov stability.

Using these two main ideas, this part of the lectures presents parallel treatments of successively more intricate fluid theories, in the sections indicated below.

1. Two-dimensional nonrelativistic barotropic fluid dynamics.
2. Two-dimensional relativistic barotropic fluid dynamics.
3. Two-dimensional barotropic relativistic multifluid plasma.

In each section, the presentation follows the same pattern. First, we identify the Hamiltonian structure and the associated Casimirs of the fluid model under consideration. (This step recapitulates some of the results of Part I of the lectures, but is included for the sake of continuity and consistent notation). Next, we show that any nondegenerate equilibrium state (where nondegenerate is defined for each fluid theory via a Bernoulli relation) is a critical point of the sum of the hamiltonian and a Casimir for each model. Finally, we calculate the second variation of this sum and find conditions on these equilibrium states for the second variation to be positive, thereby obtaining conditions for linear Lyapunov stability, as explained above. Explicit examples are given and compared with the results of spectral analysis in section 3.

We note that each of the relativistic fluid systems treated here is a regular, structure-preserving deformation, with parameter c^{-2}, of the corresponding nonrelativistic theory. Hence, the linear Lyapunov stability results given here can be extended to provide nonlinear

Lyapunov stability conditions, by employing convexity arguments as done for the nonrelativistic cases treated in Holm et al.[10]. This extension is not done explicitly in the lecture notes, however, since it follows readily from the earlier work.

Acknowledgment. The material in Part II of these lecture notes was developed in Holm and Kupershmidt[2], which also treats the three-dimensional case.

1. TWO-DIMENSIONAL NONRELATIVISTIC FLUID DYNAMICS

1.1. Equations of Motion and Hamiltonian Structure.

In ideal fluid dynamics, the physical variables are: ρ, mass density; η, specific entropy; and $\underline{M}$, fluid momentum density. The fluid moves through a fixed domain D in Euclidean space R^n with positions x^i, $i=1,2,\ldots,n$. (Later in this section we will take n=2 and specialize to the barotropic case, where pressure is a function of mass density only). In the nonrelativistic case, the momentum density is related in the fluid velocity, $\underline{v}$, by

$$\underline{M}=\rho\underline{v} \ . \tag{11}$$

The Eulerian hydrodynamics equations for an <u>adiabatic</u> fluid are

$$\partial_t\rho=-(\rho v_i)_{,i} \ , \qquad \partial_t\eta=-v_i\eta_{,i} \ , \tag{12}$$

$$\partial_t M_i=-T_{ij,j} \ , \tag{13}$$

$$T_{ij}=M_i v_j+\delta_{ij}p \ . \tag{14}$$

where ∂_t denotes partial time derivative at fixed $\underline{x}$, repeated indices are summed over $i,j=1,2,\ldots,n$, and subscript comma followed by a Latin index denotes partial space derivative with respect to the indicated component. Eq. (12a) is the continuity equation. Eq. (12b) is the adiabatic condition for each fluid element. Eq. (13) is the fluid motion equation expressed in conservative form, with nonrelativistic stress tensor T_{ij} in (14). The fluid pressure p is determined as a function of ρ and η from a prescribed relation (equation of state) for the specific internal energy $e(\rho,\eta)$, combined with the first law of thermodynamics $de=e_{\rho}d\rho+e_{\eta}d\eta=\rho^{-2}pd\rho+Td\eta$, where T is the temperature.

In Part I of these lectures we showed that the hydrodynamic system (12)-(14) can be expressed as a Hamiltonian system $\partial_t F=\{H,F\}$, with Hamiltonian

$$H=\int_D [\,|\underline{M}|^2/2\rho+\rho e(\rho,\eta)]d^nx \quad , \tag{15}$$

which is equal to the nonrelativistic energy. The hydrodynamic Poisson bracket {H,F} for functionals H and F of the variables (ρ,η,M_i) is derived in Eq. (1.18) of Part I namely,

$$\{H,F\}=-\int d^nx\left\{ \frac{\delta F}{\delta\rho}\,\partial_j\rho\,\frac{\delta H}{\delta M_j} + \frac{\delta F}{\delta\eta}\,\eta_{,j}\,\frac{\delta H}{\delta M_j} + \right.$$

$$\left. + \frac{\delta H}{\delta M_i}\left[\rho\partial_i\,\frac{\delta H}{\delta\rho} - \eta_{,i}\,\frac{\delta H}{\delta\eta} + \left(M_j\partial_i+\partial_jM_i\right)\frac{\delta H}{\delta M_j}\right]\right\} \tag{16}$$

where the operator ∂_i operates to the right on all terms that follow it. The Lie-Poisson bracket (16) (without the entropy terms) is found in Dashen and Sharp[15], Bialynicki-Birula and Iwinski[16], Iwinski and Turski[5], Enz and Turski[17], Dzyaloshinskii and Volovick[18], ans Morrison and Greene[19]. This bracket was derived from Clebsch variables and interpreted mathematically as the Lie-Poisson bracket associated to the dual space of a semidirect-product Lie algebra by Holm and Kupershmidt[20]. This and other brakets for classical continuum physics in the Eulerian representation are derived from canonical brackets in the Lagrangian representation in Holm, Kupershmidt, and Levermore[21], Marsden et al.[22], and Marseden, Ratiu and Weinstein[23]. The general-relativistic version of this Lie-Poisson bracket is obtained heuristically in Bao, Marsden, and Walton[24] and constructed systematically in Holm[25]. For an exceptionally clear derivation of (16), see Kaufman[26]. Holm, Marseden, and Ratiu[27] review the abstract mathematical framework of Lie-Poisson brackets for various representations of ideal continuum models.

The hidrodynamic equations (12)-(14) are given by $\partial_t F=\{H,F\}$, with $F\in\{\rho,\eta,M_i\}$ and Poisson bracket (16), upon using the following variational derivatives

$$\frac{\delta H}{\delta \underline{\mathbf{M}}}=\underline{\mathbf{v}}\ , \tag{17}$$

$$\frac{\delta H}{\delta \rho}=-\frac{1}{2}|\underline{\mathbf{v}}|^2+e+p/\rho\ , \tag{18}$$

$$\frac{\delta H}{\delta \eta}=\rho\partial e/\partial\eta=\rho T\ . \tag{19}$$

Remark. The quantity $\int\rho d^n x$ (the total mass) is the kernel of the Poisson bracket (16), i.e. $\left\{\int\rho d^n x,F\right\}=0$, $\forall F$, so H in (15) can be changed to $H'=H+\alpha\int\rho d^n x$ for any constant α without affecting the hydrodynamic equations. Other functionals in the kernel of Poisson bracket (16) are, for arbitrary functions ψ,ϕ of their indicate arguments,

$$\int\rho\psi(\eta)d^n x\ ,\quad \int\rho\phi(\Omega)d^3 x\ ,\ \text{where } \Omega=\rho^{-1}\text{curl}(\underline{\mathbf{M}}/\rho)\cdot\underline{\nabla}\eta\ , \tag{20}$$

the second of which is strictly three-dimensional. The scalar quantity $\Omega=\rho^{-1}\text{curl}(\underline{\mathbf{M}}/\rho)\cdot\underline{\nabla}\eta$ is known as potential voticity, and the conserved functional $\int\rho\phi(\Omega)d^3x$ plays an important role in the nonlinear stability of ideal fluids in the three dimensional nonrelativistic case (Holm, et al.[10]). As duscussed in the introduction, functionals in the kernel of the Poisson bracket are called, in general, Casimirs.
Planar barotropic case. The nonrelativistic system (11)-(16) specializes to barotropic motion in the (x,y) plane when: n=2; specific entropy η is absent; and pressure and density are related by a prescribed function, $p=p(\rho)$, so that $\rho^{-1}dp(\rho)= =:dh(\rho)$ for a function $h(\rho)$, the specific enthalpy. Poisson bracket (16) for n=2, with absent, has Casimirs $\int\rho dxdy$, and

$$C=\int \rho\phi(\Omega)dxdy \ , \quad \text{where } \Omega :=\rho^{-1}\hat{\underline{z}}\cdot\mathrm{curl}(\underline{M}/\rho) \ , \tag{21}$$

and $\hat{\underline{z}}$ is the unit vector normal to the (x,y) plane, along z. Note that the similarity between (20) and (21) can be understood by regarding the two-dimensional barotropic flow as a special case of a three-dimensional adiabatic flow which is stratified into planes of constant η, with η=z. (This observation justifies using the same notation, Ω, for the quantities in (20) and (21)).

The two-dimensional barotropic specialization of (12)-(14) in terms of velocity, $\underline{v}$, is written as

$$\partial_t \rho=-\mathrm{div}\rho\underline{v} \tag{22}$$

$$\partial_t \underline{v}=\underline{v}\times\mathrm{curl}\underline{v}-\underline{\nabla}\left[\frac{1}{2} v^2+h(\rho)\right] , \tag{23}$$

where we have used the identity $(\underline{v}\cdot\underline{\nabla})\underline{v}=\underline{\nabla}\ \frac{1}{2}\ v^2-\underline{v}\times\mathrm{curl}\underline{v}$. Taking $\hat{\underline{z}}\cdot$curl of (23) and using (22) leads to a relation that confirms the conservation of C in (21), namely

$$\partial_t\Omega=-\underline{v}\cdot\underline{\nabla}\Omega \ , \qquad \Omega :=\rho^{-1}\hat{\underline{z}}\cdot\mathrm{curl}\underline{v} \ . \tag{24}$$

Hence, C in (21) is conserved, since Ω in (24) is conserved along flow lines.

1.2. Equilibrium Relations and Critical Points of Conserved Quantities

By (22)-(24), equilibrium states $(\rho_e,\underline{v}_e)$ satisfy

$$\mathrm{div}\rho_e\underline{v}_e=0 \ , \tag{25}$$

$$\underline{v}_e\times\mathrm{curl}\underline{v}_e-\underline{\nabla}\left[\frac{1}{2} v_e^2+h(\rho_e)\right]=0 \ , \tag{26}$$

$$\underline{v}_e\cdot\nabla\Omega_e=0, \qquad \Omega_e:=\rho_e^{-1}\hat{\underline{z}}\cdot\mathrm{curl}\underline{v}_e \ . \tag{27}$$

Scalar multiplication of (26) by $\underline{v}_e$ gives

$$\underline{v}_e \cdot \underline{\nabla}\left[\frac{1}{2} v_e^2 + h(\rho_e) \right] = 0 \ . \qquad (28)$$

Conditions (27) and (28) imply that, provided the flow is nondegenerate (i.e., $\Omega_e \neq$cost.,$\Omega_e \underline{v}_e \neq 0$) there exists a functional relationship between the quantities Ω_e and $\left[\frac{1}{2} v_e^2 + h(\rho_e) \right]$ which we assume is expressible in the form of Bernoulli's law,

$$\frac{1}{2} v_e^2 + h(\rho_e) = K(\Omega_e) \ , \qquad (29)$$

for a function K, called the Bernoulli function. Vector multiplying (26) by $\hat{\underline{z}}$ leads to

$$0 = \underline{v}_e (\hat{\underline{z}} \cdot \mathrm{curl} \underline{v}_e) - (\mathrm{curl} \underline{v}_e)(\hat{\underline{z}} \cdot \underline{v}_e) - \hat{\underline{z}} \times \underline{\nabla}\left[\frac{1}{2} v_e^2 + h(\rho_e) \right] \qquad (30)$$

Then, using $\hat{\underline{z}} \cdot \underline{v}_e = 0$ and (29) in (30) gives the relation

$$\rho_e \Omega_e \underline{v}_e = \hat{\underline{z}} \times \underline{\nabla} K(\Omega_e) \ , \qquad (31)$$

or, provided $\Omega_e \neq 0$, equivalently,

$$\rho_e \underline{v}_e = \frac{K'(\Omega_e)}{\Omega_e} \hat{\underline{z}} \times \underline{\nabla}(\Omega_e) \ , \qquad (32)$$

where prime ' denotes derivative of the function K with respect to its stated argument, Ω_e. Relations (29) and (32) will be useful in establishing the following proposition.

Proposition 1.1. For smooth solutions with velocity fields tangent to the boundary, a nondegenerate equilibrium solution $(\rho_e, \mathbf{M}_e)$ of the ideal planar barotropic fluid equations is a critical state of $H_C = H + C$ with H given in (15), C given in (21) and

$$\phi(\xi) = \xi \left[\int^{\xi} \frac{K(\xi)}{z^2} \, dz + \mathrm{const} \right] \ , \qquad (33)$$

K being the Bernoulli function in (29).

Proof. Let $(\rho_e, \underline{\mathbf{M}}_e)$ be a stationary solution of (12)-(14) for planar barotropic flow. In this case the functional H_C is defined as

$$H_C=\int_D dxdy\left[\frac{M^2}{2\rho}+\varepsilon(\rho)+\rho(\phi(\Omega)+\lambda\Omega)\right],$$

where $\varepsilon(\rho)$ is the internal energy density, $\varepsilon'(\rho)=h(\rho)$ is the specific enthalpy, and λ=const. (The linear term in ϕ has been separated in H_C for later convenience). The functional has a critical point at (ρ_e,M_e), provided

$$0=\delta H_C:=DH_C(\rho_e,\underline{M}_e)\cdot(\delta\rho,\delta\underline{M})$$

$$=\int_D dxdy\left\{\rho_e^{-1}\mathbf{M}_e\cdot\delta\underline{M}-\frac{1}{2}\rho_e^{-2}M_e^2\delta\rho+\right.$$

$$\left.+\varepsilon'(\rho_e)\delta\rho+[\phi(\Omega_e)+\lambda\Omega_e]\delta\rho+\rho_e[\phi'(\Omega_e)+\lambda]\delta\Omega\right\}. \tag{34}$$

Upon using the formula

$$\delta\Omega=-\rho_e^{-1}\Omega_e\delta\rho+\rho_e^{-1}\hat{\underline{z}}\cdot\mathrm{curl}(\rho_e^{-1}\delta\underline{M}-\rho_e^{-2}\underline{M}_e\delta\rho) \tag{35}$$

in (34), we obtain

$$\delta H_C=\int_D dxdy\left\{\delta\rho\left[-\frac{1}{2}M_e^2/\rho_e^2+\varepsilon'(\rho_e)+\phi(\Omega_e)-\Omega_e\phi'(\Omega_e)\right]+\right.$$

$$\left.+\rho_e^{-1}\underline{M}_e\cdot\delta\underline{M}+(\phi'(\Omega_e)+\lambda)\hat{\underline{z}}\cdot\mathrm{curl}(\delta\underline{M}/\rho_e-\delta\rho\underline{M}_e/\rho_e^2)\right\}.$$

Integrating the last term by parts, using the divergence theorem and the vector identity $\underline{A}\cdot\mathrm{curl}\underline{B}=\mathrm{curl}\underline{A}-\mathrm{div}(\underline{A}\times\underline{B})$ gives the extremal condition

$$0=\delta H_C=\int_D dxdy\left\{\delta\rho\left[-\frac{1}{2}M_e^2/\rho_e^2+\varepsilon'(\rho_e)+\phi(\Omega_e)-\Omega_e\phi'(\Omega_e)-\right.\right.$$

$$\left.\left.-\rho_e^{-2}\underline{M}_e\cdot\mathrm{curl}\phi'(\Omega_e)\hat{\underline{z}}\right]+\rho_e^{-1}\delta\underline{M}\cdot[\underline{M}_e-\phi''(\Omega_e)\hat{\underline{z}}\times\nabla\Omega_e]\right\}-$$

$$-\oint_{\partial D} (\phi'(\Omega_e)+\lambda)\hat{\underline{z}}\times(\delta\underline{M}/\rho_e-\delta\rho\underline{M}_e/\rho_e^2)\cdot\hat{\underline{n}}ds \ , \tag{36}$$

where $\hat{\underline{n}}$ is the unit vector normal to the boundary ∂D and ds is the line element on ∂D. Now, $\Omega_e|_{\partial D}$=const, by (27) and the tangency of $\underline{v}_e$ on ∂D. Thus, the boundary term in (36) vanishes for equilibrium states, provided we choose the constant λ so that

$$\phi'(\Omega_e)|_{\partial D}+\lambda=0 \ . \tag{37}$$

The $\delta\underline{M}$ coefficient of (36) vanishes by (32) for equilibrium states, provided

$$\phi''(\Omega_e)= \frac{K'(\Omega_e)}{\Omega_e} \ , \tag{38}$$

or, solving (38),

$$\phi(\Omega_e)=\Omega_e\left[\int^{\Omega_e} \frac{K(z)}{z^2}\, dx+\text{const}\right] \ , \tag{39}$$

which is relation (33) of Proposition 1.1. Note that the integration constant in (39) can be chosen to satisfy (37). using the relation

$$\rho_e\underline{v}_e=-\text{curl}(\phi'(\Omega_e)\hat{\underline{z}}) \tag{40}$$

arising from vanishing of the δM coefficient of (36), the vanishing of the $\delta\rho$ coefficient in (36) requires

$$\frac{1}{2}\, M_e^2/\rho_e^2 + \varepsilon'(\rho_e) + \phi(\Omega_e) - \Omega_e\phi'(\Omega_e) = 0 \tag{41}$$

Since $M_e^2/\rho_e^2 = v_e^2$ and $\varepsilon'(\rho_e) = h(\rho_e)$, condition (41) becomes, by Bernoulli's law (29),

$$K(\Omega_e) + \phi(\Omega_e) - \Omega_e\phi'(\Omega_e) = 0, \tag{42}$$

which is satisfied for $\phi(\Omega_e)$ given in (39). Thus, we have proved Proposition 2.1: H_C has a critical point for equilibrium states satisfyng (29) and (32) when the function ϕ in C is determined by (33).

1.3. Second Variation and Linear Lyapunov Stability Condition.

Calculation of the second variation $\delta^2 H_C = D^2 H_C(\rho_e, \underline{M}_e)\cdot(\delta\rho, \delta\underline{M})^2$ gives, using (34)

$$\delta^2 H_C = \int_D dxdy\{\rho_e^{-1} \, |\delta\underline{M}|^2 - \rho_e^{-2}\underline{M}_e\cdot\delta\underline{M}\delta\rho + \rho_e^{-3}M_e^2(\delta\rho)^2$$

$$- \rho_e^{-2}\underline{M}_e\cdot\delta\underline{M}\delta\rho + \varepsilon''(\rho_e)(\delta\rho)^2 + \rho_e\phi''(\Omega_e)(\delta\Omega)^2$$

$$+ 2(\phi'(\Omega_e) + \lambda)\delta\rho\delta\Omega + \rho_e(\phi'(\Omega_e) + \lambda)\delta^2\Omega\} \tag{43}$$

To compute $\delta^2\Omega$, we use (35):

$$\delta^2\Omega = \rho_e^{-2}\Omega_e(\delta\rho)^2 - \rho_e^{-1}\delta\Omega\delta\rho - \rho_e^{-2}\delta\rho\hat{\underline{z}}\cdot\text{curl}(\rho_e^{-1}\delta\underline{M} - \rho_e^{-2}\underline{M}_e\delta\rho)$$

$$+\rho_e^{-1}\hat{\underline{z}}\cdot\text{curl}[\, - \rho_e^{-2}\delta\rho\delta\underline{M} + 2\rho_e^{-3}\underline{M}_e(\delta\rho)^2 - \rho_e^{-2}\delta\underline{M}\delta\rho] \tag{44}$$

Using (35) again to collect terms in (44), we arrive at

$$\delta^2\Omega = - \rho_e^{-1} \, \delta\rho[\delta\Omega - \rho_e^{-1} \, \Omega_e\delta\rho + \rho_e^{-1}\hat{\underline{z}}\cdot\text{curl} \, (\rho_e^{-1}\delta\underline{M} - \rho_e^{-2}\underline{M}_e\delta\rho)]$$

$$- 2\rho_e^{-1}\hat{\underline{z}}\cdot\text{curl}[\rho_e^{-2}\delta\rho\delta\underline{M} - \rho_e^{-3}\underline{M}_e(\delta\rho)^2]$$

$$= -2\rho_e^{-1}\delta\rho\delta\Omega - 2\rho_e^{-1}\hat{\underline{z}}\cdot\text{curl}[\rho_e^{-2}\delta\rho\delta\underline{M} - \rho_e^{-3}\underline{M}_e(\delta\rho)^2]. \tag{45}$$

Using (45), we rewrite the last two terms in (43) as

$$- \int_D dxdy^2[\phi'(\Omega_e) + \lambda]\hat{\underline{z}}\cdot\text{curl}[\rho_e^{-2}\delta\rho\delta\underline{M} - \rho_e^{-3}\underline{M}_e(\delta\rho)^2]. \tag{46}$$

Integrating by parts using the vector identity appearing just before (36) and invoking the divergence theorem, we transform expression (46) into

$$2\int_{\partial D} [\phi'(\Omega_e) + \lambda]\hat{\underline{z}} \times [\rho_e^{-2}\delta\rho\delta\underline{M} - \rho_e^{-3}\underline{M}_e(\delta\rho)^2]$$

$$-2\int_D dxdy \, [\rho_e^{-2}\delta\rho\delta\underline{M} - \rho_e^{-3}\underline{M}_e(\delta\rho)^2] \cdot \text{curl} \, [\phi'(\Omega_e) + \lambda]\hat{\underline{z}} \tag{47}$$

By (37), the first summand in (47) vanishes, and by (40) the second summand is equal to

$$2\int_D dxdy\ [\rho_e^{-2}\delta\rho\delta\underline{M} - \rho_e^{-3}\underline{M}_e(\delta\rho)^2] \cdot \rho_e\underline{v}_e \tag{48}$$

Substituting this into (43) gives

$$\delta^2 H_C = \int_D dxdy\ \{\rho_e^{-1}|\delta\underline{M}|^2 + (\delta\rho)^2[\varepsilon''(\rho_e) - \rho_e^{-3}M_e^2]+$$

$$+ \rho_e\phi''(\Omega_e)\ (\delta\Omega)^2\} \tag{49}$$

The quadratic form (49) is positive definite and, thus, as explained in the introduction the equilibrium state $(\rho_e,\underline{M}_e)$ is linearly Lyapunov stable, if

$$\rho_e > 0 \tag{50}$$

and either

$$\varepsilon''(\rho_e) - M_e^2/\rho_e^3 > 0, \qquad \phi''(\Omega_e) > 0, \tag{51}$$

or

$$\varepsilon''(\rho_e) - M_e^2/\rho_e^3 > 0, \qquad \phi''(\Omega_e) = 0. \tag{52}$$

Stability condition (50) is self - explanstory: the mass density must be positive. Using $\varepsilon''(\rho_e) = h'(\rho_e) = \rho_e^{-1}p'(\rho_e) =: \rho_e^{-1}c_e^2$ where c_e is the sound speed in the equilibrium state, condition (51) requires

$$\rho_e^{-1}\ (c_e^2 - v_e^2) \geq 0,$$

i.e., the equilibrium state must be subsonic, or, at most, sonic, and $\phi''(\Omega_e)>0$. The condition $\phi''(\Omega_e)>0$ can be expressed geometrically in terms of fluid quantities by scalar multiplying (32) by $\hat{\underline{z}}\times\underline{\nabla}\Omega_e$, solving for $K'(\Omega_e)/\Omega_e$ and using (38):

$$\phi''(\Omega) = \frac{K'(\Omega_e)}{\Omega_e} = \frac{\rho_e\underline{v}_e\cdot\hat{\underline{z}}\times\underline{\nabla}\Omega_e}{|\underline{\nabla}\Omega_e|^2} > 0. \tag{53}$$

Thus, $\phi''(\Omega_e)>0$ when the vectors $\underline{v}_e$, $\hat{\underline{z}}$ and $\underline{\nabla}\Omega_e$ form a right - handed triad. In the case (52), we have $\phi''(\Omega_e)=0$, which means that $K'(\Omega_e)=0$, so that $(1/2)v_e^2+h(\rho_e)$ is constant in D, by (29). But then, from (26) it follows that $\underline{v}_e \times \mathrm{curl}\,\underline{v}_e=0$, which implies in two dimensions that either $\underline{v}_e=0$, or $\underline{v}_e=\mathrm{const}$, in which case $\Omega_e=0$, and we have excluded this case earlier. Thus, the linear Lyapunov stability conditions for two - dimensional, nonrelativistic fluid equilibria are (50) and (51).

Remark. A nearly identical analysis of the second variation $\delta^2 H_C$ in terms of the variables $(\rho, \underline{v})$, instead of the variables $(\rho, \underline{M})$, appears in Holm et al.[9,10]. In addition, convexity arguments were given in those references which establish stability conditions not only for infinitesimal, but also for finite - amplitude perturbations. We should point out, however, that due to the presence of not algebraic, but differential $(\delta\Omega)^2$ - terms in the second variation $\delta^2 H_C$, the sufficient stability conditions one finds in different coordinates would, in general, not be equivalent. In the present case they do turn out to be equivalent. The $(\rho, \underline{M})$ basis illustrated here, as opposed to the $(\rho, \underline{v})$ basis, is especially convenient for stability analysis of relativistic fluids, to which we turn next.

2. TWO - DIMENSINAL RELATIVISTIC FLUID DYNAMICS.

2.1. Equations of Motion and Hamiltonian Structure.

Relativistic fluid dynamics can be considered as a regular structure - preserving deformation (with parameter c^{-2}, where c is the speed of light) of the nonrelativistic theory, see Holm and Kupershmidt.[6] The structure preserved by this deformation is the Hamiltonian structure: in a fixed reference frame, the Poisson bracket for the relativistic theory retains the same form as for the nonrelativistic theory, although the Hamiltonian and dynamical variables change. Since the Casimirs are associated only to the Poisson bracket, they also retain their form; but now expressed in the relativistic variables. This regular behavior and form - invariance under relativisation allows us to extend the Hamiltonian theory of Lyapunov stability illustrated in the previous section to the case of relativistic fluids. In particular, the nonrelativistic stability criteria, being a set of inequalities, are regularly deformed under relativisation.

The equations of relativistic hydrodynamics in covariant form are (see, e.g., Landau and Lifshitz)[28],

$$\partial_\mu T^\mu_\nu = 0 \tag{54}$$

$$\partial_\mu (\rho_0 u^\mu) = 0. \tag{55}$$

where Greek indices μ, ν range over $0,1,\ldots,n$, and $x^0=ct$ is the real time coordinate. Thermodynamic quantities such as density ρ_0 and specific entropy η_0 are evaluated in the proper frame of a moving fluid element. The metric tensor is given by the expression $-d\tau^2=g_{\mu\nu}dx^\mu dx^\nu$ for the proper time interval and $g_{\mu\nu}$ has signature n-1(=+2, for n=3). The equations of motion are contained in (54) which expresses conservation of energy and momentum, while (55) is the relativistic continuity equation. The energy - momentum tensor $T_{\mu\nu}$ is given by

$$T_{\mu\nu} = \rho_0 wc^2 u_\mu u_\nu + \rho_0 g_{\mu\nu}, \tag{56}$$

where

$$w = 1+(e_0 + p_0/\rho_0)/c^2, \tag{57}$$

and ρ_0, p_0, and $e_0:=e(\rho_0,\eta_0)$ are all evaluated in the comoving frame of the fluid. Together with the equation of state, equations (54) and (55) imply the relativistic adiabatic condition,

$$d\eta_0/d\tau = u^\mu \partial_\mu \eta_0 = 0. \tag{58}$$

For the purpose of the Hamiltonian formalism, we express the covariant relativistic fluid equations (54) - (57) as a dynamical system, by rewriting them <u>in the laboratory frame</u>, with respect to which the fluid velocity is denoted by $\underline{v}$. In this frame, the relativistic fluid system returns to the <u>same</u> form as (12) - (14) (see, e.g., Landau and Lifshitz[28]), but where ρ and η are now <u>relativistic laboratory frame quantities</u>, and we reinterpret the variables $\underline{\mathbf{M}}$ and T_{ij} as

$$\underline{\mathbf{M}} = \vartheta \underline{\mathbf{v}}, \tag{59}$$

$$\vartheta = \rho\gamma w, \tag{60}$$

$$\gamma = (1-v^2/c^2)^{1/2}, \tag{61}$$

$$T_{ij} = \vartheta v_i v_j + \delta_{ij} p_0 \tag{62}$$

The laboratory and proper frame quantities are related as follows:

$$\rho = \gamma\rho_0, \qquad \eta = \eta_0 , \tag{63}$$

$$u_\mu = \gamma(-1, \underline{v}/c), \tag{64}$$

$$p_0 = \rho_0^2 \partial e_0(\rho_0, \eta_0)/\partial\rho_0. \tag{65}$$

Proposition 2.1. (Iwinski and Turski[5], Holm and Kupershimdt[6]). _The relativistic hydrodynamic system_ (54) - (58) _with_ $\underline{M}$ _and_ T_{ij} _defined in_ (59) - (62) _is expressible in the laboratory frame as a Hamiltonian system, with Hamiltonian_

$$H = \int d^n x [c^2(\vartheta-\rho)-p_0], \tag{66}$$

equal to the relativistic energy minus the rest mass energy, and with Poisson bracket {H,F} _for functionals_ H _and_ F _defined to be_ (16), _which is the same as in the nonrelativistic case. The relativistic hydrodynamic system is then identical to_ $\partial_t F = \{H,F\}$ _with_ $F \in \{\rho, \eta, \underline{M}\}$.

Proof of this proposition is based on showing the following variational equalities,

$$\frac{\delta H}{\delta M_j} = v_j, \tag{67}$$

$$\frac{\delta H}{\delta \rho} = c^2(w/\gamma-1), \tag{68}$$

$$\frac{\delta H}{\delta \eta} = c^2\rho_0 w_{,\eta_0} - p_{0,\eta_0} \tag{69}$$

Substituting equations (67) - (69) into $\partial_t F=\{H,F\}$ with Poisson bracket (16) and Hamiltonian (66) immediately yields the relativistic hydrodynamic system, in the form (12) - (14), but with M_i and T_{ij} reinterpreted according to (59) - (62).

Equalities (67) - (69) are derived in Part I in detail. Their derivation is sketched again as follows. First, since

$$de_0 = \rho_0^{-2} p_0 d\rho_0 + T_0 d\eta_0 . \tag{70}$$

from (57) we find

$$c^2 dw = \rho_0^{-1} \frac{\partial p_0}{\partial \rho_0} d\rho_0 + \left[\rho_0^{-1} \frac{\partial p_0}{\partial \eta_0} + T_0 \right] d\eta_0 \tag{71}$$

Second, from (59) - (61) we have

$$1 - \frac{w\rho^2}{\vartheta^2} = 1 - \gamma^{-2} = \frac{v^2}{c^2} = \frac{1}{c^2} \frac{M^2}{\vartheta^2} , \tag{72}$$

so that

$$\vartheta = \sqrt{(\rho w)^2 + M^2/c^2} . \tag{73}$$

Using formulae (74), (73) and (63) yields:

$$1) \quad \frac{\delta H}{\delta \underline{M}} = c^2 \frac{\partial \vartheta}{\partial \underline{M}} - \frac{\partial p_0}{\partial \underline{M}} = \frac{c^2}{\vartheta} \left[\underline{M}/c^2 + \rho^2 w \frac{\partial w}{\partial \rho_0} \frac{\partial p_0}{\partial \underline{M}} \right] -$$

$$+ \frac{\partial p_0}{\partial \rho_0} \frac{\partial p_0}{\partial \underline{M}} = \underline{M}/\vartheta + \frac{\partial \rho_0}{\partial \underline{M}} \frac{c^2 \rho}{\gamma} \frac{\partial w}{\partial \rho_0} - \frac{\partial p_0}{\partial \rho_0} = \underline{M}/\vartheta = \underline{v},$$

which proves (67). Next,

$$2) \quad \frac{\delta H}{\delta \rho} = \frac{c^2}{\vartheta} \rho w \frac{\partial (\rho w)}{\partial \rho} - \frac{\partial p_0}{\partial \rho} - c^2 =$$

$$= \frac{c^2}{\gamma} \left[w + \rho \frac{\partial w}{\partial \rho} \right] - \frac{\partial p_0}{\partial \rho_0} - c^2 = c^2 \left[\frac{w}{\gamma} - 1 \right] +$$

$$+ \left[c^2 \rho_0 \frac{\partial w}{\partial \rho_0} - \frac{\partial p_0}{\partial \rho_0} \right] \frac{\partial p_0}{\partial \rho}$$

$$= c^2 (w/\gamma - 1),$$

which is (68). Finally,

$$3)\ \frac{\delta H}{\delta \eta} = \frac{c^2}{\vartheta}\rho w \frac{\partial(\rho w)}{\partial \eta} - \frac{\partial p_0}{\partial \eta} = \frac{c^2}{\gamma}\rho \frac{\partial w}{\partial \eta_0} - \frac{\partial p_0}{\partial \eta_0} =$$

$$= c^2\rho_0 \frac{\partial w}{\partial \eta_0} - \frac{\partial p_0}{\partial \eta_0},$$

which is (69). Notice that, by (71), $\delta H/\delta\eta=\rho_0 T_0$. This demonstrates formulas (67) - (69), which prove the proposition.

Remark. As $c^{-2}\to 0$, we have $\gamma\to 1$, $\vartheta\to\rho$, $\eta_0\to\eta$, $\rho_0\to\rho$, $T_0\to T$, and the quantities M and T_{ij} each tend to their nonrelativistic counter - parts. Thus, the relativistic hydrodynamic equations expressed as (12) - (14) in laboratory variables also tend to their nonrelativistic conterparts. As $c^{-2}\to 0$, the Hamiltonian density in (66) limits as follows.

$$c^2(\vartheta-\rho) - p_0 = \rho c^2(\gamma w-1) - p_0 = \rho c^2\left[\left(1+\frac{v^2}{2c^2}\right)(1 +\right.$$

$$\left. + (e_0+p_0/\rho_0)/c^2)-1\right] - p_0+O(c^{-2})$$

$$= \rho c^2[1+c^{-2}(v^2/2+e_0+p_0/\rho_0)-1] - p_0+O(c^{-2})$$

$$= \frac{1}{2}\rho v^2 + \rho e+O(c^{-2}).$$

So the Hamiltonian (66) tends to (15) as $c^{-2}\to 0$. Likewise, the variational derivatives (67) - (69) tend to their nonrelativistic counterparts. This is evident for (67) and (69), while for (68) we hawe

$$\frac{\delta H}{\delta \rho} =c^2(w/\gamma-1) = c^2[(1+(e_0+p_0/\rho_0)c^2)\left[1-\frac{v^2}{2c^2}\right]-1]+O(c^{-2})$$

$$=- \frac{1}{2}v^2 + (e+p/\rho)+O(c^{-2}), \tag{74}$$

which has nonrelativistic counterpart (18). Thus, ieal relativistic fluid dynamics is, indeed, a regular deformation, with parameter c^{-2}, of its nonrelativistic counterpart. Moreover, the Hamiltonian strucuture of the nonrelativistic system retains its form under relativisation. As a corollary of this structure preservation, the Casimirs for relativistic fluids retain the form (20) but with $\underline{\mathbf{M}}$ given by (59) - (61). Likewise, the two - dimensional barotropic specialization of the relativistic system has Casimirs $\int \rho\, dxdy$ and (21), i.e.,

$$C = \int \rho\ \phi(\Omega)dxdy, \text{ where } \Omega=\rho^{-1}\hat{\underline{z}}\cdot\text{curl }(\gamma w\underline{v}) \tag{75}$$

and we have used $\underline{\mathbf{M}}/\rho=\gamma w\underline{v}$.

2.2. Equilibrium Relations and Critical Points of Conserved Quantities

Now we derive a few useful relations for analysis of relativistic two - dimensional barotropic equilibrium states, by following an analogous procedure to that in section 1. The two - dimensional barotropic specialization of equations (12) - (14) in relativistic laboratory variables is expressible as

$$\partial_t\rho =- \text{div}\rho\underline{v} \tag{76}$$

$$\partial_t M_i = - \partial_j(\vartheta v_i v_j + \delta_{ij}p_0). \tag{77}$$

To analyze equilibrium states, we transform (77) into

$$\partial_t(\gamma w\underline{v}) = \underline{v}\times\text{curl}(\gamma w\underline{v}) - \underline{\nabla}(\gamma w c^2), \tag{78}$$

which can be obtained as follows. First, using $M=\vartheta\underline{v}$ and (76) we rewrite (77) as

$$\partial_t(\gamma w v_i) =- v_j\partial_j(\gamma w v_i) - \gamma^{-1}\partial_i(wc^2), \tag{79}$$

Then, by the identity

$$- v_j\partial_i(\gamma w v_i) = (\underline{v}\times\text{curl}\gamma w\underline{v})_i - v_j\partial_i(\gamma w v_j), \tag{80}$$

we rewrite (79) as

$$\partial_t(\gamma w v_i) = (\underline{v}\times\text{curl}\gamma w v)_i - v_j\partial_i(\gamma w v_j)- \gamma^{-1}\partial_i(wc^2). \tag{81}$$

Then (78) is found by observing that

$$-\ v_j \partial_i (\gamma w v_j) - \gamma^{-1} \partial_i (w c^2) = -\ \partial_i (\gamma w c^2). \tag{82}$$

Taking $\hat{\underline{z}}\cdot$curl of (78) and using (76) leads to

$$\partial_t \Omega = -\ \underline{v}\cdot\underline{\nabla}\Omega, \text{ where } \Omega := \rho^{-1}\hat{\underline{z}}\cdot \text{curl}(\gamma w \underline{v}). \tag{83}$$

Thus, by (76), (78), and (83), equilibrium states $(\rho_e, \underline{v}_e)$ satisfy

$$\text{div } \rho_e \underline{v}_e = 0, \tag{84}$$

$$\underline{v}_e \times \text{curl } \gamma_e w_e \underline{v}_e - \underline{\nabla}(\gamma_e w_e c^2) = 0, \tag{85}$$

$$\underline{v}_e \cdot \underline{\nabla}\Omega_e = 0, \quad \Omega_e := \rho_e^{-1}\hat{\underline{z}}\cdot \text{curl } \gamma_e w_e \underline{v}_e \tag{86}$$

Scalar multiplication of (85) by $\underline{v}_e$ gives

$$\underline{v}_e \cdot \underline{\nabla}(\gamma_e w_e c^2) = 0, \tag{87}$$

which is the relativistic exstension of (28). Conditions (86) and (87) imply that, provided $\Omega_e \neq$ const and $\Omega_e \underline{v}_e \neq 0$, there exists a functional relationship between the quantities Ω_e and $\gamma_e w_e$ which we assume is expressible in the form of a relativistic Bernoulli law,

$$c^2(\gamma_e w_e - 1) = K(\Omega_e), \tag{88}$$

for a function K, the Bernoulli function.

Vector multiplyng (85) by $\hat{\underline{z}}$ leads to

$$0 = \underline{v}_e(\hat{\underline{z}}\cdot \text{curl } \gamma_e w_e \underline{v}_e) - (\text{curl}\gamma_e w_e \underline{v}_e)(\hat{\underline{z}}\cdot \underline{v}_e) - \hat{\underline{z}}\times\underline{\nabla}\gamma_e w_e c^2.$$

Then, using $\hat{\underline{z}}\cdot\underline{v}_e = 0$ and (88) gives the relation at equilibrium

$$\rho_e \Omega_e \underline{v}_e = \hat{\underline{z}}\times\nabla K(\Omega_e)$$

or, provided $\Omega_e \neq 0$,

$$\rho_e \underline{v}_e = \frac{K'(\Omega_e)}{\Omega_e} \hat{\underline{z}} \times \nabla \underline{\Omega}_e , \tag{89}$$

which is identical in form to (32). Relations (88) and (89) will be useful in establishing the following proposition.

Proposition 2.2. For smooth solutions with velocity tangent to the boundary, a nondegenerate equilibrium state $(\rho_e \underline{\mathbf{M}}_e)$ of the ideal planar relativistic barotropic fluid equations is a critical point of H_C=H+C with H given in (66) and C given in (75), provided

$$\phi(\xi) = \xi \left[\int^{\xi} \frac{K(z)}{z^2} \, dz + \text{const} \right], \tag{90}$$

K being the relativistic Bernoulli function in (88).

Proof. Let $(\rho_e \underline{\mathbf{M}}_e)$ be a stationary solution of (12) - (14) for relativistic planar barotropic flow. The functional

$$H_C = \int_D dxdy[c^2(\vartheta - \rho) - p_0 + \rho(\phi(\Omega) + \lambda\Omega)]$$

(with λ=const) has a critical point at $(\rho_e, \underline{M}_e)$, provided the first variation

$$\delta H_C = \int_D dxdy\{\underline{v}_e \cdot \delta \underline{M} + c^2(w_e/\gamma_e - 1)\delta\rho$$

$$+ (\phi(\Omega_e) + \lambda\Omega_e)\,\delta\rho + \rho_e\,|\phi'(\Omega_e) + \lambda)\delta\Omega_e| \tag{91}$$

vanishes, where we have used (67) and (68) in (91). Substituting (35) (which is form invariant under relativization) into (91) gives

$$\delta H_C = \int_D dxdy\{\delta\rho[c^2(w_e/\gamma_e - 1) + \phi(\Omega_e) - \Omega_e\phi'(\Omega_e)]$$

$$+ \vartheta_e^{-1} \underline{M}_e \cdot \delta\underline{M} + (\phi'(\Omega_e)) + \lambda)\hat{\underline{z}} \cdot \text{curl}(\delta\underline{M}/\rho_e - \delta\rho\underline{M}_e/\rho_e^2)]$$

Integrating the last term by parts yields

$$\delta H_C = \int_D dxdy\{\delta\rho[c^2(w_e/\gamma_e - 1) + \phi(\Omega_e) - \Omega_e\phi'(\Omega_e)$$

$$- \rho_e^{\rho_2} \underline{M}_e \cdot \text{curl}\phi'(\Omega_e)\hat{\underline{z}}] + \rho_e^{-1}\delta\underline{M}\cdot[\frac{e}{\vartheta_e}\underline{M}_e - \phi''(\Omega_e)\hat{\underline{z}}(\Omega_e)\hat{\underline{z}}\times\nabla\Omega_e]$$

$$-\oint_{\partial D}[\phi'(\Omega_e))+\lambda]\ \hat{\underline{z}} \times (\delta\underline{M}/\rho_e - \delta\rho\underline{M}_e/\rho_e^2)\cdot\hat{\underline{n}}ds, \qquad (92)$$

which is a relativistic exstension of (36). Retracing the arguments of the nonrelativistic case, the boundary term in (92) vanishes for equilibrium states if (37) is satisfied ad the $\delta\underline{M}$ coefficient in (92) vanishes for equilibrium states by (89) if (38) holds for Ω_e given in (86). Then, the relation

$$\rho_e\underline{v}_e = -\,\text{curl}(\phi'(\Omega_e)\hat{\underline{z}}) \qquad (93)$$

(arising from vanishing of the $\delta\underline{M}$ coefficient) implies that the $\delta\rho$ coefficient in (92) vanishes, provided

$$\rho_e^{-1}\underline{v}_e\cdot\underline{M}_e + c^2(w_e/\gamma_e - 1) + \phi(\Omega_e) - \Omega_e\phi'(\Omega_e) = 0. \qquad (94)$$

Substituting the relativistic Bernoulli law (88) into (94) yields, as before,

$$\phi(\Omega_e) = \Omega_e\left(\int^{\Omega_e}\frac{K(z)}{z^2}\,dz + \text{const}\right),$$

which is relation (90) of Proposition 2.2. The relation (94) trasforms into (42) by using the identity

$$\rho^{-1}\underline{v}\cdot\underline{M} + c^2(w/\gamma - 1) = c^2(w\gamma - 1) \qquad (95)$$

and relation (88). So (94) is satisfied for $\phi(\Omega_e)$ given by (90). Thus, relativistic equilibria satisfying (88) and (89) are critical states of H_C when the function ϕ in C is determined by (90). This proves Proposition 2.2.

2.3. Second Variation and Linear Lyapunov Stability Conditions

Now we compute the second variation of H_C. Using (91), we get

$$\delta^2 H_C = \int_D dxdy\{\delta\underline{v}\cdot\delta\underline{M} + \delta\rho\cdot\delta(c^2 w/\gamma) + \rho_e\phi''(\Omega_e)(\delta\Omega)^2 +$$

$$+ 2[\phi'(\Omega_e) + \lambda]\delta\rho\delta\Omega + \rho_e[\phi'(\Omega_e) + \lambda]\delta^2\Omega\}. \tag{96}$$

Exactly as in nonrelativistic case (43), we use the relativistic versions of (35) and (45) to transform the last two terms in (96), noticing that (93) is identical in form to (40). Thus, we arrive again at expression (48) for the last two terms in (96). Hence, (96) becomes

$$\delta^2 H_C = \int_D dxdy\{\delta\underline{v}\cdot\delta\underline{M} + \delta\rho\delta(c^2 w/\gamma) + \rho_e\phi''(\Omega_e)(\delta\Omega)^2$$

$$+ 2\vartheta_e^{-1}\underline{M}_e\cdot[\rho_e^{-1}\,\delta\rho\delta\underline{M} - \rho_e^{-2}\,\underline{M}_e(\delta\rho)^2]\}. \tag{97}$$

To transform (97), we first notice that ϑ, w, and γ all depend upon $\underline{\mathbf{M}}$ through M^2 only, as is clear from (72) and (73). Therefore,

$$\delta\underline{v}\cdot\delta\underline{M} = \delta(\underline{M}/\vartheta)\cdot\delta\underline{M} = \vartheta_e^{-1}\,|\delta\underline{M}|^2 - (\underline{M}_e\cdot\delta\underline{M})\vartheta_e^{-2}\,\delta\vartheta$$

$$=\vartheta_e^{-1}\,|\delta\underline{M}|^2-(\underline{M}_e\cdot\delta\underline{M})\vartheta_e^{-2}\left[\left(\frac{\partial\vartheta}{\partial\rho}\right)_e\delta\rho+\left(\frac{\partial\vartheta}{\partial M^2}\right)_e 2(\underline{M}_e\cdot\delta\underline{M})\right], \tag{98}$$

where we have used the relation $\delta(M^2)=2(\underline{M}_e\cdot\delta\underline{M})$ in the last term in (98). Similarly,

$$\delta\rho\cdot\delta(c^2 w/\gamma)=\delta\rho\left[\delta\rho\left(\frac{\partial(c^2 w/\gamma)}{\partial\rho}\right)_e+2(\underline{M}_e\cdot\delta\underline{M})\left(\frac{\partial(c^2 w/\gamma)}{\partial M^2}\right)_e\right] \tag{99}$$

Substituting (98) and (99) into (97) yields

$$\delta^2 H_C = \int_D dxdy\left\{\rho_e\phi''(\Omega_e)(\delta\Omega^2)+(\delta\rho)^2\left[\frac{\partial(c^2 w/\gamma)}{\partial\rho} - \frac{2M^2}{\rho^2\vartheta}\right]_e +\right.$$

$$+ 2\delta\rho(\underline{M}_e\cdot\delta\underline{M})\left[\frac{1}{\rho\vartheta} + \frac{\partial(c^2 w/\gamma)}{\partial M^2} - \frac{1}{2\vartheta^2}\frac{\partial\vartheta}{\partial\rho}\right]_e$$

$$\left.+ \vartheta_e^{-1}\,|\delta\underline{M}|^2 + (\underline{M}_e\cdot\delta\underline{M})^2\left[-2\vartheta^{-2}\frac{\partial\vartheta}{\partial M^2}\right]_e\right\} \tag{100}$$

We pause for a moment to compare the nonrelativistic limit of (100) with expression (49). Using (74), we get the following: 1) for the $(\delta\rho)^2$ - coefficient,

$$\frac{\partial}{\partial\rho}(c^2w/\gamma) - \frac{2M^2}{\rho^2\vartheta} = \frac{\partial}{\partial\rho}\left[-\frac{M^2}{2\rho^2} + (e+p/\rho) + 0(c^{-2})\right] -$$

$$-\frac{2M^2}{\rho^3} + 0(c^{-2}) = \frac{-M^2}{\rho^3} + \varepsilon''(\rho) + 0(c^{-2}), \tag{101}$$

which limits to the $(\delta\rho)^2$ - coefficient in (49); 2) for the $\delta\rho(\underline{M}_e \cdot \delta\underline{M})$-coefficient in (100),

$$\frac{1}{\rho\vartheta} + \frac{\partial(c^2w/\gamma)}{\partial M^2} - \frac{1}{2\vartheta^2}\frac{\partial\vartheta}{\partial\rho} = \frac{1}{\rho^2} - \frac{1}{2\rho^2} - \frac{1}{2\rho^2} + 0(c^{-2}) =$$

$$= 0(c^{-2}), \tag{102}$$

in agreement with the absence of $\delta\rho(\underline{M}_e \cdot \delta\underline{M})$-terms in (49); 3) for the $(\underline{M}_e \cdot \delta\underline{M})^2$-term in (100), we obtain

$$-2\vartheta^{-2}\frac{\partial\vartheta}{\partial M^2} = -2\rho^{-2}(0+0(c^{-2})) = 0(c^{-2}), \tag{103}$$

in agreement with the absence of $(\underline{M}_e \cdot \delta\underline{M})^2$-terms in (49). The $(\delta\Omega)^2$ - and $|\delta\underline{M}|^2$-terms in (100) limit, evidently, to their respective non-relativistc cases. Thus, we see how the diagonal form of the second variation (49) changes to a nondiagonal form under relativisation.

Now we derive sufficient conditions for the positive definiteness of the relativistic second variation (100). From physical considerations, the mass deinsity ρ_e must be positive,

$$\rho_e > 0, \tag{104}$$

which results in $\vartheta_e = \rho_e \gamma_e w_e > 0$, and we require

$$\phi''(\Omega_e) > 0, \tag{105}$$

as in the nonrelativistic case. Next, for the remaining quadratic form in (100) to be nonnegative, the $(\delta\rho)^2$-coefficient must be nonnegative, i.e.,

$$\left[\partial\frac{\partial(c^2w/y)}{\partial\rho} - \frac{2M^2}{\rho^2\vartheta}\right]_e \geq 0 \tag{106}$$

Then, if $\delta\underline{M}$ is perpendicular to $\underline{M}_e$ the quadratic form in (100) becomes diagonal and nonnegative, since $\vartheta_e^{-1}>0$. So, let $\delta\underline{M}$ be parallel to $\underline{M}_e$: $(\delta\underline{M}\cdot\underline{M}_2)^2 = M_e^2|\delta\underline{M}|^2$. Hence, the last two terms in (100) combine into $(\underline{M}_e\cdot\delta\underline{M})^2$ times the quantity

$$\frac{1}{\vartheta M^2} - \frac{2}{\vartheta^2}\frac{\partial\vartheta}{\partial M^2} = \frac{1}{\vartheta}\left[\frac{1}{M^2} - \frac{\partial}{\partial M^2}\ln\vartheta^2\right] =$$

$$= \frac{1}{\vartheta}\frac{\partial}{\partial M^2}\ln\frac{M^2}{\vartheta^2} = \frac{\vartheta}{M^2}\frac{\partial v^2}{\partial M^2},$$

where we have supressed subscipt e in the computation. Thus, we get

$$\left[\frac{\partial v^2}{\partial M^2}\right]_e \geq 0, \tag{107}$$

as the requirement for the two terms in (100) to be nonnegative. Finally, the determinant of the quadratic form in $(\delta\rho, M_e\cdot\delta M)$-variables in (100) must be nonnegative:

$$\left[\frac{1}{\vartheta}\frac{\partial v^2}{\partial M^2}\frac{\partial(c^2w/\gamma)}{\partial\rho} - \frac{2M^2}{\rho^2\vartheta}\right]_e \geq \left[\frac{1}{\vartheta\rho} + \frac{\partial(c^2w/\gamma)}{\partial M^2} - \right.$$

$$\left. - \frac{1}{2\vartheta^2}\frac{\partial\vartheta}{\partial\rho}\right]_e^2. \tag{108}$$

The relativistic stability conditions for a two - dimensional barotropic fluid are (104) - (108). Notice that according to (101) the condition (106) is a relativistic version of the condition for the flow to be either subsonic, or sonic. Notice also that the $\phi''(\Omega_e)>0$ condition (105) does not change its form (53); one simply interprets Ω_e relativistically.

3. TWO - DIMENSIONAL BAROTROPIC RELATIVISTIC MULTIFLUID PLASMA.

3.1 Equations of Motion and Hamiltonian Structure.

The multifluid plasma (MFP) system describes ideal, charged fluids interacting together via selfconsistent electromagnetic forces. The fluid species are labeled by superscript s (Note: no summation convention is imposed on superscripts in this and the next section.); each species is composed of particles of mass m^s and q^s, with charge - to -

mass ratio $a^s=q^s/m^s$. Dynamical variables are: fluid velocity $\underline{v}^s$; proper mass density ρ_0^s; proper specific entropy η_0^s; electric field $\underline{E}$; and magnetic field $\underline{B}$. The proper specific internal energy is given by $e_0^s = e^s(\rho_0^s, \eta_0^s)$, where e^s is a prescribed function and the proper partial pressure is given by $p_0^s=(\rho_0^s)^2\partial e_0^e/\partial\rho_0^s$.

The MFP equations consist of Maxwell's equations for the electromagnetic fields, a continuity equation for each species, anf the MFP motion equations. In relativistic covariant form in (3+1) dimensions, these equations are

$$\partial_\mu T^{\mu\nu} = 0, \tag{109}$$

$$\partial_\mu(\rho_0^s u^{s\mu}) = 0, \tag{110}$$

$$\partial_\mu F^{\mu\nu} = j^\nu = \sum_s a^s \rho_0^s u^{s\nu}, \tag{111}$$

$$\partial_{[\sigma} \tilde{F}^{\mu\nu]} = 0, \tag{112}$$

where $u^{s\rho}$ is the 4-velocity of species s, $F^{\mu\nu}$ is the Maxwell tensor, and $\tilde{F}^{\mu\nu}$ is its dual. We use the same relativistic conventions as in the previous section. The square bracket in (112) indicates antisymmetric sum over the indices $\delta, \mu, \nu=0,1,2,3$. The energy momentum tensor, $T^{\mu\nu}$, is given by

$$T^{\mu\nu} = T^{\mu\nu}_{MAT} + T^{\mu\nu}_{EM} \tag{113}$$

$$T^{\mu\nu}_{MAT} = \sum_s [\rho_0^s w^s c^2 u^{s\mu} u^{s\nu} + \rho_0^s g^{\mu\nu}], \tag{114}$$

$$T^{\mu\nu}_{EM} = -\frac{1}{4} g^{\mu\nu} F_{\alpha\beta} F^{\alpha\beta} + F^{\mu\alpha} F^\nu_{\ \alpha}, \tag{115}$$

where w^s is the relativistic specific enthalpy,

$$w^s = 1 + (e_0^s + p_0^s/\rho_0^s)/c^2, \tag{116}$$

and ρ_0^s, p_0^s, e_0^s are all evaluated in the frame moving with the fluid at velocity $\underline{v}^s$. Together with the equation of state for each species, (109) and (110) imply the relativistic adiabatic conditions,

$$u^{s\mu}\,\partial_\mu \eta_0^s = 0. \tag{117}$$

In the laboratory frame, the relativistic MFP equations (109) - (117) can be written as a dynamical system

$$\partial_t M_i^s = -\,(M_i^s v_j^s + \delta_{ij} p_0^s)_{,j} + a^s \rho^s (E_i + \varepsilon_{ijk} v_j^s B_k), \tag{118}$$

$$\partial_t \rho^s = -\,(\rho^s v_j^s)_{,j}, \tag{119}$$

$$\partial_t \eta^s = -\,v_j^s \eta^s_{,j}, \tag{120}$$

$$\partial_t \underline{E} = \text{curl } \underline{B} - \sum_s a^s \rho^s \underline{v}^s, \tag{121}$$

$$\partial_t \underline{B} = -\,\text{curl } \underline{E}, \tag{122}$$

where i, j, k=1, 2, 3, are Cartesian components and ε_{ijk} is the totally antisymmetric tensor. This systemm preserves the nondynamical Maxwell condition,

$$\text{div } \underline{E} - \sum_s a^s \rho^s = 0, \tag{123}$$

$$\text{div } \underline{B} = 0, \tag{124}$$

which, hence, can be taken as initial conditions.

The variables in system (118) - (124) are defined in terms of previous quantities as follows.

$$M_i^s = \vartheta^s v_i^s \tag{125}$$

$$\vartheta^s = \rho^s \gamma^s w^s \tag{126}$$

$$\gamma^S = (1-|\underline{v}^S|^2/c^2)^{-1/2} \tag{127}$$

$$\rho^S = \gamma^S \rho_0^S \tag{128}$$

$$\eta^S = \eta_0^S \tag{129}$$

$$E_i = F_{0i} \tag{130}$$

$$B_i = \frac{1}{2} \varepsilon_{ijk} F_{jk} \tag{131}$$

$$u_\mu^S = \gamma^S(-1, \underline{v}^S/c). \tag{132}$$

Proposition 3.1.. (Iwinski and Turski[5], Holm and Kupershnidt[6]). <u>The</u> <u>relativistic multifluid plasma system</u> (118) - (122) <u>with variables defined in</u> (125) - (132) <u>is expressible in the laboratory frame as a Hamiltonian system, with Hamiltonian</u>

$$H = \sum_S \int d^3x [c^2(\vartheta^S - \rho^S) - p_0^S] + \frac{1}{2} \int d^3x(|\underline{E}|^2 + |\underline{B}|^2), \tag{133}$$

<u>equal to the total relativistic energy minus the rest mass energy, and with Poisson bracket</u> {H,F} <u>for functionals F and G defined to be</u>

$$\{H,F\} = - \sum_S \int d^3x \left[\frac{\delta F}{\delta \rho^S} \partial_j \rho^S \frac{\delta H}{\delta M_j^S} + \frac{\delta F}{\delta \eta^S} \eta^S_{,j} \frac{\delta H}{\delta M_j^S} + \right.$$

$$+ \frac{\delta H}{\delta M_i^S} \left\{ \rho^S \partial_i \frac{\delta H}{\delta \rho^S} \eta^S_{,i} \frac{\delta F}{\delta \eta^S} + (M_j^S \partial_i + \partial_j M_i^S) \frac{\delta H}{\delta M_j^S} \right\}$$

$$- a^S \rho^S \left(\frac{\delta F}{\delta \underline{M}^S} \cdot \frac{\delta H}{\delta \underline{E}} - \frac{\delta H}{\delta \underline{M}^S} \cdot \frac{\delta F}{\delta \underline{E}} + \underline{B} \cdot \frac{\delta F}{\delta \underline{M}^S} \cdot \frac{\delta H}{\delta \underline{M}^S} \right) \Bigg]$$

$$- \int d^3x \left(\frac{\delta F}{\delta \underline{B}} \cdot \mathrm{curl} \frac{\delta H}{\delta \underline{E}} - \frac{\delta F}{\delta \underline{E}} \cdot \mathrm{curl} \frac{\delta H}{\delta \underline{B}} \right) \tag{134}$$

<u>The relativistic</u> MFP <u>system is then identical to</u> $\partial_t F=\{H,F\}$ <u>with</u> $F\in\{\rho^s,\eta^s,M_i^s,E_i,B_i\}$.

Proof. The Hamilton H in (133) is a sum of N copies of the single fluid Hamiltonian (66), plus the quadratic electronic piece. Consequently, by reasoning similar to that for deriving relations (67) - (69), the following variational identifies hold:

$$\frac{\delta H}{\delta M_j^s} = v_j^s, \tag{135}$$

$$\frac{\delta H}{\rho^s} = c^2(w^s/\gamma^s - 1), \tag{136}$$

$$\frac{\delta H}{\eta^s} = c^2\rho_0^s w^s_{,\eta_0^s} - p^s_{0,\eta_0^s} \tag{137}$$

In addition, we evidently have

$$\frac{\delta H}{\delta \underline{E}} = \underline{E}, \tag{138}$$

$$\frac{\delta H}{\delta \underline{B}} = \underline{B}. \tag{139}$$

Substituting identities (135) (139) into $\partial_t F=\{H,F\}$ with Poisson bracket (134) and Hamiltonian (133) yields the relativistic MFP system (118) - (122), which proves the proposition.

Remark. The Casimirs for the Poisson bracket (134) in three dimensions are given by

$$\int \rho^s\psi^s(\eta^s)d^3x, \quad \int \rho^s\phi^s(\tilde{\Omega}^s)d^3x,$$

$$\tilde{\Omega}^s = (\rho^s)^{-1}\, \underline{\nabla}\eta \cdot \left[\mathrm{curl}\left(\frac{M^s}{\rho^s}\right) + a^s\, \underline{B}\right] \tag{140}$$

This is seen by noticing that the bracket (134) is equivalent to the sum of a Poisson bracket in canonically conjugate variables $(\underline{E},\underline{A})$, where $\underline{A}$ is the magnetic vector potential ($\underline{B}$=curl $\underline{\mathbf{A}}$), plus N brackets of the form (16), but with the variables $\underline{M}^s$ replaced by $\underline{\tilde{M}}^s=\underline{\mathbf{M}}^s+a^s\rho^s\,\underline{\mathbf{A}}$. (See Holm and Kupershmidt[6]. Thus, Casimirs (140) are of the same form as (20) but with $\underline{\mathbf{M}}$ replaced by $\underline{\tilde{M}}^s$, and (140) results upon using

$\mathrm{curl}(\tilde{\underline{M}}^s/\rho^s) = [\mathrm{curl}\ (\underline{M}^s/\rho^s)+a^s\underline{B}]$. The canonical $(\underline{E},\underline{A})$ piece of the brackets, having no nontrivial Kernel, contributes no additional Casimirs beyond those in (140).

Specialization to Two Dimensional MFP. In the two dimensional barotropic case (i.e. $p_0^s=p_0^s(\rho_0^s)$) with magnetic field $\underline{B}=\underline{B}(x,y,t)\hat{\underline{z}}$ normal to the xy plane, and electric field $\underline{E}(x,y,t)$ in the plane, the Poisson bracket (134) specializes to

$$\{H,F\} =- \sum_s \int dxdy \left[\frac{\delta F}{\delta\rho^s}\, \mathrm{div} \left(\rho^s \frac{\delta H}{\delta \underline{M}^s} \right) + \frac{\delta F}{\delta M_i^s} \left(M_j^s\partial_i + \right.\right.$$

$$\left. + \partial_j M_i^s \right) \frac{\delta H}{\delta M_j^s} - a^s\rho^s B\hat{\underline{z}} \cdot \frac{\delta F}{\delta \underline{M}^s} \times \frac{\delta H}{\delta \underline{M}^s} + a^s\rho^s \frac{\delta F}{\partial \underline{E}} \cdot \frac{\delta H}{\delta \underline{M}^s}$$

$$\left. + \frac{\delta F}{\delta \underline{M}^s} \cdot \left(\rho^s \underline{\nabla} \frac{\delta H}{\delta\rho^s} - a^s\rho^s \frac{\delta F}{\partial \underline{E}} \right) \right] + \int dxdy \left(\frac{\delta F}{\partial \underline{E}} \cdot \right.$$

$$\left. \cdot\ \mathrm{curl}\ \frac{\delta H}{\partial B} \hat{\underline{z}} - \frac{\delta F}{\delta B} \hat{\underline{z}} \cdot \mathrm{curl}\ \frac{\delta H}{\partial \underline{E}} \right) \qquad (141)$$

The Casimirs corresponding to (141) are

$$C^s \int \rho^s \phi^s(\bar{\Omega}^s)\ dxdy,\quad \bar{\Omega}^s=(\rho^s)^{-1}\ [\hat{\underline{z}}\cdot \mathrm{curl}(\underline{M}^s/\rho^s)+a^s B] \qquad (142)$$

or, upon using $\underline{M}^s/\rho^s=\underline{v}^s\vartheta^s/\rho^s=\gamma^s w^s \underline{v}^s$,

$$\bar{\Omega}^s = (\rho^s)^{-1}\ [\ \hat{\underline{z}}\cdot \mathrm{curl}(\gamma^s w^s \underline{v}^s)+a^s B] =: \Omega^s+(a^s/\rho^s)B, \qquad (143)$$

where Ω^s, the purely fluid part, is defined by (75) for each species s.

3.2. Equilibrium Relations and Critical Points of Conserved Quantities

The two - dimensional barotropic specialization of the relativistic MFP system (118) - (122) in laboratory - frame variables is expressed as

$$\partial_t\rho^s =- \mathrm{div}\ \rho^s\underline{v}^s, \qquad (144)$$

$$\partial_t M_i^s = -\partial_j(\vartheta^s v_i^s v_j^s + \delta_{ij} p_0^s) + a^s \rho^s (E_i + \varepsilon_{ij3} v_j^s B), \tag{145}$$

$$\partial_t \underline{E} = \nabla B \times \hat{\underline{z}} - \sum_s a^s \rho^s \underline{v}^s, \tag{146}$$

$$\partial_t \mathbf{B} = -\hat{\underline{z}} \cdot \mathrm{curl}\ \underline{E}, \tag{147}$$

and the nondynamical Maxwell conditions remain (123) (124). Following the same procedure as in (76) - (82), the motion equation (145) can be cast into a convenient form analogous to (78):

$$\partial_t(\gamma^s w^s \underline{v}^s) = \rho^s \underline{v}^s \times \hat{\underline{z}} \bar{\Omega}^s + a^s \underline{E} - \underline{\nabla} c^2 w^2 \gamma^2. \tag{148}$$

Taking $\hat{\underline{z}}\cdot$curl of (148) and using (144), (147), and (143), leads to

$$\partial_t \bar{\Omega}^s = -\underline{v}^s \cdot \underline{\nabla} \bar{\Omega}^s. \tag{149}$$

Thus, by (144) and (146) (149), the equilibrium states $(\rho_e^s, \underline{v}_e^s, \underline{E}_e, B_e)$ satisfy the following relations,

$$\mathrm{div}\ \rho_e^s \underline{v}_e^e = 0, \tag{150}$$

$$\underline{\nabla} B_e \times \hat{\underline{z}} = \sum_s a^s \rho_e^s v_e^s \tag{151}$$

$$\hat{\underline{z}} \cdot \mathrm{curl}\ E_e = 0, \tag{152}$$

$$\rho_e^s \underline{v}_e^s \times \hat{\underline{z}} \bar{\Omega}_e^s + a^s \underline{E}_e - \underline{\nabla} c^2 w_e^s \gamma_e^s = 0, \tag{153}$$

$$\underline{v}_e^s \cdot \underline{\nabla} \bar{\Omega}_e^s = 0. \tag{154}$$

Equilibrium relation (152) implies

$$\underline{E}_e = -\underline{\nabla}\phi(x,y), \tag{155}$$

for an electrostatic potential, ϕ. Substituting (155) into (153) and scalar multiplying by $\underline{v}_e^s$ gives

$$\underline{v}_e^s \cdot \underline{\nabla}(c^2 w_e^s \gamma_e^s + a^s \phi) = 0. \tag{156}$$

Provided $\bar{\Omega}_e^s \neq$ const and $\bar{\Omega}_e^s \underline{v}_e^s \neq 0$, conditions (154) and (156) imply a functional relationship between the quantities $\bar{\Omega}_e^s$ and $(c^2 w_e^s \gamma_e^s + a^s \phi)$ which we assume is expressible as a set of Bernoull laws

$$c^2(w_e^s \gamma_e^s - 1) + a^s \phi = K^s(\bar{\Omega}_e^s), \tag{157}$$

for functions K^s, the Bernoulli functions.

Vector multiplying (153) by $\hat{\underline{z}}$ and using (155) and (157) gives the relation

$$\rho_e^s \bar{\Omega}_e^s \underline{v}_e^s = \hat{\underline{z}} \times \underline{\nabla} K^s(\bar{\Omega}_e^s), \tag{158}$$

or, since we have assumed $\bar{\Omega}_e^s \neq 0$,

$$\rho_e^s \underline{v}_e^s = \frac{K^{s\prime}(\bar{\Omega}_e^s)}{\bar{\Omega}_e^s} \hat{\underline{z}} \times \underline{\nabla} \bar{\Omega}_e^s, \tag{159}$$

which is identical in form to (32) and (89). Finally, substitution of (159) into (151) and vector multiplication by $\hat{\underline{z}}$ gives

$$\underline{\nabla} B_e = - \sum_s a^s \frac{K^{s\prime}(\bar{\Omega}_e^s)}{\bar{\Omega}_e^s} \underline{\nabla} \bar{\Omega}_e^s. \tag{160}$$

Relations (157), (159), and (160) will be useful in establishing the following proposition.

Proposition 3.2. <u>For smooth flows with boundary conditions</u>

$$\underline{v} \cdot \hat{\underline{n}}|_{\partial D} = 0, \qquad \underline{E} \times \hat{\underline{n}}|_{\partial D} = 0,$$

<u>a nondegenerate equilibrium state</u> $(\rho_e^s, \underline{M}_e^s, B_e, \underline{E}_e)$ <u>of the ideal barotropic planar relativistic</u> MFP <u>equations is a critical point of the functional</u> $H_C = H + \sum_s C^s$ <u>with</u> H <u>given in</u> (133) <u>and</u> C^s <u>given in</u> (142), <u>provided</u>

$$\phi^s(\xi) = \xi \left[\int^{\xi} \frac{K^s(z)}{z^2}\, dz + \text{const}\right],$$

K^s being the relativistic Bernoulli function in (157) for the s-th species.

Remark. For nonrelativistic MFP the correspond proposition is given in Holm[29].

Proof. Let $(\rho_e^s, \underline{M}_e^s, B_e, \underline{E}_e)$ be an equilibrium state of (144) - (147). The functional

$$H_c = \sum_s \int_D dxdy[(c^2(\vartheta^s-\rho^s)-p_0^s+\rho^s(\phi^s(\bar{\Omega}^s)+\lambda^s\bar{\Omega}^s+\mu a^s)] \tag{161}$$

$$+ \frac{1}{2}\int_D dxdy(|E|^2+B^2),$$

with μ=const, λ^s=const, has a critical point at $(\rho_e^s, \underline{M}_e^s, B_e, \underline{E}_e)$ provided the first variation

$$H_c = \sum_s \int_D dxdy\{\underline{v}_e^s\cdot\delta\underline{M}^s+c^2(w_e^s/\gamma_e^s-1)\delta\rho^s$$

$$+ [\phi^s(\bar{\Omega}_e^s)+\lambda^s\bar{\Omega}_e^s+\mu a^s]\delta\rho^s+\rho_e^s(\phi^s,(\bar{\Omega}^s)+\lambda^s)\delta\bar{\Omega}^s\}$$

$$+ \int_D dxdy(\underline{E}_e\cdot\delta\underline{E}+B_e\delta B) \tag{162}$$

vanishes, where we have used variational derivatives (135) - (139) in (162). By writing

$$\bar{\Omega}_e^s = \Omega_e^s+(a^s/\rho_e^s)B_e,\quad \Omega_e^s = (\rho_e^s)^{-1}\hat{\underline{z}}\cdot\text{curl}(\underline{M}_e^s/\rho_e^s),$$

we have

$$\delta\bar{\Omega}^s = \delta\Omega^s + (a^s/\rho_e^s)\delta B - (a^s/(\rho_e^s)^2)B_e\delta\rho \tag{163}$$

$$= (\rho_e^s)^{-1}\,\bar{\Omega}_e^s\delta\rho^s + (\rho_e^s)^{-1}\,\hat{\underline{z}}\cdot\text{curl}[(\rho_e^s)^{-1}\delta\underline{M}^s-(\rho_e^s)^{-2}\,\underline{M}_e^s\delta\rho^s]+$$

$$+ (a^s/\rho_e^s)\delta B,$$

$$\delta^2\bar{\Omega}^s = \delta^2\Omega^s - 2(a^s/(\rho_e^s)^2)\delta H\delta\rho^s + 2(a^s/(\rho_e^s)^3)B_e(\delta\rho^s)^2, \tag{164}$$

where $\delta\Omega^s$ and $\delta^2\Omega^s$ are given by the formulae analogous to (35) and (45), respectively. In particular, one finds

$$2\delta\rho^s\delta\bar{\Omega}^s + \rho_e^s\delta^2\bar{\Omega}^s = 2\delta\rho^s\delta\Omega^s + \rho_e^s\delta^2\Omega^s. \tag{165}$$

Substituting (163) into (162) yields

$$\delta H_c = \sum_s \int_D dxdy\{(\vartheta_e^s)^{-1}\underline{M}_e^s\cdot\delta\underline{M}^s + [c^2(w_e^s/\gamma_e^s-1) + \mu a^s + \phi^s(\bar{\Omega}_e^s) -$$

$$+ \bar{\Omega}_e^s\phi^s(\bar{\Omega}_e^s)\delta\rho^s + [\phi^{s\prime}(\bar{\Omega}_e^s)+\lambda^s][a^s\delta B+\hat{\underline{z}}\cdot\mathrm{curl}(\delta\underline{M}^s/\rho_e^s-\delta\rho^s\underline{M}_e^s/(\rho_e^s)^2)]\}$$

$$+ \int_D dxdy\ (\underline{E}_e\cdot\delta\underline{E}+B_e\delta B). \tag{166}$$

Substituting $\underline{E}_e=-\underline{\nabla}\phi$ into (160), integrating the terms in $\delta\underline{E}$ by parts, and using div $\delta\underline{E} = \Sigma^s a^s\delta\rho^s$ gives

$$\delta H_c = \sum_s \int_D dxdy\{(\vartheta_e^s)^{-1}\underline{M}_e^s\circ\delta\underline{M}^s + [c^2(w_e^s/\gamma_e^s-1) + a^s\phi + \phi^s(\bar{\Omega}_e^s) -$$

$$+ \bar{\Omega}_e^s\phi^s(\bar{\Omega}_e^s)]\delta\rho^s + [\phi^{s\prime}(\bar{\Omega}_e^s)+\lambda^s]\hat{\underline{z}}\cdot\mathrm{curl}\ (\delta\underline{M}^s/\rho_e^s-\delta\rho^s\underline{M}_e^s/(\rho_e^s)^2)\}$$

$$+ \int_D dxdy\{B_e + \sum_s a^s[\phi^{s\prime}(\bar{\Omega}_e^s)+\lambda^s]\}\delta B-(\phi|_{\partial D}-\mu)\oint_{\partial D}\delta\underline{E}\cdot\hat{\underline{n}}ds. \tag{167}$$

Integrating the curl term in the first integrand of (167) by parts yields,

$$\delta H_c = \sum_s \int_D dxdy\ \Big\{\delta\rho^s[c^2(w_e^s/\gamma_e^s-1) + a^s\phi + \phi^s(\bar{\Omega}_e^s) - \bar{\Omega}_e^s\phi^s(\bar{\Omega}_e^s)$$

$$- (\rho_e^s)^{-2}\underline{M}_e^s\cdot\mathrm{curl}\ \phi^s(\bar{\Omega}_e^s)\hat{\underline{z}}] + (\rho_e^s)^{-1}\delta\underline{M}^s\cdot\Big[\frac{\rho_e^s}{\vartheta_e^s}\underline{M}_e^s - \phi^{s\prime\prime}(\bar{\Omega}_e^s)\hat{\underline{z}}\times\underline{\nabla}\Omega_e^s\Big]\Big\}$$

$$+ \int_D dxdy\left\{B_e + \sum_s a^s[\phi^{s\prime}(\bar{\Omega}^s_e)+\lambda^s]\}\delta B\right\} \tag{168}$$

$$- (\phi|_{\partial D} - \mu)\oint_{\partial D} \delta \underline{E}\cdot\hat{\underline{n}}ds - \Sigma\, [\phi^{s\prime}(\bar{\Omega}^s_e)|_{\partial D} + \lambda^s]\, \oint_{\partial D} \hat{\underline{z}}\times(\delta \underline{M}^s/\rho^s_e$$

$$- \delta\rho^s \underline{M}^s_e(\rho^s_e)^2)\cdot\hat{\underline{n}}ds.$$

Since each connected component of the boundary, ∂D, is a curve of constant ϕ and $\bar{\Omega}^s_e$, the constant terms have been taken outside the coundary integrals as coefficients. These coefficients vanish upon choosing

$$\phi|_{\partial D} - \mu = 0, \tag{169}$$

$$\phi^{s\prime}(\bar{\Omega}^s_e)|_{\partial D} + \lambda^s = 0. \tag{170}$$

By (125) and (159), the $\delta\underline{M}^s$ coefficient of (168) vanishes for equilibrium states, provided

$$\frac{K^{s\prime}(\bar{\Omega}^s_e)}{\bar{\Omega}^s_e} = \phi^{s\prime\prime}(\bar{\Omega}^s_e), \tag{171}$$

which implies the relation

$$\rho^s_e \underline{v}^s_e = -\ \mathrm{curl}\ \phi^{s\prime}(\bar{\Omega}^s_e)\hat{\underline{z}}. \tag{172}$$

Substitution of (172) into (168) reduces the $\delta\rho^s$ coefficient to

$$(\rho^s_e)^{-1}\ \underline{v}^s_e\cdot\underline{M}^s_e + c^2(w^s_e/\gamma^s_e - 1) + \phi^s(\bar{\Omega}^s_e) - \bar{\Omega}^s_e\phi^{s\prime}(\bar{\Omega}^s_e),$$

which vanishes provided, using (157),

$$K^s(\bar{\Omega}^s_e) + \phi^s(\bar{\Omega}^s_e) - \bar{\Omega}^s_e\phi^{s\prime}(\bar{\Omega}^s_e) = 0, \tag{173}$$

which is a first integral of (171) viewed as an ordinary differential equation in $\bar{\Omega}^s_e$, for each s. Thus, both relations (171) and (173) are satisfies for ϕ^s given by

$$\phi^S(\xi) = \xi \left[\int^{\xi} \frac{K^S(z)}{z^2}\, dz + \text{const}\right],$$

which is the relation of Proposition 3.2. Finally, relation (171) implies, via (160), that for equilibrium states,

$$\underline{\nabla}[B_e + \sum_S a^S\phi^{S\prime}(\bar{\Omega}^S_e)] = 0.$$

Thus,

$$B_e + \sum_S a^S\phi^{S\prime}(\bar{\Omega}^S_e) = k$$

for some constant k. The δB coefficient in (168) vanishes for equilibrium states, upon choosing

$$K + \sum_S a^S\lambda^S = 0. \tag{174}$$

Hence, relativistic MFP equilibria satisfying (157), (159) and (160) are critical points of H_C in (161), when the function ϕ^S in C^S are determined as in the statement of Proposition 4.1 and the constants μ, λ^S satisfy (169), (170), and (174). This proves Proposition 3.2.

3.3. Second Variation and Linear Lyaponov Stability Condition.

The second variation $\delta^2 H_C$ at equilibrium is expressed using (162) as

$$\delta^2 H_C = \sum_S \int_D dxdy\{\delta\underline{v}^S\cdot\delta\underline{M}^S + \delta\rho^S\delta(c^2 w^S/\gamma^S) + \rho^S_e\phi^{S\prime\prime}(\bar{\Omega}^S_e)(\delta\bar{\Omega}^S)^2$$

$$+ 2\,(\phi^{S\prime}(\bar{\Omega}^S_e) + \lambda^S)\delta\rho^S\bar{\Omega}^S + \rho^S_e[\phi^{S\prime}(\bar{\Omega}^S_e) + \lambda^S\,]\delta^2\,\bar{\Omega}^S\}$$

$$+ \int_D dxdy[\,|\delta\underline{E}^2| + (\delta B)^2]\,. \tag{175}$$

The similarity of this equation with (96) and the identity (165) allows us immediately to write the formula analogous to (97) for the MFP case:

$$\delta^2 H_C = \sum_S \int_D dxdy\{\delta\underline{v}^S\cdot\delta\underline{M}^S + \delta\rho^S\delta(c^2 w^S/\gamma^S) + \rho^S_e\phi^{S\prime\prime}(\bar{\Omega}^S_e)(\delta\bar{\Omega}^S)^2$$

$$+ 2(\vartheta_e^S)^{-1}\ \underline{M}_e^S[(\rho_e^S)^{-1}\delta\rho^S\delta\underline{M}^S - (\rho_e^S)^{-2}\underline{M}_e^S(\delta\rho^S)^2]\}$$

$$+ \int_D dxdy[\,|\delta\underline{E}|^2 + (\delta B)^2] \ . \tag{176}$$

Transforming (176) exactly as in (98) - (100) for each species, one obtains

$$\delta^2 H_C = \sum_S \int_D dxdy\ \{\rho_e^S \phi^{S\prime\prime}(\bar\Omega_e^S)(\delta\bar\Omega^S)^2 + (\vartheta_e^S)^{-1}\ |\delta\underline{M}^S|^2 +$$

$$+ (\delta\rho^S)^2 \left[\frac{\partial(c^2 w^S/\gamma^S)}{\partial\rho^S} - \frac{2|M^S|^2}{(\rho^S)^2\vartheta^S} \right]_e 2\delta\rho^S \underline{M}_e^S\cdot\delta\underline{M}^S \left[(\rho^S\vartheta^S)^{-1} + \right.$$

$$\left. + \frac{\partial(c^2 w^S/\gamma^S)}{\partial|\underline{M}^S|^2} - \frac{1}{2}(\vartheta^S)^{-2}\frac{\partial\vartheta^S}{\partial\rho^S} \right]_e + (\underline{M}_e^S\cdot\delta\underline{M}^S)^2\ [-2(\vartheta^S)^{-2}\cdot$$

$$\frac{\partial\vartheta^S}{\partial|\underline{M}^S|^2}\ ]_e\} + \int_D dxdy\ |\delta\underline{E}|^2 + (\delta B)^2. \tag{177}$$

Sufficient conditions for positivity of the second variation, $\delta^2 H_C$, in (177) can be obtained, as in the fluid case, by decomposing $\delta\underline{M}^S$ into components perpendicular and parallel to $\underline{M}_e^S$:

$$\delta\underline{M}^S = \delta\underline{M}^S_\perp + \delta\underline{M}^S_{||},$$

$$\delta\underline{M}^S_\perp \cdot \underline{M}_e^S = 0, \ \delta\underline{M}^S \cdot \underline{M}_e^S = |\delta\underline{M}^S_{||}|\,|\underline{M}_e^S|$$

Hence, $|\delta\underline{M}^S|^2 = |\delta\underline{M}^S_\perp|^2 + |\delta\underline{M}^S_{||}|^2$ and (177) becomes

$$\delta^2 H_C = \sum_S \int_D dxdy\ \{\rho_e^S \phi^{S\prime\prime}(\bar\Omega_e^S)(\delta\bar\Omega^S)^2 + (\vartheta_e^S)^{-1}\ |\delta\underline{M}_\perp^S|^2 +$$

$$+ Q^S\ (\delta\rho^S_{,}\ |\delta\underline{M}^S_{||}|)\} + \int_D dxdy\ [\,|\delta E|^2 + (\delta B)^2], \tag{178}$$

where the matrix of the quadratic form Q^S in (178) is $\begin{bmatrix} a_{11} & a_{12} \\ a_{21} & a_{22} \end{bmatrix}$

where

$$a_{11}=\left[\frac{\partial(c^2w^S/\gamma^S)}{\partial\rho^S}-\frac{2|M^S|^2}{(\rho^S)^2\vartheta^S}\right]_e$$

$$a_{12}=a_{21}=\left[\frac{1}{\rho^S\vartheta^S}+\frac{\partial(c^2w^S/\gamma^S)}{\partial|\underline{M}^S|^2}-\frac{1}{2}(\vartheta^S)^{-2}\frac{\partial\vartheta^S}{\partial\rho^S}|\underline{M}^S|\right]_e$$

$$a_{22}=\left[(\vartheta^S)^{-1}|\underline{M}^S|^2\frac{\partial|v^S|^2}{\partial|\underline{M}^S|^2}\right]_e \tag{179}$$

The second variation (178) is positive definite and, hence, the corresponding equilibrium flow is linearly Lyapunov stable, when the following sufficient conditions are satisfied by the equilibrium flow variables.

$$\rho^S_e > 0, \tag{180}$$

$$\phi^{S\prime\prime}(\bar{\Omega}^S_e) = \frac{\rho^S_e \underline{v}^S_e \cdot \hat{\underline{z}}\times\underline{\nabla}\bar{\Omega}^S_e}{|\underline{\nabla}\bar{\Omega}^S_e|^2} > 0, \tag{181}$$

$$\left(\frac{\partial(c^2w^S/\gamma^S)}{\partial\rho^S}-\frac{2|M^S|^2}{(\rho^S)^2\vartheta^S}\right)_e \geq 0, \tag{182}$$

$$\left(\frac{\partial|v^S|^2}{\partial|\mathbf{M}^S|^2}\right)_e \geq 0, \tag{183}$$

$$\left[|\underline{M}^S|^2(\vartheta^S)^{-1}\frac{\partial|v^S|^2}{\partial|\underline{M}^S|^2}\left(\frac{\partial(c^2w^S/\gamma^S)}{\partial\rho^S}-\frac{2|M^S|^2}{(\rho^S)^2\vartheta^S}\right)\right]_e \geq$$

$$\geq\left[\left(\frac{1}{\rho^S\vartheta^S}+\frac{\partial(c^2w^S/\gamma^S)}{\partial|\underline{M}^S|^2}-\frac{1}{2}(\vartheta^S)^{-2}\frac{\partial\vartheta^S}{\partial\rho^S}\right)^2|\underline{M}^S|^2\right]_e \tag{184}$$

Condition (180) is the physical condition of positive mass density, which implies $\vartheta^S_e>0$, as well. Condition (181) is the relativistic MFP version of (53) and (105), which we shall discuss further in the examples belowe. Condition (182) is the relativistic subsonic (or sonic) requirement for each species. Condition (183) requires that the magnitude of each species velocity be an increasing function of the magnitude of the corresponding momentum density. Lastly, condition (184) represents an additional, essentially - relativistic requirement for stability, which is not present in the nonrelativistic limit (in that limit, (184) reduces to the trivial inequality $0 \geq 0$).

3.4. Applications to Zonal Flows and Circular Shear Flows.

Two applications of the stability conditions (180) - (184) follow: for zonal and circular relativistic MFP planar shear flows.

ZONAL FLOWS. A zonal flow is an equilibrium of the relativistic MFP equations (144) - (147) and (123) - (124) in the strip $\{(x,y)\in R^2 | Y_1 \leq y \leq Y_2\}$. This is a plane - parallel flow along x, additing arbitrary velocity profiles $v_e^s(x,y)=(\bar{v}^s(y),0)$ and densities $\rho_e^s(x,y)=\bar{\rho}^s(y)$; with electrostatic potential $\phi(x,y)=\bar{\phi}(y)$ and magnetic field $B_e(x,y)=\bar{B}(y)$ determined by the requirement that B_e satisfies (151) and $\underline{E}_e=-\underline{\nabla}\phi$ satisfies (155). The domain of x can be either periodic, or on the entire real line. Zonal flows model relativistic, non - neutral, warm plasma currents in a planar diode with ideally conduction walls, cf. Davidson, Tsang, and Swegle[4].

To determine the meaning of the stability condition (181) for the zonal case, one first finds $\bar{\Omega}_e^s$ from its definition (143)

$$\bar{\Omega}_e^s := (\rho_e^s)^{-1} \left(\hat{\underline{z}}\cdot \text{curl}\ (\gamma_e^s w_e^s \underline{v}_e^s) + a^s B_e\right)$$

$$= (\bar{\rho}^s)^{-1} \left[-\frac{d}{dy} (\bar{\gamma}^s \bar{w}^s \bar{v}^s) + a^s \bar{B}(y) \right] =: \bar{\Omega}^s(y),$$

where $\bar{Y}^s(y) := \gamma_e^s$ and $\bar{w}^s(y) := \bar{w}_e^s$. Thus, denoting y - derivatives by primes, e.g., $\bar{v}^{s\prime} = d\bar{v}^s(y)/dy$, condition (181) becomes

$$\phi^{s\prime\prime}(\bar{\Omega}_e^s) = \frac{\bar{\rho}^s \bar{v}^s}{\{((\bar{\gamma}^s \bar{w}^s \bar{v}^s)' - a^s \bar{B})/\bar{\rho}^s\}} > 0. \tag{185}$$

We consider several cases that elucidate the meaning of (185) for zonal flows that are admissible according to the other stability crtiteria (180), (182) - (184), and velocities are nonvanishing in the domain of flow.

Zonal Case A. In the case of neutral fluids, we have $a^S=0$, as in the previous section. The positivity condition (185) then reduces to

$$\bar{\rho}^S\bar{v}^S/[(\gamma^S\bar{w}^S\bar{v}^S)'/\bar{\rho}^S]'>0. \tag{186}$$

If, further, the zonal flow has constant density $(\bar{\rho}^S)'(y)=0)$, we find a relativistic generalization of Rayleigh's inflection - point criterion for stability of shear flows: all admissible zonal flows in the constant - density case having no inflection point in the profile of $(\gamma^S\bar{w}^S\bar{v}^S)$ are stable, when $\bar{v}^S$ is of a single sign satisfying $\bar{v}^S/(\gamma^S\bar{w}^S\bar{v}^S)''>0$ throughout the domain of flow, D. In the nonrelativistic limit $\gamma^S\bar{w}^S\longrightarrow 1$ and we recover Rayleigh's inflection point criterion for the velocity profile: $\bar{v}^S(\bar{v}^{S}{}''(y)>0$; in particular $\bar{v}^{S}{}''(y)\neq 0$, for stability of constant - density, barotropic zonal flows.

Zonal case B. In the case of charged fluid $(a^S\neq 0)$ at constant density $(\bar{\rho}^{S}{}'(y)=0)$, positivity in condition (185) reduces to

$$\frac{(\bar{\rho}^S)^2\bar{v}^S}{(\bar{\gamma}^S\bar{w}^S\bar{v}^S)'-a^S\bar{B}'} > 0. \tag{187}$$

All admissable zonal flows in this case having no critical points in the profile of the quantity $((\bar{\gamma}^S\bar{w}^S\bar{v}^S)'-a^S\bar{B})$ are stable, when $\bar{v}^S$ is of a single sign satisfying (187) throughout domain D. In particular, inflection points in the profile $(\gamma^S\bar{w}^S\bar{v}^S)(y)$ could occur without loss of stability, in general, provided a compensating profile exists for the magnetic field.

Zonal Case C. In the case of relativistic MFP with charged compressible fluids (i.e., $a^S\neq 0$, $\bar{\rho}^{S}{}'\neq 0$), the stability condition (185) implies that all admissible zonal flows having no extremum in the profile of the quantity $\{[(\gamma^S\bar{w}^S\bar{v}^S)'-a^S\bar{B}]/\bar{\rho}\}$ are stable, when $\bar{v}^S$ is of a single sign satisfying (185) troughout domain D. When $\bar{v}^S(y)$ is positive in D, (185) requires simply

$$((\gamma^{s}\bar{w}^{s}\bar{v}^{s})' - a^{A}\bar{B}^{s})/\bar{\rho}^{s})'(y)>0, \tag{188}$$

for stability of zonal reltivistic MFP equilibra that are admissible under the other stability criteria (180), (182) - (184).

CIRCULAR FLOWS. A circular flow is an equilibrium solution of the relativistic MFP equations (144) - (147) and (123) - (124) in the planar annulus $\{(r,\vartheta)\,|\,0<a\leq r\leq b,\ 0\leq\vartheta<2\pi\}$ in (r,ϑ) polar coordinates. Such circular flows model relativistic plasma currents in a coaxial waveguide with ideally conduction walls. This is the basic, "E-layer" configuration for a number of devices, such as the Astron, gyrotron, orbitron, heliotron, etc., as discussed, e.g., in Chernin and Lau[3]. The circular flow admits arbitrary velocity profiles $\underline{v}_{e}^{s}=\bar{v}^{s}(r)\hat{\vartheta}$ and densities $\rho_{e}^{s}=\bar{\rho}^{s}(r)$, with electric and magnetic fields, $\underline{\mathbf{E}}_{e}=\bar{E}(e)\hat{\underline{r}}$ and $B_{e}=\bar{B}(r)$, determined at equilibrium by (151) and (155), respectively. In this case, the equilibrium motion equations (153) become

$$0 = -(\bar{Y}^{s})^{-1}\,\frac{d}{dr}\,(c^{2}\bar{w}^{s}) + \frac{\bar{\gamma}^{s}\bar{w}^{s}(\bar{v}^{s})^{2}}{r} + a^{s}(\bar{E} + \bar{v}^{s}\bar{B}). \tag{189}$$

Upon using the barotropic relation $d(c^{s}w^{s})/d\rho_{0}^{s}=(\rho_{0}^{s})^{-1}(c_{0}^{s})^{2}$ (cf. (71)), the equilibrium equation (189) becomes

$$(\gamma^{s}\bar{\rho}^{s})^{-1}(c_{0}^{s})^{2}\bar{\rho}^{s\prime} = (\bar{\rho}^{s})^{-1}\bar{\gamma}^{s}(c_{0}^{s})^{2}\bar{v}^{s}\bar{v}^{s\prime} + \frac{\gamma^{s}\bar{w}^{s}(\bar{v}^{s})^{2}}{r}$$

$$+ a^{s}(\bar{E} + \bar{v}^{s}\bar{B}), \tag{190}$$

where $\bar{\gamma}^{s}(r):=\gamma_{e}^{s},\ \bar{w}^{s}(r):=w_{e}^{s}$, and prime $'$ denotes d/dr for circular flows. Relation (190) dfetermines $\bar{\rho}^{s\prime}(r)$ in terms of the other equilibrium profiles. To express the stability condition (181) for circular flows, we first calculate from (143) that

$$\bar{\Omega}_{e}^{s} = (\bar{\rho}^{s})^{-1}\,[\frac{1}{r}\,(\bar{\gamma}^{s}\bar{w}^{s}\bar{v}^{s}r)' + a^{s}\bar{B}] =: \bar{\Omega}^{s}(r) \tag{191}$$

The stability condition (181) then becomes

$$0 < \phi''(\bar{\Omega}_{e}^{s}) = \frac{\bar{\rho}^{s}\bar{v}^{s}}{\bar{\Omega}^{s\prime}(r)} \tag{192}$$

$$= \frac{(\bar{\rho}^s)^2\bar{v}^s}{\{[(\bar{\gamma}^s\bar{w}^s\bar{v}^s r)'/r + a^s\bar{B}]/\bar{\rho}^s\}'} . \quad (193)$$

For the special case of "relativistically rigidly rotating" flows, defined by the relation

$$\bar{\gamma}^s\bar{w}^s\bar{v}^s = \tilde{\omega}^s r/2, \ \tilde{\omega}^s = \text{const} > 0, \quad (194)$$

we have

$$\bar{\Omega}^s(r) = [\tilde{\omega}^s + a^s\bar{B}]/\bar{\rho}^s \quad (195)$$

and condition (192) or (193) requires

$$[(\tilde{\omega}^s + a^s\bar{B})/\bar{\rho}^s]'(r) > 0, \quad (196)$$

if $\bar{v}^s > 0$ throughut the domain of flow.

Thus, the quantity $(\tilde{\omega}^s + a^s\bar{B})/\bar{\rho}^s$ must be monotonically incressing with radius for stability of such "rigid" circular MFP flows.

Remark. The Lyapunov stability criteria (188) and (192) (193), for admissible zonal and circular relativistic MFP equilibria, respectively, can be compared with the spectral analyses of Davidson et al.[4] and Chernin and Lau[3], upon taking the "cold plasma" limit. The cold plasma limit is obtained by setting w^s=1 everywhere from the onset (where w^s is defined in (116)). For <u>zonal equilibria</u>, this means that after the dendities $\rho^s_e = \bar{\rho}^s(y)$ are used to determine the equilibrium electric fields $\underline{E}_e$ through Gauss's law (123), then the motion equation (151) and Maxwell equation (153) determine $\underline{v}^s_e$ and B_e according to

$$\bar{B}'(y) = \sum_s a^s\bar{\rho}^s(y)\bar{v}^s(y) \quad (197)$$

$$\bar{E}(y) + \bar{v}^s(y)\bar{B}(y) = 0. \quad (198)$$

Relation (198) implies that the species velocities for zonal equilibria in the cold plasma limit are all equal. The magnetic field is then determined by substituting $\bar{v}^s(y) = -\bar{E}(y)/\bar{B}(y)$ into (197). To determine the stability criterion corresponding to (181) for relativistic zonal flows of a cold plasma, we set $\bar{w}^s=1$ and then suppress the species label s in (188) to find

$$\{[(\bar{\gamma}\bar{v})' - a\bar{B}]/\bar{\rho}\}'(y) > 0, \tag{199}$$

when $\bar{v}(y)$ is positive in D, or, equivalently,

$$[(\bar{\gamma}^3\bar{v}' - a\bar{B})/\bar{\rho}]'(y) > 0, \tag{200}$$

for stability. Stability criteria (188) for relativistic zonal flows of warm plasmas and (199) (200) for cold plasmas generalize the result of Davidson et al.[4]; their criterion is

$$(\bar{\gamma}/\bar{\rho})'(y) > 0, \tag{201}$$

for spectral stability of relativistic cold plasmas, in the <u>guiding center</u> approximation.

To obtain the Lyapunov stability criterion for circular flows of a relativistic cold plasma, we set $\bar{w}^s=1$ in (193) to find, in the case that $\bar{v}^s>0$ throught the domain of flow, the stability condition

$$\{[(\bar{\gamma}^s\bar{v}^s r)'/r+a^s\bar{B}]\rho^s\}'(r) > 0 \tag{202}$$

In this case, the equilibrium densities $\bar{\rho}^s$ determine the equilibrium electric field as usual by Gauss's law, after which the equil ibrium velocities and magnetic field for the cold plasma are determined from (151) and (153) with $\bar{w}^s\equiv 1$, i.e.,

$$- \bar{B}'(r) = \sum_s a^s\bar{\rho}^s(r)\bar{v}_s(r), \tag{203}$$

$$- (\bar{v}^s)^2\bar{\gamma}^s/r+a^s[\bar{E}(r) + \bar{v}^s(r)\bar{B}(r)] = 0 \tag{204}$$

The Lyapunov stability criterion (202) for circular flows of relativistic cold plasma applies to varions cases in the E - layer configuration studied in Chernin and Lau[3].

4. CONCLUSION.

For two - dimensional barotropic fluid and plasma equilibria, we have investigated Lyapunov stability conditions in both the nonrelativistic and relativistic cases. The method used relies on the general fact that critical points of $H_C=H+C$ where H is the energy and C is a Casimir are always equilibrium points of flows generated by H. Vice versa, we have shown in each case that nondegenerate equilibrium flows are critical points of energy plus the appropriate Casimirs (and, in general, other conserved quantities). To find stability conditions for these flows, we have computed the second variation of H_C at equilibrium and found sufficient conditions for which this second variation is positive definite. The equilibrium flows that satisfy these conditions are linearly Lyapunov stable, as explained in the introduction.

We have also shown how the relativistic Lyapunov stability conditions reduce to their nonrelativistic conterparts. Some of these relativistic stability conditions such as (105) are form - invariant relativisations of corresponding nonrelativistic conditions. This is due to form invariance of the Casimirs, which is, in turn, associated to the common form of the Poisson bracket structure for both the relativistic and nonrelativistic theories. However, other relativistic stability conditions such as (108) have no nonrelativistic counterparts, since they reduce to the trivial requirement $0\geq0$ in the nonrelativistic limit.

In the three dimensional adiabatic case, Holm and Kupershmidt[2] show that nondegenerate equilibrium flows of multifluid plasmas are critical points of H_C, the energy plus appropriate Casimirs, just as in the two - dimensional cases. In contrast to the two - dimensional cases, the second variation of H_C at equilibrium for three dimensional fluids and plasmas involves nonalgebraic expressions, in particular, expressions containing gradients of the specific entropy variations. Such nonalgebraic terms in δ^2H_C preclude deriving **unconditional** stability conditions by the energy - Casimr method. However, **conditional** stability results are derivable in three dimensions, e.g., when the variations of specific entropy for each MFP species remain sufficiently smooth. Hence, one can consider monitoring the development of disturbance of an initial equilibrium state, expecting the onset of instability to occur some time after gradients (or wave numbers) of the specific entropy variations leave the stable range for that inizial state.

REFERENCES (Part II).

1. K.M.Watson, S.A.Bludman, and M.N. Rosenbluth, **Phys. Fluid 3,** 741 (1960).
2. D.D. Holm and B.A.Kupershmidt, Phys. Fluid **29,** 49 (1986).
3. D. Chernin and Y.Y. Lau, Phys. **Fluids 27,** 2319 (1984).
4. R.C.Davidson, K.T. Tsang, J.A. Swegle, **Phys. Fluids 27,** 2332 (1984).
5. Z.R.Iwinski and L.A.Turski, **Lett. in Appl. and Sci. Eng. 4,** 179 (1976).
6. D.D. Holm and B.A.Kupershmidt, **Phys. Lett. 101A,** 23 (1984).
7. V.I.Arnold, **Sov.Math.Dokl.162** (5), 773 (1965); **Prikl. Mat Mekh. 29,** 846 (1965). (Translated in **J. Appl. Math. Mech. 29,** 1002.)
8. V.I.Arnold, **Am.MathSoc.Transl.79,** 267 (1969).
9. D.D.Holm J.E.Marsden, T.Ratiu, and A.Weinstein, **Phys.Lett. 98A,** 15 (1983).
10. D.D:Holm, J.E.Marsden, T.Ratiu, A.Weinstein, **Physics Reports 123,** 1(1985).
11. W.A.Newcomb, Appendix in I.B.Bernstein, E.A. Frieman, M.D. Kruskal and R.M.Kulsrud, **Proc. Roy. Soc.** (London) **244,**, 17 (1958).
12. M.C.Kruskal and C.R.Oberman, **Phys. Fluids, 1** 275 (1958).
13. B.A. Kupershmidt, Phys. Lett. A **114,** 231 (1986).
14. H.D.I.Abarbanel, D.D.Holm, J.E.Marsden, and T.Ratiu, Phil. Trans. Roy. Soc. (London) A **318,** 349 (1986).
15. R.F.Dashen and D.H.Sharp, **Phys.Rev.165,** 1857 (1968).
16. I.Bialynicki - Birula and Z.Iwinski, **Rep.Math.Phys.4,** 139 (1973).
17. C.P. Enz and L.A.Turski, **Physica** (Utrecht) **96A,** 369 (1979).
18. I.E.Dzyaloshinskii and G.E.Volovick, **Ann. of Phys. (NY) 125,** 67 (1980).
19. P.J.Morrison and J.M.Green, **Phys.Rev.Lett.45,** 790 (1980).; errata, **Ibid. 48,** 569.
20. D.D.Holm and B.A.Kupershmidt, **Physica** D **6,** 347 (1983).
21. D.D.Holm, B.A.Kupershmidt, and C.D. Levermore, **Phys.Lett 98A,** 389 (1983).
22. J.E.Marsden, A.Weistein, T.Ratiu, R. Schmid, and R.G.Spencer, **Atti della Academia della Scienze di Torino, 117,** 289 (1983).
23. J.E.Marsden, A.Weistein, and T.Ratiu, **Trans.Am.Math.Soc. 281,** 147 (1984).

24. D.Bao, J.E.Marsden and R.Walton, Comm. Math. Phys, **99**, 319 (1985).

25. D.D. Holm, Physica D **17**, 1 (1985).

26. A.N.Kaufman, Phys. Fluids 25, 1993 (1982).

27. D.D. Holm, J.E.. Marsden, and T.Ratiu, **Hamiltonian Structure and Lyapunov Stability for Ideal Continuum Dynamics**, Univ.Montreal Press: Montreal (1986).

28. L.D.Landau and E.M.Lifshitz, **Fluid Mechanics**, Pergamon Press: New York (1959).

29. D.D. Holm, **Comtemp. Math. AMS 28**, 25 (1984).

PART III

RELATIVISTIC CHROMOHYDRODYNAMICS AND YANG-MILLS VLASLOV PLASMA

ABSTRACT

The equations of a special-relativistic, classical, ideal quark-gluon fluid plasma interacting self-consistently with a Yang-Mills field are obtained by starting from the single-particle description and using the cold plasma limit of the relativistic Yang-Mills Vlasov kinetic equations. The Hamiltonian structures are given for the single-particle, kinetic theory, and fluid descriptions.

INTRODUCTION

The core of a neutron star at relativistically hygh energy density is sometimes postulated to be a quark-gloun plasma, i.e., a classical fluid composed of quarks and gluons interacting via a self-consistent Yang-Mills field. In ultra-relativistic heavy ion collisions, nuclear matter is also expected to undergo a deconfinement phase transition to a quark-gluon plasma. Both of these situations involve dense matter having a very large number of degrees of freedom and are approximately describable by a fluid dynamics model, seem e.g., Clare and Strottam[1].

In this part of the lectures, we use the techniques and language developed earlier for the Hamiltonian description of special-relativistic fluid plasmas to describe an ideal classical quark-gluon plasma, including the effects of special relativity and Yang-Mills internal degrees of freedom. Previously, nonrelativistic chromohydrodynamics (CHD) (that is, the theory of nonrelativistic ideal fluids with gauge internal degrees of freedom interacting self-consistently with a Yang-Mills field) has been derived and given a Hamiltonian description in Gibbons, Holm, and Kupershmidt[2,3]. Moreover, these nonrelativistic CHD euqtions, together with their Hamiltonian structure are obtained in that reference by taking the cold-plasma limit of the corresponding equations and hamiltonian structure for a Yang-Mills Vlasov plasma (YMV). The YMV equations for a relativistic plasma are formulated in Heinz[4,5]. Here (following Holm and Kupershmidt[6]), the relativistic CHD equations are formulated by starting from the single-particle description, passing to the YMV kinetic theory description, and using the cold plasma limit of the relativistic YMV equations. The Hamiltonian structure is given in each case. As noted earlier in the lectures and for each case discussed here, the Hamiltonian structure remains unchanged in passing from the relativistic to the nonrelativistic description, while the physical variables and the Hamiltonian functional for each relativistic system reduce regularly (in the parameter c^{-2}) to their nonrelativistic forms.

The plan of Part III is as follows. First, we discuss the YM single-particle description. Then we pass to the kinetic equations for relativistic YMV and cast them into Hamiltonian form. Upon taking the cold plasma limit and introducting thermodynamic internal energy, we finally derive the relativistic CHD equations and their associated Hamiltonian structure.

1. YANG-MILLS PARTICLE DYNAMICS

We first show that the noncanonical Poisson bracket given in Gibbons, Holm, and Kupershmidt[2,3] produces the correct relativistic YM particle dynamics (in the laboratory frame) from the relativistic single-particle Hamiltonian. The single-particle classical phase space for a Yang-Mills particle in a fixed, laboratory frame consists of the particle's spatial coordinates $\underline{\mathbf{x}}$, canonically conjugate momentum components $\underline{\mathbf{p}}$, and gauge charge g. The YM single-particle Poisson bracket between functions of $\underline{\mathbf{x}}$, $\underline{\mathbf{p}}$, and g is given by

$$\{J,K\}_1=-[(\partial J/\partial\underline{\mathbf{p}})\cdot(\partial K/\partial\underline{\mathbf{x}})-(\partial J/\partial\underline{\mathbf{x}}\cdot(\partial K/\partial\underline{\mathbf{p}})+\langle g,[\partial J/\partial g,\partial K/\partial g]\rangle] \qquad (1)$$

This is the direct sum of a canonical bracket in $\underline{\mathbf{p}}$ and $\underline{\mathbf{x}}$, with the Lie algebraic $\tilde{g}^*$-braket. Here the charge g belongs to the dual $\tilde{g}^*$ of the gauge-symmetry Lie algebra $\tilde{g}$. The relativistic single-particle Hamiltonian (Routhian) is

$$H_1=c^2[(|\underline{\mathbf{p}}-\langle g,A\rangle|^2/c^2+m^2)^{1/2}-m]-\langle g,A_0\rangle\ , \qquad (2)$$

where $A_\mu=(A_0,A)$ is the Yang-Mills vector potential taking values in $\tilde{g}$, m the particle mass, and c the speed of light. greek indices $\mu,\nu,\sigma,\alpha,\beta$ run from 0 to n (=3), Latin indices i,j,k from 1 to n and a,b,r from 1 to $N=\dim\tilde{g}$; summation over repeated indices is implied; ad: $\tilde{g}\longrightarrow \mathrm{End}\,\tilde{g}$ denotes multiplication in $\tilde{g}$: $ad(y)z=[y,z]$; a map $\tilde{g}\rightarrow\tilde{g}^*$, $\tilde{g}\ y\longrightarrow {}^*y\in\tilde{g}^*$ is defined by the rule $\langle y^*,z\rangle=(y,z)$ where (,) is an invariant symmetric nondegenerate from on $\tilde{g}$ (e.g., the Killing form for $\tilde{g}$ semi-simple); $t^r_{a,b}$ are the structure constants of $\tilde{g}$ in a basis with elements e_b; $g=ge^a$, where e^a are the elements of the dual basis.

From the Poisson bracket (1) and Hamiltonian (2) one may derive

$$m\ddot{x}_i=\langle g,E_i\rangle-\langle g,\dot{x}_jB_{ji}\rangle \ , \tag{3}$$

wich is Yang-Mills analog of the Lorentz force. The Yang-Mills fields E_i and B_{ij} are defined in terms of potentials $\underline{\mathbf{A}}$ and A_0 by

$$E_i=-\partial A_i/\partial t+A_{0,i}-[A_i,A_0], \quad B_{ij}=A_{i,j}-A_{j,i}+[A_i,A_j] \ . \tag{4}$$

In addition,

$$\dot{g}=-ad(\partial H_1/\partial g)^*g=ad(A_0+\dot{\underline{\mathbf{x}}}\cdot\underline{\mathbf{A}})^*g$$

$$[\text{in components: } \dot{g}_a=t^r_{a,b}(A^b_0+\dot{\underline{\mathbf{x}}}\cdot\underline{\mathbf{A}}^b)g_r] \ . \tag{5}$$

Eqs. (3) and (5) appear in Wong[7] for the motion of a gauge-charge particle in a Yang-Mills field.

2. YANG-MILLS VLASOV PLASMA

The Poisson bracket for YMV in single-particle phase space is defined for any two functionals J[f] and K[f] depending on the distribution function f on phase space to be[2,3]

$$\{J[f],K[f]\}_f=\int f\{\delta J/\delta f,\delta K/\delta f\}_1 d^nx\, d^np\, d^Ng \ . \tag{6}$$

The full Hamiltonian structure of relativistic YMV is exactly the same as the nonrelativistic one derived in Gibbons, Holm, and Kupershmidt[2,3], namely, the direct sum of (6) with a canonical structure for the fields ${}^*\underline{\mathbf{E}}$ and $\underline{\mathbf{A}}$:

$$\partial f/\partial t=\{H,f\}_f(=-\{\delta H/\delta f,f\}_1), \ \partial\underline{\mathbf{A}}/\partial t=-\delta H/\delta{}^*\underline{\mathbf{E}}, \ \delta{}^*\mathbf{E}/\partial t=\delta H/\delta\underline{\mathbf{A}} \ . \tag{7}$$

The Hamiltonian H in (7) is

$$H=\int H_1 f d^n x d^n p d^N g+\int(<{}^*E^i,A_{0,i}-[A_i,A_0]>+\frac{1}{2}<{}^*E^i,E_i>$$

$$+\frac{1}{2}<{}^*B^{ij},B_{ij}>)d^n x\ , \tag{8}$$

where A_0 in the section is considered as a Lagrange multiplier for the Yang-Mills version of Gauss's Law. Transforming Eqs. (7a) to kinetic momentum $\underline{p}$-<g, A> produces the relativistic YMV equations in Heinz[4,5] when written in the laboratory frame.

Remark. When $\tilde{g}$ is abelian, this Hamiltonian structure has a counterpart in the space of variables where the vector potential $\underline{\mathbf{A}}$ is replaced by the magnetic field strenght B_{ij}. The latter structure was derived in Holm and Kupershmidt[8].

Passing to the cold plasma limit by considering only the first moment ρ=<1>, $\underline{\mathbf{M}}$=<$\underline{p}$>, G=<g>, where $<\ \cdot\ >=\int\cdot f d^n p d^N g$, yelds the barotropic CHD equations with pressure depending only on mass density. Adding entropy dependence to the internal energy, we get the full CHD system, treated in the next section.

3. CHROMOHYDRODYNAMICS

The relativistic CHD equations express conservation of energy - momentum and particle number, combined with the Yang-Mills field equations.

$$\partial_\mu T^{\mu\nu}=0\ ,\quad \partial_\mu(\rho_0 u^\mu)=0,\quad D_\mu F^{\mu\nu}=j^\nu,\quad D[_\sigma \tilde{F}^{\mu\nu}]=0\ , \tag{9}$$

where we use the following additional notation: We work in a flat space-time with coordinates x^μ, wher x^0=ct is the real time coordinate. The signature of the metric tensor $g_{\alpha\beta}$ is n-1(=2). derivatives are denoted $\partial_\mu=\partial/\partial x^\mu$. The energy momentum tensor is $T^{\mu\nu}$,u^μ is the particle velocity vector, ρ_0 is proper density in the frame of a moving fluid element, $F^{\mu\nu}$ is the Yang-Mills field strenght tensor, $j^\nu=G_0 u^\nu$ is current density, where G_0 is the gauge charge density in the proper frame; $\tilde{F}^{\mu\nu}$ is the tensor dual to $F^{\mu\nu}$,$\tilde{F}_{\mu\nu}=\frac{1}{2}\vartheta_{\mu\nu\alpha\beta}F^{\alpha\beta}$, and D_μ is the covariant derivative, $D_\mu=\partial_\mu-\mathrm{ad}(A_\mu)$. In terms of $\underline{\mathbf{A}}$, $\mathrm{ad}(F_{\nu\mu})=[D_\mu,D_\nu]$.

The expression for energy-momentum is $T^{\mu\nu}_{YM}=T^{\mu\nu}_{MAT}+T^{\mu\nu}_{YM}$, where $T^{\mu\nu}_{MAT}=$

$=\rho_0 wu^\mu u^\nu + p_0 g^{\mu\nu}$ and $T^{\mu\nu}_{YM} = -\frac{1}{4} g^{\mu\nu} \langle {}^*F_{\alpha\beta}, F^{\alpha\beta} \rangle + \langle {}^*F^{\mu\alpha}, F^\nu_\alpha \rangle$ where $w=1+(e_0+p_0/\rho_0)c^{-2}$, p_0 is pressure, and e_0 is specific internal energy, all evaluated in the proper frame of the fluid.

In the nonrelativistic limit, eqs. (9) reduce to the CHD equations. As a consequence of the antisymmetry of $F^{\mu\nu}$ and Eqs. (9c), charge is conserved: $D_\nu j^\nu=0$.

To treat a covariant system by the Hamiltonian formalism, one rewrites the covariant equations in a fixed frame, cf. Dirac [1962]. For our CHD system we choose the laboratory frame (to get the Eulerian description). With $u^\mu=\gamma(1,v^i/c)$, and $\gamma=(1-v^2/c^2)^{-1/2}$, eq. (9) become

$$\partial_t N_i = -\partial_j (N_i v^j + \delta^j_i p_0) + G(E_i + B_{ij} v^j) \ ,$$

$$\partial_t \rho = -\partial_i (\rho v^i), \quad \partial_t \eta = -\eta_{,i} v^i \ , \tag{10}$$

$$D_t({}^*E^i) = -Gv^i + {}^*(D_k B^{ki}), \quad D_t A_i = -E_i, \quad D^*_t G = -D^*_i (Gv^i) \ , \tag{11}$$

wher $\underline{D}^* = \underline{\nabla} + \mathrm{ad}(A)^*$, $D^*_t = \partial/\partial t + \mathrm{ad}(A_0)^*$, $B^{ki} = -F^{ki}$, and the quantities $\underline{N}, \rho, \eta$, and G are, respectively, the hydrodynamic momentum density, mass density, specific entropy, and gauge charge density in laboratory frame. The velocity of the fluid relative to the laboratory frame is denoted by $\underline{v}$. The laboratory frame quantities are related to the proper frame quantities by $\rho=\gamma\rho_0$, $\underline{N}=\gamma^2\rho_0 \underline{v} w$, $\eta=\eta_0$, $G=\gamma G_0$. Eqs. (10) and (11) preserve Gauss' law [the zero-component of (9c)]

$$D^*_i({}^*E^i) + G = 0 \tag{12}$$

To make eqs. (10) and (11) into a dynamical system, we choose the radiation gauge $A_0=0$. Then eqs. (11) become

$$\partial_t {}^*E^i = -Gv^i + {}^*(D_k B^{ki}), \quad \partial_t A_i = -E_i, \quad \partial_t G = -D^*_i (Gv^i) \ . \tag{13}$$

The main result of this section is that the relativistic equations (10) and (13) are Hamiltonian with the following Poisson bracket

$$\{H,F\}=-\int d^nx\Big\{\Big[\frac{\delta F}{\delta\rho}\partial_i\rho+\frac{\delta F}{\delta\eta}\eta_{,i}+\frac{\delta F}{\delta N_j}(N_i\partial_j+\partial_i N_j-\langle G,B^{ij}\rangle)$$

$$-\langle G,\frac{\delta F}{\delta^{*}E^i}\rangle\Big]\frac{\delta H}{\delta N_i}+\langle D_i^{*}G\frac{\delta H}{\delta N_i},\frac{\delta F}{\delta G}\rangle+\langle G,\Big[\frac{\delta H}{\delta G},\frac{\delta F}{\delta G}\Big]\rangle$$

$$+\frac{\delta F}{\delta N_j}\Big[\langle G,D_j\frac{\delta H}{\delta G}\rangle+\rho\partial_j\frac{\delta H}{\delta\rho}-\eta_{,j}\frac{\delta H}{\delta\eta}+\langle G,\frac{\delta F}{\delta^{*}E^j}\rangle\Big]+$$

$$+\frac{\delta F}{\delta A_i}\frac{\delta H}{\delta^{*}E^i}-\frac{\delta H}{\delta A_i}\frac{\delta F}{\delta^{*}E^i}\Big\}\tag{14}$$

and Hamiltonian given by

$$H=\int T^{00}d^nx=\int d^nx\Big\{c^2[(|\mathbf{N}|^2/c^2+(\rho w)^2)^{1/2}-\rho]-p_0+\frac{1}{2}\langle{}^{*}E^i,E_i\rangle$$

$$+\frac{1}{4}\langle{}^{*}B^{ij}B_{ij}\rangle\Big\}.\tag{15}$$

This result follows immediately by substituting the identities (proven by the same technique as in the earlier lectures)

$$\delta H/\delta\underline{N}=\underline{v},\quad \delta H/\delta\rho=c^2(w\gamma^{-1}-1),\quad \delta H/\delta\eta=c^2\rho_0 w_{,\eta}-p_{0,\eta},$$

$$\delta H/\delta^{*}E^i=E_i,\quad \delta H/\delta A_i={}^{*}(D_kB^{ki}),\quad \delta H/\delta G=0\,,\tag{16}$$

into the equations $\partial_t F=\{H,F\}$ for $F\in\{\underline{N},\rho,\eta,G,{}^{*}\underline{E},\underline{A}\}$ and using (14). The Poisson bracket (14) satisfies the Jacobi identity because this bracket transforms to the semidirect-product Lie-Poisson bracket derived for nonrelativistic CHD in Gibbons, Holm, and Kupershmidt[2,3] by changing variables to $\underline{M}$, the local momentum density, instead of the hydrodynamic momentum density $\underline{N}=\underline{M}-\langle G,\underline{A}\rangle$. Notice that when $c^{-2}\to 0$, the motion equations (10), (13), the Hamiltonian (15), and the functional derivatives (16) all reduce to their nonrelativistic counterparts. Thus, relativistic CHD is a regular, structure preserving deformation (with parameter c^{-2}) of nonrelativistic CHD.

For multiple gauge-charge species (flavors) all variables except ${}^{*}\underline{E}$

and $\underline{A}$ acquire species labels, and one sums over species in Gauss' law (12), in the Poisson bracket (14), and in the Hamiltonian (15). The resulting relativistic multi-species CHD system has a number of limiting cases. In particular, for the Abelian, single-species case, Eqs. (9) reduce to the equations for a classical relativistic electromagnetically charged fluid discussed in Part II of these lectures.

REFERENCES (Part III)

1. R.B.Clare and D.Strottman, Relativistic hydrodynamics and heavy ion reactions, Phys. Rep. (1986).

2. J.Gibbons, D.D.Holm and B.A.Kupershmidt, Phys. Lett. A 90, 281 (1982).

3. J.Gibbons, D.D.Holm and B.A.Kupershmidt, Physica 6, 179 (1983).

4. U.Heinz, in Quark matter formation and heavy ion collisions, eds. M.Jacob and H.Satz (World Scientific, Singapore, 1982) p.439.

5. U.heinz, Phys. Rev. Lett. 51, 351 (1983).

6. D.D.Holm and B.A.Kupershmidt, Phys. Lett. A 105, 225 (1984).

7. S.K.Wong, Nuovo Cimento 65A, 689 (1970).

8. D.D.Holm and B.A.Kupershmidt, Phys. Lett. A 101, 23 (1984).

9. P.A.M.Dirac, in Recent developments in general relativity, A collection of papers dedicated to L.Infeld (Pergamon, New York, 1962).

Lectures delivered at C.I.M.E. Session on "Relativistic Fluid Dynamics"
(ed. A.M. Anile and Y. Choquet-Bruhat)
at Centro Studi Noto, Noto (Syracuse) May 25 - June 3, 1987

COVARIANT FLUID MECHANICS AND THERMODYNAMICS: AN INTRODUCTION

Werner Israel*
Theoretical Physics Institute, Avadh Bhatia Physics Laboratory
University of Alberta, Edmonton, T6G 2J1, Canada

ABSTRACT

These lectures are intended as a pedagogical introduction to the covariant formulation of nonequilibrium and statistical thermodynamics and kinetic theory. The various formulations of nonequilibrium thermodynamics and their individual difficulties - with causality, stability, shock structure - are critically reviewed. As an illustration of the general formalism, a fairly detailed treatment is presented of the covariant statistical thermodynamics of superfluids.

*Work supported by Canadian Institute for Advanced Research and by Natural Sciences and Engineering Research Council of Canada

CONTENTS

INTRODUCTION

These notes are intended as a brief introduction to the covariant thermomechanics of fluids and as background for some of the seminars on special topics that will come later. For fluids out of thermal equilibrium, the covariant phenomenological theory is still unsettled and currently exists in a variety of formulations; I shall therefore attempt to lay a framework that is sufficiently broad to encompass most of them.

As far as I am aware, there has been only one previous School devoted exclusively to relativistic fluid mechanics. That was organized by the late Professor Cattaneo [1], also under the auspices of C.I.M.E., and held at the Alpine resort of Bressanone 17 years ago. Courses offered then included: shock waves in relativistic magnetohydrodynamics (A. Lichnerowicz), variational principles (A.H. Taub), relativistic kinetic theory (J. Ehlers) and problems connected with the propagation of heat in relativistic theories (G. Boillat). If some of these titles have a disconcertingly topical ring today, it is not that there has been no movement in the meantime, rather that a reasonable doubt exists whether all of it has been forwards! At worst, even if little has been gained on some fronts beyond a better appreciation of the difficulties, that must be accounted a form of progress.

By contrast, the scope for applications of the theory has expanded almost beyond recognition. Neutron star superfluids and pion condensates, false vacuum states in the early universe and other exotic fluid-like media scarcely imagined a decade or two ago are now objects of everyday discussion. Several of the new problems opened up by these developments have the added fascination of being inseparably bound up with the subtleties of general relativity.

A case in point is the curious general-relativistic instability first noted by Chandrasekhar [2] about the time of the Bressanone School and since then studied extensively [3]. Rotating stars made of perfect fluid are unstable to emission of gravitational waves in a bar-like mode with azimuthal dependence $e^{im\varphi}$. The source of emission is a counter-rotating hydrodynamic wave which is itself driven by radiation-reaction. An attractive idea, which has received some support from detailed calculations [4,5], is that this instability is the factor which limits the angular velocities of accreting neutron stars and which might make the fastest of these objects promising candidates for sources of gravitational waves.

Viscosity tends to suppress the instability, and in neutron stars is effective below temperatures of about 10^7 K [4]. It was while studying this effect a few years ago [6] that Hiscock and Lindblom [7] came upon the unexpected result that the standard (Eckart, Landau-Lifshiftz) relativistic theories of heat flow and viscosity predict instabilities on time-scales

$\sim$ (mean collision time) $\times$ (speed of sound/c)^4 ,

about 10^{-34} seconds for water at room temperature! This is a further embarrassment for theories already notorious for violating causality. These drawbacks and the astrophysical exigencies highlight the need for a stable, hyperbolic phenomenological theory of irreversible processes that is compatible with causality and clearly correlated with microscopic theory where their domains of validity overlap.

The following pages will discuss these and other issues, and attempt to place them in a coherent perspective. I hope they may be of service in initiating non-relativists into the elegance of thermodynamics in its covariant guise. A good case can be made for the thesis that thermodynamics becomes simpler when formulated covariantly. As we shall see, there are many reasons for this; the most obvious is that, through its fusion of energy and momentum, relativistic covariance welds their balance laws into a single compact equation.

In Chapter I the covariant formalism for thermal equilibrium is built up by induction, starting with the familiar example of a simple fluid. The once-controversial relativistic formulation of the second law ("Does a moving body appear cool?") is briefly dealt with in §2. Then follows an excursion (§§3,4) into the statistical foundations of the theory and the thermodynamics of relativistic superfluids, an instructive illustration of the effectiveness and some of the subtleties of the covariant formulation. These two sections can be omitted without affecting the continuity of the exposition.

Chapter II outlines the attempts, past and present, to formulate a covariant thermodynamics of off-equilibrium processes, their successes and their failures. The freedom to choose the rest-frame arbitrarily (within certain limits) in relativistic theories - which arises from the equivalence of mass and energy and has no exact classical counterpart - is explained together with its implications (§2). In §3 a general covariant framework for off-equilibrium thermodynamics is set up based on little more than the requirement of compatibility with equilibrium theory in the limit of vanishing dissipation. The two following sections review the "standard" theories of Eckart and Landau-Lifshitz and their difficulties with causality and stability. In §6 we look at some proposals for improved theories - "nonlocal thermodynamics" - entailing memory effects and frequency-dependent transport coefficients - and "extended thermodynamics" in which entropy acquires dependence on heat flux and viscous stress. The extended theories are hyperbolic and exhibit a satisfying interrelationship between causality and stability. Finally, in §7, we discuss a less satisfactory aspect of the extended theories: their inability to give a satisfactory account of the structure of shocks above a certain (quite moderate) strength.

The concluding Chapter develops the rudiments of covariant kinetic theory up to the point where its close links to the phenomenology of Chapter II becomes apparent.

Because of the pedagogical character of the lectures, I have supplied only a modicum of references that might serve for further study and, apart from sporadic comments, there is no attempt to provide historical perspective. These omissions are handsomely redressed in a number of recent reviews, in particular those collected in the 1984 Barcelona Proceedings [34].

CHAPTER I. COVARIANT THERMODYNAMICS OF EQUILIBRIUM

§1. Energy Tensor

In textbooks, the stress-energy-momentum tensor of a continuous medium is often introduced by specifying the physical significance of its components in the rest-frame. I need to emphasize at the outset that this basic object has a direct significance, totally independent of the choice, or even existence, of a "rest-frame".

The direct definition is this [9]. Let $d\Sigma_\beta$ be an arbitrarily oriented element of 3-area. Then the 4-momentum dp^α that crosses $d\Sigma_\beta$ in the sense of its normal is

$$dp^\alpha = c^{-1}\epsilon T^{\alpha\beta} d\Sigma_\beta \quad . \qquad (1)$$

The sign factor ϵ is +1 if the normal $d\Sigma_\alpha$ is spacelike and -1 if it is timelike (the null case is dealt with by a continuity argument); it ensures the positivity of $\epsilon A^\alpha d\Sigma_\alpha$ in all cases where A^α and $d\Sigma^\alpha$ are co-directed. By applying Gauss's divergence theorem

$$\int \epsilon A^\beta d\Sigma_\beta = \int \partial_\beta(\sqrt{-g}\, A^\beta)\, d^4x$$

to a small 4-dimensional tetrahedron, it is easy to show that the object $T^{\alpha\beta}$ defined by (1) is independent of $d\Sigma_\beta$ if the material 4-momentum created in a small 4-dimensional region (for example, by interaction with an external field) is proportional to the 4-volume.

The conventional interpretation of $T^{\alpha\beta}$ of course follows readily from (1). If, in an arbitrary local Lorentz frame

$$x^\mu = (ct,x,y,z) \quad , \quad g_{\mu\nu} = \eta_{\mu\nu} = \mathrm{diag}(-1,1,1,1)$$

we choose a 3-surface perpendicular to the x-axis, we obtain $\delta p^\alpha = T^{\alpha 1}dydzcdt$ as the 4-momentum transferred across area dydz in the time dt. Thus, cT^{o1} is identified as energy flux in the positive x-direction and T^{11} as the x-component of stress on the yz plane (a pressure if it is positive). Again, if we choose $d\Sigma^\alpha = \delta^\alpha_o dxdydz$ tangent to the 3-space t=0, we find that $\delta p^\alpha = c^{-1}T^{\alpha o}dxdydz$ is the 4-momentum that transits across the 3-area dxdydz from past to future, i.e. the 4-momentum present in dxdydz at time t=0. Therefore T^{oo} and $c^{-1}T^{ao}$ (a=1,2,3) are densities of energy and momentum.

In general, a local observer having an arbitrary 4-velocity u^α ($u^\alpha u_\alpha = -c^2$) will measure energy density $c^{-2}T^{\alpha\beta}u_\alpha u_\beta$ and energy flux $-T^{\alpha\beta}u_\alpha n_\beta$ along the direction of a unit vector n_α in his rest-frame.

Conservation of energy and momentum for a system isolated from forces other than gravity is expressed by the condition that the integral of 4-momentum flux (1) over a small closed 3-surface should vanish to an order higher than the enclosed

4-votume, when evaluated in locally inertial co-ordinates (i.e. co-ordinates such that $\partial_\gamma g_{\alpha\beta}=0$ at a selected enclosed point). The result has the generally covariant form

$$\nabla_\beta T^{\alpha\beta} = 0 .$$

On the other hand, if $T^{\alpha\beta}$ describes only a part of an interacting system, there is no requirement that it be either conserved or symmetric. For instance, it may be useful to distinguish, at some level of description, between orbital and spin angular momentum. For a polarized medium one would then introduce an angular momentum flux $S^{\lambda\mu\beta}$ (skew in λ,μ) by a formula analogous to (1), which defined $T^{\alpha\beta}$ as flux of <u>linear</u> 4-momentum. The resulting ponderomotive equations for an isolated gravitating system [10]

$$\nabla_\beta T^{\alpha\beta} = -\frac{1}{2} R^\alpha{}_{\beta\lambda\mu} S^{\lambda\mu\beta} \quad , \quad \nabla_\beta S^{\lambda\mu\beta} = 2T^{[\lambda\mu]}$$

show a characteristic spin-curvature coupling. (The square brackets indicate antisymmetrization.) The second equation expresses conservation of angular momentum, i.e. the vanishing of the total (spin plus orbital) angular momentum flux

$$\int (S^{\lambda\mu\beta} - 2x^{[\lambda}T^{\mu]\beta})\epsilon d\Sigma_\beta = 0$$

through a small closed 3-surface (using locally inertial co-ordinates). A symmetric, conserved energy tensor can be derived from $T^{\alpha\beta}$ by the well-known Belinfante-Rosenfeld procedure; it is this which couples to the Einstein tensor in the gravitational field equations and represents the total density of <u>gravitating</u> energy and momentum. However, the unsymmetrized $T^{\alpha\beta}$ is more immediately linked to microscopic physics at the molecular level. I shall not pursue these complications here, but merely note that gravitational effects of polarized spin have sometimes been considered in a cosmological context [11], and that spin-orbital coupling is a basic dynamical feature of the anisotropic neutron-proton superfluids believed to be present in neutron stars (§4).

§2. <u>Thermal Equilibrium: Phenomenology</u>

Fully covariant thermodynamics tends to make a somewhat remote first impression on a non-relativist (even on relativists). It is best approached in a familiar setting: the thermal equilibrium of a simple fluid.

Working in the rest-frame of the fluid, we denote the particle density by n, specific volume by $V=(1/n)$ and entropy per particle by $s=kS/n$ where k is Boltzmann's constant. The energy density is

$$\rho c^2 = n(mc^2 + \epsilon) \tag{2}$$

where ϵ is the mean internal (thermal plus chemical) energy per particle. From the equation of state $s=s(\epsilon,V)$, the temperature and pressure emerge as partial derivatives:

$$ds(\epsilon,V) = T^{-1}(d\epsilon + PdV) . \tag{3}$$

In relativistic theory it is convenient to include the rest-mass energy when defining the chemical potential per particle:

$$\mu \equiv mc^2 + \mu_{class} , \quad \mu_{class} = \epsilon + PV - Ts . \tag{4}$$

It follows from (4), (2) and (3) that

$$\mu n = \rho c^2 + P - kTS \tag{5}$$

$$kTdS = d(\rho c^2) - \mu dn . \tag{6}$$

This brings out the meaning of μ as an "injection energy", the energy required to inject one extra particle reversibly into a fixed volume.

Next we introduce the "thermal potential" α and inverse-temperature β by the definitions

$$\alpha \equiv \mu/kT \quad , \qquad \beta \equiv c^2/kT . \tag{7}$$

Then equations (5), (6) take the form

$$S = \beta(\rho + P/c^2) - \alpha n \tag{8}$$

$$dS = \beta d\rho - \alpha dn . \tag{9}$$

It is amusing to note that all of thermodynamics can be boiled down to the single equation

$$d(SX) = \beta d(\rho X) - \alpha d(nX) + \beta(P/c^2)dX \tag{10}$$

in which X is arbitrary. For example, X=1 gives (9), X=V gives (3).

We move on to the kinematics of the fluid. Let u^α be the hydrodynamical 4-velocity, so that $u^\alpha u_\alpha = -1$ (from here on we set c=1) and let

$$\Delta^{\alpha\beta} \equiv g^{\alpha\beta} + u^\alpha u^\beta$$

be the unit spatial tensor that projects onto the 3-flat orthogonal to u^α, i.e., in the rest-frame we have

$$u^\alpha \stackrel{*}{=} \delta^\alpha_o \quad , \qquad \Delta^{\alpha\beta} \stackrel{*}{=} \mathrm{diag}(0,1,1,1) .$$

Since for a perfect fluid the stress-energy tensor should reduce to

$$T^{\alpha\beta} \stackrel{*}{=} \mathrm{diag}(\rho,P,P,P)$$

in the rest-frame, we infer the generally covariant expression

$$T^{\alpha\beta} = \rho u^\alpha u^\beta + P\Delta^{\alpha\beta} . \tag{11}$$

The particle and entropy current 4-vectors are

$$J^\alpha = nu^\alpha \quad , \qquad S^\alpha = Su^\alpha . \tag{12}$$

Everything is now in place for a last transmutation of the thermodynamical relations into a form that involves the covariant objects S^μ, J^μ and $T^{\lambda\mu}$ directly as basic variables. In (10) set $X=u^\mu$, and note that $\rho u^\mu = -u_\lambda T^{\lambda\mu}$ and that $Pdu^\mu = P\Delta^{\lambda\mu}du_\lambda$ since du_λ is orthogonal to u^λ. The result is

$$dS^\mu = -\alpha dJ^\mu - \beta_\lambda dT^{\lambda\mu} \tag{13}$$

where

$$\beta_\lambda \equiv \beta u_\lambda = u_\lambda/kT \tag{14}$$

defines the inverse-temperature 4-vector. From (8) we have immediately

$$S^\mu = P\beta^\mu - \alpha J^\mu - \beta_\lambda T^{\lambda\mu} . \tag{15}$$

It follows from (13) and (15) that

$$d(P\beta^{\mu}) = J^{\mu}d\alpha + T^{\lambda\mu}d\beta_{\lambda} \ . \tag{16}$$

The last two relations show how the basic variables J^{μ}, $T^{\lambda\mu}$, S^{μ} can all be generated from partial derivatives of the "fugacity" 4-vector $\Phi^{\mu}(\alpha,\beta_{\lambda})\equiv P\beta^{\mu}$, once the equation of state is known.

At first glance this transition to covariant form is nothing more than a trivial and unnecessary complication of the simple relations (8) and (9). Its advantages will become more apparent as we proceed. I list some of them here:

(i) In the covariant formulation the hydrodynamical velocity is co-opted in a natural way as an extra thermodynamical variable.

(ii) The thermodynamical relations remain valid in their covariant form even when a rest-frame (i.e. a timelike 4-velocity) no longer exists, e.g., for a pencil of radiation issuing from a pinhole in the wall of a heated cavity [12], or for Hawking radiation on the horizon of a black hole. In both cases, the "rest-frame" has the speed of light.

(iii) Since the covariant equations are expressed directly in terms of the basic conserved variables J^{μ}, $T^{\lambda\mu}$, they can be manipulated that much more easily and transparently.

(iv) Above all, (15) remains valid (at least to first order in deviations) for states that deviate from equilibrium. It is more now than just a transcription of (8), since the vectors S^{μ}, J^{μ}, ... are not parallel. As we shall see (§3 of Chapter II), the extra information in (15) is precisely the standard linear relation between entropy flux and heat flux.

To extend these considerations to a mixture of fluids is straightforward. There are now N components or conserved charges (e.g., baryon number, electric charge, ...) with densities n_A and currents $J^{\mu}_A = n_A u^{\mu}_A$. If the components are free to exchange momentum without constraint - as we shall assume here for simplicity, although it is not always true (compare §4, where we discuss superfluids) - all macroscopic velocities must be equal in thermal equilibrium: $u^{\mu}_A = u^{\mu}$. An equilibrium state is then characterized by (N+4) parameters,

$$\alpha_A = \mu_A/kT \ (A=1,\dots N) \ , \qquad \beta^{\mu} = u^{\mu}/kT \ . \tag{17}$$

From these all other quantities are derivable by partial differentiation of the fugacity $P\beta = \Phi(\alpha_A,\beta)$ or its equivalent covariant form $P\beta^{\mu} = \Phi^{\mu}(\alpha_A,\beta_{\lambda})$:

$$d(P\beta) = \sum_A n_A d\alpha_A - \rho d\beta \quad ; \quad d(P\beta^{\mu}) = \sum_A J^{\mu}_A d\alpha_A + T^{\lambda\mu}d\beta_{\lambda} \tag{18}$$

$$S = \beta(\rho+P) - \sum_A \alpha_A n_A \ ; \quad S^{\mu} = P\beta^{\mu} - \sum_A \alpha_A J^{\mu}_A - \beta_{\lambda}T^{\lambda\mu} \tag{19}$$

These equations imply the Gibbs relation

$$dS = \beta d\rho - \sum_A \alpha_A dn_A \qquad ; \quad dS^{\mu} = - \sum_A \alpha_A dJ^{\mu}_A - \beta_{\lambda}dT^{\lambda\mu} \ . \tag{20}$$

As already indicated, (19) continues to hold off-equilibrium to first order. In Chapter II we discuss this in detail. Anticipating the result here for a moment, it is then easy to infer the constraints that global thermal equilibrium imposes on gradients of the thermodynamical parameters α_A, β_λ.

We have

$$\delta S^\mu = - \sum_A \alpha_A \delta J^\mu_A - \beta_\lambda \delta T^{\lambda\mu} \tag{21}$$

for a perturbation (δS^μ, δJ^μ_A, $\delta T^{\lambda\mu}$) from an equilibrium state parametrized by (α_A, β_λ). If no charge is injected into the system, and if gravity is the only external influence,

$$\nabla_\mu \delta J^\mu_A = \nabla_\mu \delta T^{\lambda\mu} = 0 \ , \tag{22}$$

$\delta T^{\lambda\mu}$ is symmetric, and (21) yields

$$\nabla_\mu \delta S^\mu = - \sum_A (\partial_\mu \alpha_A)\delta J^\mu_A - (\nabla_\mu \beta_\lambda)\delta T^{\lambda\mu} \quad . \tag{23}$$

Since the unperturbed state is one of thermal equilibrium, the integral

$$\int (\nabla_\mu \delta S^\mu)\sqrt{-g}\, d^4x = \left[\int -S^\mu d\Sigma_\mu\right]_{t=-\infty}^{t=+\infty} \tag{24}$$

which gives the resultant change of entropy, must vanish to first order for locally arbitrary perturbations δJ^μ_A, $\delta T^{\lambda\mu}$. It follows that

$$\partial_\mu \alpha_A = 0 \quad , \quad \nabla_{(\mu}\beta_{\lambda)} = 0 \tag{25}$$

where the parentheses denote symmetrization.

At least the first of these conditions is modified in the presence of additional, nongravitational forces. Consider, for instance, a simple charged medium in an external electromagnetic field. The second of (22) becomes

$$\nabla_\mu \delta T^{\lambda\mu} = F^{\lambda\mu}\delta J_\mu \ , \tag{26}$$

and (25) is replaced by

$$\partial_\mu \alpha + \beta_\lambda F^\lambda{}_\mu = 0 \ , \qquad \nabla_{(\mu}\beta_{\lambda)} = 0 \ . \tag{27}$$

The conditions (25) or (27) have a simple physical meaning. One easily verifies that

$$2\nabla_{(\mu}\beta_{\lambda)} = \beta^\nu \partial_\nu g_{\lambda\mu} + 2(\partial_{(\lambda}\beta^\nu)g_{\mu)\nu} = \mathcal{L}_\beta g_{\lambda\mu} \tag{28}$$

where $\mathcal{L}$ denotes what differential geometers call the Lie derivative. Inspection of (28) shows that $\partial g_{\lambda\mu}/\partial t=0$ in co-ordinates adapted to β^μ, i.e. such that $\beta^a=0$ ($a=1,2,3$), $\beta^o=1/kT_o$, with T_o constant. It follows that the spacetime geometry and the gravitational field appear stationary to observers with 4-velocity $u^\mu=kT\beta^\mu$ at rest in the medium, who therefore also see a constant spatial separation between neighbouring world-lines. This means that the motion is rigid in the sense of Born [9], ensuring absence of viscous effects in global equilibrium.

Moreover,

$$-1 = g_{\mu\nu}u^\mu u^\nu = g_{oo}(T/T_o)^2$$

so that locally measured temperature varies with gravitational potential in accordance with Tolman's law

$$T(-g_{oo})^{1/2} = T_o \tag{29}$$

becoming higher at deeper levels. This is easy to understand. Radiation emanating from these levels is redshifted and cooled as it climbs, so they must be warmer to compensate for this and maintain thermal equilibrium.

In a (purely) gravitational field, all thermal potentials α_A are constant by (25). The chemical potential $\mu_A = \alpha_A kT$ therefore varies with depth in the same way as temperature (Klein's law [14]). In the presence of an electromagnetic field with (stationary) vector potential A_λ, Klein's law holds for the augmented chemical potential $\mu - u^\lambda A_\lambda$ (per unit charge), i.e.

$$\alpha - e\,\beta^\lambda A_\lambda = (kT_o)^{-1}[\mu(-g_{oo})^{1/2} + e\varphi] = \text{const.} \tag{30}$$

(where $\varphi = -A_o$ is the electrostatic potential), as one readily verifies from (27).

As a further illustration of the economy of the covariant thermodynamical relations (15) or (19) (extended off equilibrium), I present a simple derivation of the second law of thermodynamics in its discrete form.

Consider a set of components or subsystems that interact only weakly, so that it becomes meaningful to assign an individual energy and entropy to each, and to set

$$T^{\lambda\mu} = \sum_A T_A^{\lambda\mu} \quad , \qquad S^\mu = \sum_A S_A^\mu \quad .$$

The subsystems might in particular be physically distinct bodies in marginal or momentary thermal contact, and having arbitrary 4-velocities u_A^μ. It is furthermore consistent to assume each subsystem close to its own internal equilibrium, characterized by

$$\alpha_A = \mu_A/T_A \quad , \qquad \beta_A^\lambda = u_A^\lambda/T_A \quad .$$

Then it follows from (13), applied to each component separately, that

$$\nabla_\mu S^\mu = -\sum_A \alpha_A \nabla_\mu J_A^\mu - \sum_A \beta_{A\lambda} \nabla_\mu T_A^{\lambda\mu} \quad .$$

Integrating this result over the spacetime region of interaction, we immediately obtain the total change of entropy

$$\boxed{\delta s = -\sum_A \; (\alpha_A \delta n_A + \beta_{A\lambda} \delta p_A^\lambda),} \tag{31}$$

in which $\delta p_{A\lambda}$, δn_A represent the increments of 4-momentum and particle number (or charge) acquired by component A in the interaction. The left-hand side must, of course, be nonnegative for any spontaneous process and must vanish in thermal equilibrium.

Equation (31) is the covariant form of the second law, first demonstrated convincingly by van Kampen [15] for thermal exchanges. The appearance of the exchanged 4-momentum δp_A^λ in place of exchanged energy is an explicit recognition that in relativity heat carries momentum. The old cliché, "heat flows spontaneously from warmer to cooler bodies" is not necessarily true in situations where momentum exchange plays a significant role.

Neglect of this point has led in the past to a great deal of fruitless discussion on the pseudo-question: "Does a moving body appear cool?" One could attempt to put this on an operational footing: Two bodies, A and B, have relative speed v. What must be the relation between the temperatures T_A, T_B measured in their rest-frames if they are to be in thermal equilibrium? The answer is quite clear. If there is no constraint on their momentum exchanges, the two bodies can never be in thermal equilibrium as long as they retain any relative motion, since any exchange of heat must entail a change of momentum for at least one of them. At best, one could hope to contrive a partial equilibrium by constraining the momentum exchanges in some way.

For example, if A is somehow contrained so that it can only emit and absorb heat isotropically - i.e., if A is unable to change its 3-momentum, so that we have to set*

$$\delta p_A^\mu = u_A^\mu \delta E = - \delta p_B^\mu \tag{32}$$

in (31) - then the condition for thermal equilibrium is $T_B = \gamma T_A$ where

$$\gamma = -u_{A\mu} u_B^\mu = (1 - v^2/c^2)^{-1/2}$$

is the dilation factor associated with their relative speed. It then also follows from (31) that the chemical potentials are related by $\mu_B = \gamma \mu_A$. A constraint of this type is operative if A is the superfluid component of a superfluid (§4), or the central part of a wheel. If the wheel is set spinning, heat will flow outwards to warm up the rim.

§3. Statistical Thermodynamics†

A covariant statistical theory for thermal equilibrium can be developed in a formalism that dovetails perfectly with the phenomenology of the previous section. In the special case of a gas this was already shown nearly fifty years ago by van Dantzig [13].

In the general case, one considers a grand canonical ensemble with density matrix $\hat{\rho}$, and maximizes the entropy $s = -\mathrm{Tr}(\hat{\rho}\, \ell n\, \hat{\rho})$ subject to the usual constraints

$$\mathrm{Tr}\, \hat{\rho} = 1 \quad , \quad \mathrm{Tr}(\hat{\rho}\hat{n}_A) = n_A \, , \quad \mathrm{Tr}(\hat{\rho}\hat{p}^\lambda) = p^\lambda \quad . \tag{33}$$

This leads to

---

*In setting $\delta(p_A^\mu + p_B^\mu) = 0$ we have assumed that there is no gravitational or other potential difference between A and B.

†§3 and §4 may be omitted without breaking the continuity of the exposition.

$$\hat{\rho} = Z^{-1}\exp(\sum_A \alpha_A\hat{n}_A + \beta_\lambda\hat{p}^\lambda) \tag{34}$$

where α_A, β_μ are Lagrange multipliers.

Now suppose that the members of the ensemble are small (macroscopic) subregions of an extended body in thermal equilibrium, whose world-tubes intersect an arbitrary spacelike hypersurface in small 3-areas $\Delta\Sigma$. Then one can express the extensive quantities n, p^μ and ℓn Z as flux integrals over $\Delta\Sigma$:

$$n_A = \int_{\Delta\Sigma} - J^\mu_A d\Sigma_\mu \ , \quad p^\lambda = \int_{\Delta\Sigma} - T^{\lambda\mu}_{(mat)} d\Sigma_\mu \ , \quad \ell n\ Z = \int_{\Delta\Sigma} - \Phi^\mu d\Sigma_\mu \ . \tag{35}$$

Taking the differential of

$$Z = \mathrm{tr}\ \exp(\sum \alpha_A\hat{n}_A + \beta_\lambda\hat{p}^\lambda) \tag{36}$$

one sees readily that all thermodynamical quantities can be derived from the fugacity 4-vector $P\beta^\mu = \Phi^\mu(\alpha,\beta_\lambda)$:

$$\begin{aligned} J^\mu_A &= \partial\Phi^\mu/\partial\alpha_A \ , \qquad T^{\lambda\mu}_{(mat)} = \partial\Phi^\mu/\partial\beta_\lambda \ , \\ S^\mu &= \Phi^\mu - \sum_A \alpha_A J^\mu_A - \beta_\lambda T^{\lambda\mu}_{(mat)} \ . \end{aligned} \tag{37}$$

For a system subject to external influences in addition to gravity, it is important to distinguish between the total (conserved, symmetric) stress energy tensor $T^{\lambda\mu}$ and its (in general, nonsymmetric) material or canonical part $T^{\lambda\mu}_{(mat)}$. Only the material part enters (37), ensuring that reversible flows of field energy (e.g., circulating Poynting currents in crossed electric and magnetic fields) are not accompanied by an entropy flux.

A simple gas in an external field provides an elementary concrete illustration of the general formalism. A state of the system is specified by giving the occupation numbers n_r of each of the 1-particle states labelled by r and having 4-momentum $p^\mu_{(r)}$. This state has total particle number and 4-momentum

$$n = \sum_r n_r \ , \qquad p^\mu = \sum_r n_r p^\mu_{(r)} \tag{38}$$

in the absence of interactions. The partition function

$$Z = \sum_{n_1,n_2,\ldots} \exp(\alpha n + \beta_\lambda p^\lambda) \tag{39}$$

therefore splits into a product

$$Z = \prod_r Z_r \ , \quad Z_r = \sum_{\nu=0,1,\ldots} \exp[\nu(\alpha + \beta_\lambda p^\lambda_{(r)})] \tag{40}$$

where ν ranges over all admissible occupation numbers of a 1-particle state. This yields

$$\ell n\ Z = -\epsilon^{-1} \sum_r \ell n[1 - \epsilon\ \exp(\alpha + \beta_\lambda p^\lambda_{(r)})] \tag{41}$$

with ϵ=+1 for bosons and -1 for fermions. (Classical Boltzmann statistics is recoverable in the limit $\epsilon\to 0$.) If the discrete levels are approximated by a continuum, we can make the replacement

$$\sum_r \longrightarrow h^{-3}\int d^3x\, d^3p_{\cdot} = h^{-3}\int d\Sigma_\mu d\omega\, v^\mu \ , \tag{42}$$

where

$$d\omega = \int (-g)^{-1/2} d^4p_{\cdot}\ \delta(H + m) \tag{43}$$

is the element of 3-area on the mass-shell $H=-m$ and $H(x^\alpha, p_\beta)$ is the one-particle Hamiltonian. The particle 4-velocity $v^\mu = \partial H/\partial p_\mu$ is also the normal to the mass-shell; it is generally not parallel to p^μ if external fields are present (apart from gravity). From (35) and (37) we can read off

$$\begin{aligned} \Phi^\mu &= (\epsilon h^3)^{-1}\int \ell n(1 + \epsilon h^3 N)d\omega \ , \\ J^\mu &= \int N v^\mu d\omega \ , \quad T^{\lambda\mu}_{(mat)} = \int N p^\lambda v^\mu d\omega \ , \end{aligned} \tag{44}$$

where

$$N(x^\alpha, p_\beta) = h^{-3}[\exp(-\alpha - \beta_\lambda p^\lambda) - \epsilon]^{-1} \tag{45}$$

is the one-particle distribution function.

In the general case of a system with self-interaction in (classical) external fields, $\hat{n}$ and $\hat{p}^\lambda$ are expressible as integrals of field operators:

$$\hat{n} = \int_\Sigma -\hat{J}^\mu(x)d\Sigma_\mu \ , \quad \hat{p}^\lambda = \int_\Sigma -\hat{T}^{\lambda\mu}_{(mat)}(x)d\Sigma_\mu \ . \tag{46}$$

(I have assumed for simplicity that there is just one conserved current.) Specializing to stationary co-ordinates (t, x^a), and defining the constant $\mu_o = \alpha/\beta^o = \mu\sqrt{-g_{oo}}$ (μ_o may be considered to be the chemical potential at infinity for a bounded, self-gravitating system), we can set

$$\alpha\hat{n} + \beta_\lambda\hat{p}^\lambda = \int_{\Sigma\times\beta^\mu} (\mu_o\hat{J}^o + \hat{T}^{\ o}_{o(mat)})(-g)^{1/2}d^4x = -\beta^o\hat{H}(\mu) \ , \tag{47}$$

where I have introduced the μ-modified Hamiltonian operator

$$\hat{H}(\mu) = -\int_\Sigma (\mu_o\hat{J}^o + \hat{T}^{\ o}_{o(mat)})(-g^{oo})^{-1/2} d\Sigma \tag{48}$$

Let us assume for simplicity that the quantized fields describing the system, which we denote generically by $\hat{\varphi}(x)$, are ungauged boson fields. Feynman's path-integral prescription for the transition amplitude is

$$\varphi''|e^{-i\hat{H}(t''-t')}|\varphi'\rangle = \langle\varphi'', t''|\varphi', t'\rangle = \int_{\varphi'}^{\varphi''} d[\varphi]\exp\left(i\int_{t'}^{t''} L dt\right) \ . \tag{49}$$

Now extend this analytic formula into the complex t-plane by setting $t=-i\tau$, $t''-t'=-i\beta^o$, identify end-configurations $\varphi''=\varphi'$ and functionally integrate over φ'. The result is Feynman's path-integral representation for the partition function

$$\begin{aligned} Z &= \mathrm{tr}[\exp(\alpha\hat{n} + \beta_\lambda\hat{p}^\lambda)] = \mathrm{tr}[\exp(-\beta^o\hat{H}(\mu))] \\ &= \int d[\varphi]\ \exp\int_{\Sigma\times\beta^\mu} \mathcal{L}(\varphi, \partial_\mu\varphi, g_E^{\mu\nu}, \mu)g_E^{1/2}d^4x \end{aligned} \tag{50}$$

The domain of the action integral in (50) is an oblique cylindrical segment $\Sigma\times\beta^\mu$ in the Euclidean sector, generated by sliding Σ, a slice of constant $x^o=\tau$, through an interval $0<x^o<\beta^o$. The functional integral is over all fields $\varphi(x)$ on R^4 satisfying the periodicity condition

$$\varphi(x^\mu + \beta^\mu) = \varphi(x^\mu) . \tag{51}$$

(Fermionic fields would be anti-periodic.) The metric $g_E^{\mu\nu}(x)$ in the Euclidean sector is derived from its value $g_L^{\mu\nu}(x)$ in the Lorentzian sector by formally applying the co-ordinate transformation law $x_E^o=ix_L^o$. The same formal rule is applied to any other stationary external fields that might be present; however, the functional integration variables φ and their gradients $\partial_\mu\varphi$ are not transformed, and x^μ, β^μ are always taken as real. (This peculiar recipe is the essence of what is meant by a Wick rotation, which is <u>not</u> a co-ordinate transformation [17]). Finally, the integrand is, of course, that Lagrangian density $\mathcal{L}(\varphi, \partial_\mu\varphi, g^{\mu\nu}, \mu)$ which generates, via the standard prescription in the Lorentzian sector, the μ-modified Hamiltonian functional $H(\varphi(x), \pi(x), \mu]$ defined by (48).

An example of how this (Matsubara) prescription for calculating the partition function works in practice is provided in the next section, where we consider the thermodynamics of a simple scalar-field model for a relativistic superfluid.

Extension of the path integral expression (50) to gauge fields requires a Faddeev-Popov ansatz to eliminate spurious gauge degrees of freedom [18].

Cosmological developments, in particular the dynamics of phase transitions in the GUT era, have focussed current interest on the extension to nonequilibrium statistical field theory. This presents a much more formidable problem, now under attack from several approaches: closed-time path integrals [19], thermofield dynamics [20] and the Schrödinger picture [21]. Comparative reviews of the various formalisms of thermofield statistics (focussing on the equilibrium case) have been given by van Hove [22] and Landsman and van Weert [23].

§4. An Example: Covariant Superfluid Mechanics

The relativistic generalization [24] of Landau's two-fluid theory of superfluids affords an instructive example of the economy of the covariant formalism and of the interplay between the phenomenological and macroscopic levels of description [25,17].

Since the superfluid component is supposed to be ideal and frictionless in the two-fluid model, it will retain well-defined thermodynamical properties that can be cleanly separated from the normal component, even out of equilibrium.

A necessary condition for thermal equilibrium is then that the chemical potentials μ_n, μ_s of the two components should be related by

$$\mu_n = \gamma\mu_s \quad , \quad \gamma = - u_n^\mu u_{s\mu} = (1 - v^2/c^2)^{-1/2} \tag{52}$$

(and also $T_n=\gamma T_s$), where v is their relative speed (cf end of §2). This condition leaves an imprint even in the nonrelativistic limit: by virtue of (4),

$$\mu_n^{class} = \mu_s^{class} + \frac{1}{2} mv^2 . \tag{53}$$

This has a particularly transparent interpretation [26]: a small (but macroscopic) mass element, consisting of δN particles, that is transferred from the superfluid to the normal component, necessarily has zero momentum with respect to the superfluid. The energy $\mu_n(\delta N)$ injected (reversibly) into the normal component therefore exceeds the energy $\mu_s \delta N$ extracted from the superfluid by the relative kinetic energy $\frac{1}{2} mv^2 \delta N$.

(a) Two-fluid model: phenomenological theory

Since a nonvanishing chemical potential requires a conserved charge in the relativistic statistical version of the theory, it is convenient to consider a charged system in an external electromagnetic field. (The field can be switched off at any stage of the analysis.) The (nonequilibrium) phenomenological two-fluid model is then characterized by currents J^μ_A and stress-energy tensors $T^{\lambda\mu}_A$ (A=n,s) subject to overall conservation laws

$$\sum_A \nabla_\mu J^\mu_A = 0 \quad , \qquad \sum_A \nabla_\mu T^{\lambda\mu}_A = F^{\lambda\mu} \sum_A J_{A\mu} \ . \tag{54}$$

The ideal properties of the superfluid component are encoded in the restrictions

$$J^\mu_s = en_s u^\mu_s \quad , \qquad T^{\lambda\mu}_s = \rho_s u^\lambda_s u^\mu_s + P_s \Delta^{\lambda\mu}_s \tag{55}$$

$$\nabla_\mu T^{\lambda\mu}_s - F^{\lambda\mu} J_{s\mu} \propto u^\lambda_s \ . \tag{56}$$

I shall postulate further that the entropy of the superfluid component vanishes identically. While there is no compulsion to adopt this postulate from a strictly phenomenological point of view, it accords naturally with the perception that all thermal excitations are associated exclusively with the normal component, and is supported by microscopic theory. In that case temperature - defined generally as an integrating factor for the entropy - loses meaning for the superfluid component, and T_s drops out of the formalism.

With $S_s \equiv 0$, the equation of state $\rho_s = \rho_s(n_s)$ of the superfluid component now involves only a single independent parameter, and the Gibbs-Duhem relations (8), (9) reduce to

$$\frac{d\rho_s}{dn_s} = \frac{\rho_s + P_s}{n_s} = \mu_s \longrightarrow dP = n_s d\mu_s \ . \tag{57}$$

Substituting (55) and (57) into (56) fixes the proportionality coefficient,

$$\nabla_s T^{\lambda\mu}_s - F^{\lambda\mu} J_{s\mu} = (\mu_s/e) u^\lambda_s \nabla_\mu J^\mu_s \tag{58}$$

as well as the superfluid acceleration $\dot{u}^\lambda_s = u^\mu_s \nabla_\mu u^\lambda_s$:

$$\mu_s \dot{u}^\lambda_s = eF^{\lambda\mu} u_{s\mu} - \Delta^{\lambda\mu}_s \partial_\mu \mu_s \ . \tag{59}$$

This equation can be recast into the form

$$u^\mu_s \nabla_{[\mu} (\mu_s u_s + eA)_{\lambda]} = 0 \ , \tag{60}$$

which implies conservation of the superfluid circulation

$$\mathcal{L}_{u_s^\mu} \oint (\mu_s u_{s\lambda} + eA_\lambda) dx^\lambda = 0 , \tag{61}$$

i.e. the line integral is independent of time for a closed loop convected with the superfluid component.

The conservation law (61) guarantees that it will be consistent, within the phenomenological theory, to impose the Onsager-Feynman flux quantization condition

$$\oint (\mu_s u_{s\lambda} + eA_\lambda) dx^\lambda = nh . \tag{62}$$

But this integral must be invariant under continuous deformations of the loop. So it is necessary - and, by virtue of (60), dynamically consistent - to strengthen the condition (60) to

$$\nabla_{[\mu}(\mu_s u_{s\lambda]}) = \frac{1}{2} eF_{\lambda\mu} \tag{63}$$

which must hold everywhere except on vortex lines.

The conditions (52) for thermal equilibrium can also be readily recovered from this continuum theory. The divergence of (15), applied to the normal component, yields with the aid of (54) and (58),

$$0 \leq \nabla_\mu S_n^\mu = - J_n^\mu(e^{-1}\partial_\mu\alpha + \beta^\lambda F_{\lambda\mu}) - T_n^{\lambda\mu}\nabla_\mu\beta_\lambda + \beta(\mu_n - \gamma\mu_s)e^{-1}\nabla_\mu J_s^\mu . \tag{64}$$

(The currents have been taken to be purely convective, $J_A^\mu = en_A u_A^\mu$, to avoid uninteresting complications from Ohmic effects.) The parameters $\alpha = \mu_n/kT$, $\beta^\lambda = u_n^\lambda/kT$ refer, of course, to an equilibrium state of the normal component; neither is defined for the superfluid component. From (64) one reads off at once the thermal equilibrium conditions (27) and (52). The equation of state has the separable form

$$P = P_s(\mu_s) + P_n(\mu_n, T) , \quad n_s = dP_s/d\mu_s , \quad n_n = (\partial P_n/\partial\mu_n)_T . \tag{65}$$

(b) Statistical theory

The microscopic theory is based on the now 50-year old conception by Fritz London that superfluidity is a manifestation of Bose-Einstein condensation - in the simplest case, a breaking of U(1) symmetry (phase invariance) - in a boson system.

An appropriate model is therefore the thermal equilibrium of a self-interacting charged scalar field φ, with a condensed phase, in a stationary Einstein-Maxwell background. Accordingly, suppose that classical fields $g_{\lambda\mu}(x)$, $F_{\lambda\mu}(x)$, $\alpha(x)$ and $\beta^\lambda(x)$ are given, subject to

$$\nabla_{(\mu}\beta_{\lambda)} = \frac{1}{2}\mathcal{L}_\beta g_{\lambda\mu} = 0 , \quad \mathcal{L}_\beta F_{\lambda\mu} = 0 \tag{66}$$

$$\partial_\mu\alpha + \beta^\lambda eF_{\lambda\mu} = 0 . \tag{67}$$

In a stationary gauge ($\mathcal{L}_\beta A_\lambda = 0$) the last condition can be recast in the form (30). Equivalently, we can define a chemical vector potential μ_λ (up to an arbitrary spatial gradient) by

$$\partial_{[\kappa}\mu_{\lambda]} = 0 , \qquad \beta^\lambda(\mu_\lambda + eA_\lambda) = \alpha , \tag{68}$$

then (67) asserts that

$$\beta^{\lambda}\mu_{\lambda} = \text{constant} \qquad (\mathcal{L}_{\beta}A_{\lambda} = 0) \; . \tag{69}$$

The partition function is then given by the path integral (50), with the periodicity condition (51), and Lagrangian

$$-\mathcal{L}(\mu) = g^{\mu\nu}(D^{\dagger}_{\mu}\varphi^*)(D_{\nu}\varphi) + (m^2 + \tfrac{1}{2}\lambda U)\varphi\varphi^* \; . \tag{70}$$

Dependence on the chemical potential has been smuggled in via a suitably modified covariant derivative:

$$D_{\lambda} = \partial_{\lambda} - i(eA_{\lambda} + \mu_{\lambda}) = \nabla_{\lambda} - i\mu_{\lambda} \quad , \quad D^{\dagger}_{\lambda} = \nabla^{\dagger}_{\lambda} + i\mu_{\lambda} \; . \tag{71}$$

Thus, D_{λ} and $D^{\dagger}_{\lambda}$ are complex conjugate on the Lorentzian sector, but not on the Euclidean sector since the time-component of the external field $\mu_{\lambda}+eA_{\lambda}$ is pure imaginary there.

The Lagrangian (70) is invariant under the restricted group of time-independent local phase transformations

$$\begin{aligned} &\varphi \longrightarrow \varphi e^{i\Theta(x)} \quad , \quad \beta^{\lambda}\partial_{\lambda}\Theta = 0 \\ &\mu_{\lambda} + eA_{\lambda} \longrightarrow \mu_{\lambda} + eA_{\lambda} + \partial_{\lambda}\Theta \; . \end{aligned} \tag{72}$$

The conserved current

$$J_{\mu}(x) = i(\varphi D^{\dagger}_{\mu}\varphi^* - \varphi^* D_{\mu}\varphi) \tag{73}$$

is, of course, associated with the unrestricted form of this invariance. (Restriction to time-independence is needed in the case of thermal equilibrium to preserve (68) and the periodicity condition $\varphi(x^{\mu}+\beta^{\mu})=\varphi(x^{\mu})$.)

For each of the class of chemical 4-potentials satisfying (68), a stationary time-function $t(x)$ exists such that

$$\mu_{\lambda} = \mu_{o}\partial_{\lambda}t \quad , \quad \mu_{o} = (\alpha/\beta^{o}) - eA_{o} = \text{constant} \; .$$

The restricted gauge transformations (72) just express the invariance of the stationary formalism under the group of space-dependent time translations

$$t \longrightarrow t + \mu_{o}^{-1}\Theta(x) \; .$$

(In fact, it is clear from (68), (71) and (72) that the chemical potential could be entirely absorbed into an electromagnetic potential $\overset{*}{A}_{\lambda}=A_{\lambda}+\mu_{\lambda}/e$ partially gauge-fixed by the condition $\beta^{\lambda}\overset{*}{A}_{\lambda}=\alpha$.)

Since the general form of the macroscopic properties may be expected to be insensitive to the details of the interaction, one can treat the self-interaction phenomenologically as an artifice that short-circuits the explicit introduction of auxiliary mediating fields. We require that U be a real, gauge-invariant functional of φ and φ^*, independent of μ. A possible form, borrowed from nonrelativistic many-body theory, is

$$U[\varphi\varphi^*;x] = \int d^4x' g^{1/2}(x')\varphi^*(x')\varphi(x')V(x',x') \tag{74}$$

where the real symmetric biscalar $V(x',x)$ is a form-factor representing the "potential of two-body interactions". In a covariant theory, such a nonlocal structure is, strictly, incompatible with the canonical analysis and justification of the choice of Lagrangian (70) that is given immediately below, and there are other, well-known, problems with nonlocal relativistic interactions (failure of microcausality at short range, etc. [27]) when considered as fundamental theories. Nevertheless, it appears that (74) encounters no obstacles when used in (70) simply as a phenomenological ansatz for the calculation of the partition function. Strict locality can, of course, be reimposed at any stage by replacing V by the pseudopotential $g^{-1/2}\delta^4(x-x')$. Then (70) would reduce to $\lambda|\varphi|^4$ theory, a model that has been used in some of the nonrelativistic path-integral analyses [28] as well as my own relativistic discussion [17]. To one-loop order, the results are sufficiently illustrative of what happens with more general interactions, although there is mounting evidence that the $\lambda|\varphi|^4$ model is actually trivial, i.e. that the renormalized coupling constant will be found to vanish when the perturbation series is summed to all orders. This has recently been proved for the nonrelativistic version of the theory [29].

To check that the chemical potential has been correctly incorporated by the prescription (71), we must examine the canonical structure of the model. The canonical momentum density is

$$\pi(x) = \partial\mathcal{L}(\mu)/\partial\dot{\varphi}^* = - g^{o\mu}D_\mu\varphi \ . \tag{75}$$

The (gauge-invariant, μ-modified) canonical stress-energy tensor is (disentangling the μ-dependence from the covariant derivatives)

$$- T_{\rho(\mathrm{mat})}^{\ \sigma}(\mu) = \frac{\partial\mathcal{L}(\mu)}{\partial(\nabla_\sigma\varphi)}\nabla_\rho\varphi + \mathrm{c.c.} - \mathcal{L}(\mu)\delta_\rho^\sigma \tag{76}$$

which yields the canonical energy density

$$\mathcal{H}(\mu) = - T_{o(\mathrm{mat})}^{\ o}(\mu) = \pi D_o^\dagger\varphi^* + \pi^* D_o\varphi - \mathcal{L}(\mu) - \mu_o J^o \tag{77}$$

where the time-component of the conserved current (73),

$$J^o(x) = i(\pi\varphi^* - \varphi\pi^*) \ ,$$

has been expressed in canonical variables using (75). Inspection of (77) and (70) shows at once that μ_o enters the canonical expression $\mathcal{H}(\pi,\varphi;\mu)$ of the Hamiltonian density only through the term

$$\mu_o J^o = \mu_\lambda J^\lambda + (\text{spatial divergence}), \tag{78}$$

since its appearances elsewhere (through $D_o\varphi$, $D_o^\dagger\varphi^*$) are supplanted by π,π^*. Thus

$$\mathcal{H}(\mu) = \mathcal{H}(\mu{=}0) - \mu_o J^o \ ,$$

whose agreement with (48) vindicates the choice (70) for the μ-modified Lagrangian.

A Wick rotation now yields the expression

$$Z = e^{-W} = \int d[\varphi]d[\varphi^*]e^{-S(\varphi)} \tag{79}$$

$$S[\varphi] = \int d^4x\cdot g^{1/2}[-\mathcal{L}(\varphi)] \tag{80}$$

for the partition function, in which it is understood, without explicit indication, that the external fields $g_{\mu\nu}$, A_μ are now defined on the Euclidean sector. The path integral is over all fields φ satisfying the Euclidean periodicity condition (51).

The superfluid phase, in which Bose-Einstein condensation has developed, is characterized by a nontrivial ground state, in which the Euclidean action S is minimized for some nonvanishing $\varphi(x)$. The properties of the condensate are then described by a macroscopic wave function $\psi(x)$ which agrees, at least approximately, with the function that minimizes the classical action. (More accurately, it can be defined as the function that minimizes the total (zero-temperature) action, including higher-loop corrections from purely quantum (non-thermal) fluctuations, after the theory has been renormalized at zero temperature.) This is not the place to enter into too many technical details, so I shall just sketch the key ideas and results.

One performs an expansion of $W = -\ell n\ Z$ in terms of a dimensionless "loop parameter" [30,28], which can be thought of formally as $\hbar^{1/2}$, by setting

$$\varphi(x) = \psi(x) + \hbar^{1/2}\ \varphi_{New}(x) \tag{81}$$

and expanding $S[\varphi]$ in powers of this parameter. The Euclidean effective action W is then given by

$$W = S[\psi] + \Delta W \quad , \quad \Delta W = \hbar W_1 + \hbar^{3/2} \ldots \tag{82}$$

where $S[\psi]$ may be thought of as describing the "classical" properties of the ground state, and ΔW, defined by

$$\begin{aligned} e^{-\Delta W} &= \int d[\varphi]d[\varphi^*]e^{-\Delta S} \\ \Delta S &= S[\psi + \hbar^{1/2}\varphi] - S[\psi] = \hbar S_1[\psi,\varphi] + \hbar^{3/2} \ldots \end{aligned} \tag{83}$$

as containing effects of quantum fluctuations and thermal excitations.

Now require that $\psi(x)$ be a solution of the classical field equations, modulo a disposable functional f of one-loop order, $\hbar$, that will be determined a posteriori when the higher-loop corrections have been calculated:

$$-\hbar f[|\psi|^2;x]\psi(x) = -\frac{\delta S[\psi,\psi^*]}{\delta\psi^*(x)} = (D^\mu D_\mu - m^2 - \lambda U[|\psi|^2;x])\psi(x) \ . \tag{84}$$

This is the covariant form of the Gross-Pitaevski equation. Imposing the further restrictions

$$\partial_o\psi(x) = 0 \quad , \quad f = f^* \quad , \tag{85}$$

ensures that the complex conjugate of (84) is also satisfied.

The higher-loop corrections can now be computed by standard methods. For example, the one-loop correction, in the effective potential approximation (gradients of $|\psi(x)|$ and of external fields neglected in the loop corrections, not in the classical term $S[\psi]$), by functionally integrating the trace of a 2×2 Green's matrix [17]:

$$\delta W_1/\delta\kappa(x) = -\frac{1}{2}\, g^{1/2}\ \mathrm{tr}\ G(x,x) \ . \tag{86}$$

This matrix is defined by

$$G(x,x') = \langle \varphi(x)\varphi^T(x')\rangle \equiv e^{W_1}\int d[\varphi]\varphi(x)\varphi^T(x')e^{-S_1[\psi,\varphi]} \quad ; \tag{87}$$

it is periodic in $x^o - x'^o$ with period β^o, and satisfies a linear differential equation (integro-differential equation in the case of nonlocal interactions) of the form

$$L_x G(x,x') = g^{-1/2}\delta^4(x-x')1_{2\times 2} \ . \tag{88}$$

(In (87), the real and imaginary parts of the complex field $\varphi(x)$ have been rewritten as a real column-vector φ.) For example, the detailed structure of L_x in the case of the $\lambda|\varphi|^4$ model is

$$L_x = \begin{bmatrix} -\partial^\mu\partial_\mu + \Gamma^\mu\Gamma_\mu + \kappa + \varepsilon & -2\Gamma^\mu\partial_\mu \\ 2\Gamma^\mu\partial_\mu & -\partial^\mu\partial_\mu + \Gamma^\mu\Gamma_\mu + \kappa - \varepsilon \end{bmatrix} \tag{89}$$

where

$$\kappa(x) = m^2 + 2\lambda|\psi|^2 \ , \quad \varepsilon(x) = \lambda|\psi|^2 \ , \quad \Gamma_\lambda = eA_\lambda + \mu_\lambda \ . \tag{90}$$

In four dimensions, the formal prescription (86) leads to a divergent W_1, the divergence arising from the singular behaviour of $G(x,x')$ when $x'\to x$ in spaces of dimension $N>2$. Since this singular behaviour is independent of boundary conditions, it is at once apparent that the divergences of the formalism are independent of β^o. This is at the root of the now well-known fact that it is possible to renormalize finite-temperature quantum field theory in accordance with a temperature-independent scheme.

Using dimensional regularization, the singularity of W_1 can be isolated as a simple pole at $N=4$. The next step is to absorb this into a redefinition of the parameters m, λ that appear in the effective action (82). (To one loop order, μ and $|\psi|$ do not require renormalization.) This brings us to a key point. As it stands, the Gross-Pitaevski equation (84), which describes the properties of the condensate, is independent of β^o. In order not to contaminate it with temperature-dependent terms, it is essential to define the renormalized parameters λ_R, m_R at zero temperature. For example, in $\lambda|\varphi|^4$ theory, one defines

$$\lambda_R = -\left.\frac{\partial^2\mathcal{L}}{\partial(|\psi|^2)^2}\right|_{\substack{\psi=0\\ \beta=\infty}} \ ; \quad m_R^2 + \Gamma^\mu\Gamma_\mu = -\left.\frac{\partial\mathcal{L}}{\partial(|\psi|^2)}\right|_{\substack{\psi=0\\ \beta=\infty}} \tag{91}$$

where $\mathcal{L}(\psi,D_\mu\psi)$ is the effective Lagrangian (the integrand of (82)). It is common to renormalize at zero temperature anyway, as a computational convenience. The present argument shows that this choice is actually mandatory if one wishes to make manifest the separability of the thermodynamical properties of the normal and superfluid components.

After renormalization, one obtains for the effective Lagrangian to one-loop order, in the effective potential approximation and weak-coupling limit, (small λ),

$$\mathcal{L} = \mathcal{L}_{class}(\psi,D_\mu\psi) + \hbar\mathcal{L}_{fluc}(|\psi|) + \hbar\mathcal{L}_{therm}(\alpha,\beta^\mu) \tag{92}$$

in which the first two terms are independent of temperature and describe the classical properties and quantum fluctuations in the condensate, and the last term gives the partition function for a hot gas of free ("quasi-") particles and antiparticles. This is the relativistic, finite-temperature generalization of Bogoliubov's famous 1947 result. To one-loop order, it would now be appropriate to choose the disposable functional f in the Gross-Pitaevski equation [84] so that $\psi(x)$ minimizes the action formed from $\mathcal{L}_{class} + \hbar\mathcal{L}_{fluc}$.

Explicitly, $\mathcal{L}_{class}(\psi)$ has the functional form (70) (where it is understood that it is now the renormalized parameters that enter all equations), and, for the $\lambda|\varphi|^4$ model,

$$-\mathcal{L}_{fluc} = (4\pi)^{-2}\{-\lambda m^2|\psi|^2 + \frac{1}{2}(m^2 + 2\lambda|\psi|^2)^2 \ell n(1 + 2\lambda|\psi|^2/m^2)\} \tag{93}$$

$$\ell n\, Z_{therm} \equiv \hbar \int_{\Sigma\times\beta^\mu} \mathcal{L}_{therm} d^4x = -V\int^{(3)} g^{-1/2} \frac{d^3p}{(2\pi)^3} \sum_{\epsilon=\pm 1} \ell n\{1-\exp(\epsilon\alpha+\beta^\mu p_\mu)\}. \tag{94}$$

By comparing with (41), the last expression is recognized as giving the partition function for a gas of bosons and antibosons with thermal potentials $\pm\alpha$. The energy $|p_o|$ in the integral is defined in terms of the 3-momentum by

$$g^{\mu\nu}p_\mu p_\nu = -\kappa = -(m^2 + 2\lambda|\psi|^2) , \tag{95}$$

which shows that the rest-masses of the quasi-particles are shifted by the interaction energy.

After renormalization, the Gross-Pitaevski equation (84) has the general form

$$[D^\mu D_\mu - m^2 - F(|\psi|^2;x)]\psi(x) = 0 \quad , \quad \partial_o\psi(x) = 0 , \tag{96}$$

where F is some real functional. This form is unaltered when we rotate back to the Lorentzian sector. I reiterate the key point: because we have chosen to renormalize at zero temperature, equation (96) contains no temperature-dependence.

It is now easy to delineate the correspondence between the microscopic theory and the phenomenology. The conservation law

$$\partial^\mu[|g|^{1/2}|\psi D_\mu\psi^* - \psi^* D_\mu\psi] = 0 , \tag{97}$$

follows from (96). Accordingly, define μ_s and a normalized 4-velocity u_s^λ ($u_{s\lambda}u_s^\lambda=-1$) by

$$\mu_s u_{s\mu} = \frac{1}{2} i(\psi D_\mu\psi^* - \psi^* D_\mu\psi) = \partial_\mu\Phi - \Gamma_\mu \quad , \tag{98}$$

where we have set

$$\psi(x) = [n(x)]^{1/2} e^{i\Phi(x)} . \tag{99}$$

Then

$$(n_s u_s^\mu)_{|\mu} = 0 \quad , \quad n_s \equiv n\mu_s \quad . \tag{100}$$

From (98) it follows that

$$\oint (\mu_s u_{s\mu} + eA_\mu)dx^\mu = \oint d\Phi = 2\pi n = nh\ , \qquad (101)$$

where the Planck constant has been restored in the last step. Now define u_n^μ, β, μ_n by

$$\beta^\mu = \beta u_n^\mu\ , \quad u_{n\mu} u_n^\mu = -1\ , \quad \mu_n = \alpha/\beta = -(-g^{oo})^{-1/2}\mu_o\ . \qquad (102)$$

Multiplication of (98) by u_n^μ, noting $\partial_o \psi = 0$, yields

$$\gamma\mu_s = \mu_n\ , \quad \gamma \equiv -u_{s\mu} u_n^\mu\ . \qquad (103)$$

Finally, (96) and (99) yield the equation of state of the superfluid component,

$$\mu_s^2 = m^2 + F[n;x] - \hbar^2 n^{-1/2} \nabla^2 (n^{1/2}) \qquad (104)$$

which accords, aside from the nonlocal dependence, with the general (one-parameter, temperature-independent) form $n_s = n_s(\mu_s)$ that we postulated in the phenomenological theory.

One may conclude that the phenomenology is fully corroborated by microscopic theory, at least for thermal equilibrium.

The extension of these considerations to the Bose-Einstein condensation of multicomponent scalar fields in external Yang-Mills fields is straightforward and has wide-ranging applications, for example, to the theory of pion condensation in nucleons [31] and phase transitions in cosmology [32].

An area of considerable complexity, as yet unexplored in a relativistic context, is the dynamics of the $^3P_2$-paired (anisotropic) neutron superfluids that might realistically be expected in neutron stars [33].

CHAPTER II. NONEQUILIBRIUM PHENOMENOLOGICAL THEORIES

§1. Introduction. The primary and auxiliary variables of covariant nonequilibrium thermodynamics

There are as many attitudes to nonequilibrium thermodynamics as there are theories [34]. They run the gamut from pragmatism to axiomatism, and are represented in forms ranging from brief papers directed toward specific experiments to monumental Handbuch articles.

This chapter aims at a middle course which, however, leans strongly toward the pragmatic and uncommitted. I attempt to draw together, in a covariant setting, the common features of the class of theories which accept the premise that the macroscopic concept of entropy (since it corresponds to a mere counting of microstates in statistical theory) retains a definite meaning for at least a sufficiently small range of states outside equilibrium.

The primary ingredients of the covariant theory are then the usual basic dynamical variables $T^{\lambda\mu}$, J^{λ}_{A}, satisfying conservation laws

$$\nabla_{\mu}T^{\lambda\mu} = \nabla_{\mu}J^{\mu}_{A} = 0 \quad , \qquad (A=1,\ldots N) \tag{1}$$

and one additional extensive variable - the entropy 4-current S^{λ}, subject to the inequality

$$\nabla_{\mu}S^{\mu} \geqslant 0 \tag{2}$$

expressing positivity of local entropy production. (In theories which incorporate memory effects or fluctuations, (2) may be relaxed to a global inequality

$$\int \nabla_{\mu}S^{\mu}\,(-g)^{1/2}\,d^{4}x \geqslant 0 \quad .) \tag{2a}$$

Other concepts familiar from nonrelativistic theory - temperature, rest-frame, hydrodynamical velocity - are subordinated in the covariant formalism to these primary covariant objects. Indeed, it is characteristic of relativistic theories that there are two or more ways, a priori equally natural, of defining a 4-velocity and rest-frame for a given nonequilibrium state. Each flux, whether of particles, energy, entropy or various kinds of charge, defines its own characteristic 4-velocity and out of equilibrium they are generally all different.

Although $T^{\lambda\mu}$, J^{λ}_{A} and S^{λ} are the objects of primary interest in the theory, it is not implied or expected that they specify a nonequilibrium state completely. Even for such a simple thing as a gas, kinetic theory teaches that a complete specification requires knowledge of at least the one-particle distribution $N(x^{\alpha},p_{\beta})$ in phase space, or, equivalently, the infinite set of moments $\int Np^{\lambda}p^{\mu}p^{\nu}\ldots d^{4}p$ at the macroscopic level.

In general, a complete theory of nonequilibrium processes must deal with an infinite set of auxiliary quantities $X^{\lambda\mu\nu\ldots}_{(i)}$ (i=1,2,...) in addition to the primary variables. An equation of state would have the form

$$S^{\mu} = F^{\mu}(T^{\lambda\mu},\ J^{\lambda}_{A},\ X^{\cdots})\ . \tag{3}$$

For equilibrium states (subscript "o") the auxiliary quantities must vanish, reducing (3) to

$$S^{\mu}_{(o)} = F^{\mu}(T^{\lambda\mu}_{(o)},\ J^{\lambda}_{A(o)},\ 0) \tag{4}$$

in accordance with the fact that all quantities in (4) are functions of the (N+4) variables α_A, β_λ in equilibrium. (These latter variables are, of course, defined only for equilibrium states.) Further, the expectation that entropy is (locally) a maximum in equilibrium for given concentrations of the charges and 4-momentum suggests that dependence on the auxiliary variables appears only at second order in deviations from equilibrium, i.e.,

$$\left.\frac{\partial F^{\mu}(T^{\lambda\mu}_{(o)},\ J^{\lambda}_{A(o)},\ X^{\cdots})}{\partial X^{\cdots}}\right|_{X^{\cdots}=0} = 0\ . \tag{5}$$

The covariant theory of "extended thermodynamics" recently elaborated by Liu, Müller and Ruggeri [35] is founded on the postulate that no auxiliary quantities enter, i.e. that a covariant equation of state of the form (4) holds for arbitrary states. It formulates the program of rational thermodynamics as the systematic extraction of the rigorous mathematical consequences of postulates (1), (2) and (4). This is not the viewpoint taken here. The general framework for covariant theories developed in this Chapter leaves entirely open the issue of auxiliary variables and nonequilibrium equations of state. In §6, however, we shall look at the phenomenological equations that result, when, as part of a pragmatic and essentially ad hoc approximation for small deviations from equilibrium, the auxiliary variables are neglected.

§2. Heat flux in relativistic theories: the Eckart and Landau-Lifshitz alternatives

Because the point is central to much of the debate concerning alternative thermodynamical formalisms, it may be well to interpolate here some elementary remarks on the way in which heat flow is described in relativistic theory.

We recall from §1 of the previous chapter that an observer with arbitrary 4-velocity u^{λ} measures energy density $\rho(u)=u_{\lambda}u_{\mu}T^{\lambda\mu}$ and energy flux $q(u,n)=-u_{\lambda}T^{\lambda\mu}n_{\mu}$ along a unit spatial vector n_{ν} in his frame.

Consider first a local observer in a simple fluid who is at rest with respect to the average motion of the particles. His 4-velocity $u^{\lambda}=u^{\lambda}_{J}$ is parallel to the particle flux J^{λ}:

$$J^{\lambda} = n_J u^{\lambda}_{J}\ . \tag{6}$$

He observes heat flow as a flux of energy $\vec{q}$ in his rest-frame:

$$-u_J^\lambda T_\lambda^\mu = \rho_J u_J^\mu + q^\mu \ . \tag{7}$$

With respect to this observer the stress-energy tensor thus decomposes as follows:

$$\begin{aligned} T^{\lambda\mu} &= \rho_J u_J^\lambda u_J^\mu + q^\lambda u_J^\mu + u_J^\lambda q^\mu + P_J^{\lambda\mu} \\ P^{\lambda\mu} &= P^{\mu\lambda} \ , \quad P_\mu^\lambda u_J^\mu = q_\mu u_J^\mu = 0 \end{aligned} \tag{8}$$

$$P_J^{\lambda\mu} = P\Delta_J^{\lambda\mu} + \pi\Delta_J^{\lambda\mu} + \pi^{\lambda\mu} \ , \qquad \Delta_J^{\lambda\mu}\pi_{\lambda\mu} = 0 \ . \tag{9}$$

Successive terms on the right-hand side of (9) represent the thermodynamical pressure, bulk stress and shear stress measured by our observer. This decomposition and choice of rest-frame is the basis of the first relativistic formalism constructed in 1940 by Eckart [36].

We pass now to a second observer, drifting slowly in the direction of heat flow with a velocity $\vec{v}_D=\vec{q}/nmc^2$, adjusted to make the mass-energy counterflow of particles relative to his frame exactly cancel the heat flux $\vec{q}$. The 4-velocity u_E^μ of this observer is defined by the condition that he measures no nett energy flux in his rest-frame: $u_E^\lambda T_\lambda^\mu n_\mu=0$ for all vectors n_μ orthogonal to u_E^μ, i.e.

$$u_E^\lambda T_\lambda^\mu = -\rho_E u_E^\mu \ . \tag{10}$$

Thus u_E^λ is the timelike eigenvector of $T^{\lambda\mu}$ - unique if $T^{\lambda\mu}$ satisfies a positive energy condition [9]. The structure of $T^{\lambda\mu}$ is now comparatively simple,

$$T^{\lambda\mu} = \rho_E u_E^\lambda u_E^\mu + P_E^{\lambda\mu} \ , \tag{11}$$

while

$$J^\lambda = n_E u_E^\lambda + j^\lambda \qquad (j_\lambda u_E^\lambda = 0) \tag{12}$$

reveals the expected counterdrift of particles $\vec{j}=-n\vec{v}_D=-\vec{q}/mc^2$. This alternative formulation of the theory is due to Landau and Lifshitz [37].

We have here formally different but completely equivalent descriptions of the same phenomenon (a flow of energy relative to the particles) from the viewpoints of two observers who decompose the same invariant objects $T^{\lambda\mu}$, J^λ differently. One sees the phenomenon as an energy flux, the other as a particle drift. In this equivalence, and its extension to fluid mixtures, one glimpses the germ of a remarkable fusing of Fourier's law of heat conduction with Fick's law of diffusion, stemming from relativistic mass-energy equivalence.

The structural difference between (6) and (12), and between (8) and (11) lends distinctive computational advantages to each of these formulations. The virtues of the Landau-Lifshitz formalism come to the fore when it is convenient to reduce the stress-energy tensor to the simplest possible form - for instance, when treating the Einstein field equations. Eckart's formalism has complementary advantages when it is desirable to have a simple first integral of the particle conservation law $\nabla_\mu J^\mu=0$ - for example, when analyzing the structure of plane shock layers. (In the case of mixtures, this advantage is largely dissipated.)

Although I have fastidiously marked the distinction between the pairs (n_J, n_E) and (ρ_J, ρ_E) - the particle and energy densities measured in the two frames - the actual differences involve Lorentz factors $(1-v_D^2/c^2)^{1/2}$ and are insignificant for most practical purposes if deviations from equilibrium are small. More precisely,

$$n_E = n_J \cosh \epsilon \quad , \quad \rho_J = \rho_E \cosh^2\epsilon + P_E \sinh^2\epsilon + \pi^{\lambda\mu} j_\lambda j_\mu / n_J^2$$

where

$$\cosh \epsilon = - u_E^\mu u_{J\mu} \quad ; \tag{13}$$

this defines the angle

$$\epsilon \sim j/n \sim v_D/c \sim q/nmc^2 \tag{14}$$

as a dimensionless measure of the deviation from equilibrium. Thus, $n_E - n_J$ and $\rho_E - \rho_J$ are each of order ϵ^2.

It is clear that we are not confined to just the Eckart and Landau-Lifshitz alternatives for defining a macroscopic 4-velocity, although these have proved the most convenient in practice. We may choose to decompose the primary variables $T^{\lambda\mu}$, J^λ in terms of any 4-velocity u^λ that falls within a cone of angle $\sim\epsilon$ containing u_J^λ and u_E^λ. Each choice u^λ leads to a particle density $n(u) = -u_\lambda J^\lambda$ and energy density $\rho(u) = u_\lambda u_\mu T^{\lambda\mu}$ that are actually independent of u^λ if one neglects terms of order ϵ^2.

The insensitivity of ρ and n to the choice of frame (within the limits prescribed) has a valuable advantage. It means that one can, without further ado, and for any such choice, read off the entropy and thermodynamical pressure of an off-equilibrium system directly from the equilibrium equation of state. Since, for given energy and particle densities, the entropy density is maximal at thermal equilibrium, it follows that the measured off-equilibrium entropy density $S(u) = -u_\mu S^\mu$ differs from the equilibrium density $S_o(\rho(u), n(u))$ only by terms of second order:

$$S(u) - S_o(\rho(u), n(u)) \sim \epsilon^2 \ . \tag{15}$$

Thus, the thermodynamical pressure $P(u)$, defined as work done in an isentropic expansion,

$$P(u) = - \left(\frac{\partial(\rho/n)}{\partial(1/n)}\right)_{S/n} \ , \tag{16}$$

is very nearly the pressure given by the equilibrium equation of state $P_o = P_o(\rho_o, n_o)$:

$$P(u) - P_o(\rho(u), n(u)) \sim \epsilon^2 \tag{17}$$

and thus is also insensitive to the choice of frame u^λ.

The correct identification of thermodynamical pressure off equilibrium is essential for separating (irreversible) bulk stress $\pi(u)$ from (reversible) pressure $P(u)$ in the decomposition of the stress-energy tensor as in (8) and (9). Since the point has often been a source of confusion in the literature, I re-emphasize that the simple procedure just outlined requires the matching, to the actual, off-equilibrium state $(T^{\lambda\mu}, J^\lambda, S^\lambda)$, of a neighbouring equilibrium state $(T_{(o)}^{\lambda\mu}, J_{(o)}^\lambda = n_o u^\lambda)$

whose 4-velocity u^λ lies anywhere within the ϵ-cone of u_J^λ and u_E^λ, and which has the same particle and energy densities:

$$(T^{\lambda\mu} - T^{\lambda\mu}_{(o)})u_\lambda u_\mu = (J^\lambda - J^\lambda_{(o)})\, u_\lambda = 0 \,. \tag{18}$$

Considering that the choice of rest-frame is an arbitrary element of the theory to within fairly wide limits, it might, from a severely formal point of view, seem attractive to avoid introducing a 4-velocity altogether. We shall see that it is possible and advantageous to proceed quite far in this direction. In the end, however, a formulation of the Navier-Stokes equations in terms of third-rank tensors simply does not carry the intuitive connotations built up from a century of experience with velocity gradients and shear stresses. The best compromise seems to be to tolerate a measure of arbitrariness in return for the formal flexibility, convenience and intuitive appeal that one wins in exchange.

In an interesting paper, demonstrating that the Chapman-Enskog expansion is systematically recoverable by a separation of fast and slow variables, van Kampen [38] has commented on the "irritating arbitrariness in the definition of hydrodynamic quantities" for relativistic gases. He suggests that an expansion scheme in which the conserved densities $T^{\lambda o}$, J^o are identified as the "slow" variables in an arbitrary fixed (global) Lorentz frame "leads to a perfectly unique definition...not Lorentz-invariant". Considering that J^o and T^{oo}, while uniquely defined in a given frame, are somewhat remote from what a comoving observer would recognize as particle and energy density, it is an amusing exercise to formulate, e.g., the Gibbs relation in terms of them. (The answer emerges most quickly from the covariant formalism, see equation (13) of the previous Chapter.) Even in the simplest case of a dissipationless flow, for which the Eckart and Landau-Lifshitz choices for u^λ agree in giving the standard covariant ideal-fluid description, the fixed-frame choice leads to a more formidable appearance, unless the fluid happens to be motionless.

One point, which plays a key role in van Kampen's argument, is worth emphasis here. Once a definite choice of u^λ is supplemented by a set of linear phenomenological laws, different choices are no longer merely different descriptions of the same physics; they are, speaking strictly, inequivalent theories. If, for instance, one translates Fourier's law $\vec{q}=-\kappa(\nabla T+c^2 T\dot{\vec{u}})$ in the Eckart frame to the Landau-Lifshitz frame, one would find the same result, provided one consistently throws away all terms nonlinear in ϵ, the deviation from equilibrium as well as gradients and time-derivatives of $\vec{v}_D$. An exact transcription would, however, reveal a term $(\vec{v}_D/c^2)\kappa\dot{T}$ and a host of other corrections of order ϵ^2, reflecting the relative drift $\vec{v}_D$. These corrections generally remain unimportant until the entire linearized formalism is on the verge of breakdown anyway. But they could be significant in special circumstances. Neglect of the corrections could, for instance, result in certain disturbances, for which the Landau-Lifshitz linear laws predict marginal stability, becoming actually unstable when analyzed in terms of the linearized

Eckart equations. Precisely this is found to happen in the stability analysis of Hiscock and Lindblom [39] (§5). I am indebted to Maria Ekiel-Jezewska for a discussion on this point.

§3. Covariant Off-Equilibrium Thermodynamics: A General Framework

From the foregoing reconnaissance we can now go on to a more concise and systematic development of a broad class of nonequilibrium theories.

An arbitrary state of a fluid system insulated from nongravitational forces is described by a set of primary variables $T^{\lambda\mu}$, J_A^λ, S^λ satisfying

$$\nabla_\mu T^{\lambda\mu} = \nabla_\mu J_A^\mu = 0 \quad , \quad \nabla_\mu S^\mu \geq 0 , \tag{19}$$

and by an unspecified set of auxiliary variables.

The subclass of thermal equilibrium states $\{T_{(o)}^{\lambda\mu}, J_{A(o)}^\lambda, S_{(o)}^\lambda\}$, characterized by $\nabla_\mu S_{(o)}^\mu=0$, forms an (N+4)-dimensional subspace $\mathcal{E}$, parametrized by N+4 parameters α_A, β^λ, where β^λ is non-spacelike. Each fluid is characterized by a vectorial fugacity function Φ^λ $(\alpha_A, \beta^\mu, g_{\lambda\mu})$ (necessarily parallel to β^λ by requirements of covariance) which defines its equilibrium equation of state:

$$P(\alpha,\beta)\ \beta^\lambda = \Phi^\lambda(\alpha, \beta^\mu) \tag{20}$$

and from which all equilibrium variables can be derived by

$$d(P\beta^\lambda) = \sum_A J_{A(o)}^\lambda d\alpha_A + T_{(o)}^{\lambda\mu} d\beta_\mu \tag{21}$$

$$S_{(o)}^\lambda = P\beta^\lambda - \sum \alpha_A J_{A(o)}^\lambda - \beta_\mu T_{(o)}^{\lambda\mu} \quad . \tag{22}$$

It follows from these relations that $T_{(o)}^{\lambda\mu}$, $J_{A(o)}^\lambda$, $S_{(o)}^\lambda$ necessarily take their standard, perfect-fluid form in terms of the 4-velocity $u^\lambda=\beta^\lambda/\beta$. Further, in the global thermal equilibrium of viscous, heat-conducting fluids (<u>not</u> superfluids), thermodynamical gradients are constrained by

$$\partial_\mu \alpha = 0 \quad , \quad \nabla_\mu \beta_\lambda + \nabla_\lambda \beta_\mu = 0 \quad . \tag{23}$$

From (21) and (22) it follows that

$$dS_{(o)}^\lambda = - \sum \alpha_A\, dJ_{A(o)}^\lambda - \beta_\mu dT_{(o)}^{\lambda\mu} \tag{24}$$

for an infintesimal displacement from equilibrium state $(\alpha_A, \beta_\lambda)$ to a neighbouring point on the equilibrium hypersurface $\mathcal{E}$.

Only one assumption is needed to extend this formalism off equilibrium. One postulates that the covariant Gibbs relation holds, not just for transitions between equilibrium states - i.e., displacements tangent to $\mathcal{E}$ - but for <u>arbitrary</u> infinitesimal displacements $(dJ_A^\lambda, dT^{\lambda\mu})$ from an equilibrium state $(\alpha_A, \beta^\lambda)$. (It is important to bear in mind that the parameters α_A, β^λ are defined only for points on $\mathcal{E}$.) Then, if the displacement $(\delta J_A^\lambda, \delta T^{\lambda\mu}, \ldots)$ from a point of $\mathcal{E}$ is small but finite,

$$\delta S^\lambda = - \sum_A \alpha_A \delta J_A^\lambda - \beta_\mu \delta T^{\lambda\mu} - Q^\lambda \tag{25}$$

where the added term Q^λ is of second order in the deviations δJ_A^λ, $\delta T^{\lambda\mu}$ and generally will also depend (quadratically) on other, "auxiliary" variables that vanish in equilibrium. It is solely in the specific form assumed for Q^μ that all viable off-equilibrium theories are distinguished from each other. For the present we shall leave the form of Q^μ entirely open.

By simple addition of (22) and (25) we now obtain

$$\boxed{S^\lambda = P(\alpha_A,\beta)\beta^\lambda - \sum_A \alpha_A J^\lambda_A - \beta_\mu T^{\lambda\mu} - Q^\lambda(\delta J^\mu_A,\ \delta T^{\lambda\mu},\ldots)} \tag{26}$$

for the entropy of an off-equilibrium state defined by the primary variables ($T^{\lambda\mu}$, J^λ_A, S^λ). The parameters α_A, β^λ that enter (26) refer to an arbitrarily chosen nearby equilibrium state. It is clear from (24) that (26) remains quite unchanged in form under infinitesimal displacements ($d\alpha_A$, $d\beta^\lambda$) of the nearby state on $\mathcal{E}$, and even under small finite displacements if the second-order term Q^λ is transformed appropriately.

What are the implications of the postulate (25) which takes us off the equilibrium surface ? Let us look first at a simple fluid with particle current J^λ. Since the fiducial equilibrium state (α,β^λ) is arbitrary, we are free to choose $\beta^\lambda = u^\lambda/kT$ parallel to the current J^λ of the given off-equilibrium state, so we are in the Eckart frame. Projecting (26) onto the 3-space orthogonal to u^λ gives

$$\sigma^\kappa \equiv \Delta^\kappa_\lambda S^\lambda = \beta\, q^\kappa - Q^\lambda \Delta^\kappa_\lambda \quad , \quad q^\kappa = -\Delta^\kappa_\lambda T^{\lambda\mu} u_\mu \tag{27}$$

so that

$$\vec{\sigma} = (\vec{q}/kT) + (\text{possible 2nd order terms}) \quad , \tag{28}$$

which, to linear order, is just the standard relation between entropy flux $\vec{\sigma}$ and heat flux $\vec{q}$.

Proceeding now to the general case of a mixture, let us choose the fiducial equilibrium state this time so that u^λ is the timelike eigenvector of $T^{\lambda\mu}$, i.e.

$$u_\mu T^{\lambda\mu} \Delta^\kappa_\lambda = 0 \ , \tag{29}$$

which brings us into the Landau-Lifshitz frame. Spatial projection of (26) gives

$$\sigma^\kappa = -\sum \alpha_A j^\kappa_A - Q^\lambda \Delta^\kappa_\lambda \tag{30}$$

or

$$\vec{\sigma} = -\sum_A (\mu_A/kT)\vec{j}_A + (\text{possible 2nd order terms}) \ , \tag{31}$$

the standard relation (at linear order) between entropy flux and diffusive flux. In this formulation, heat flow and diffusion have been subsumed into the diffusive fluxes $\vec{j}_A$ relative to the mean mass-energy flow.

Before proceeding to extract the consequences of the entropy inequality $\nabla_\mu S^\mu > 0$, it is useful to introduce the idea of a "local equilibrium field". Consider a fluid with given equilibrium equation of state $P\beta^\lambda = \Phi^\lambda(\alpha_A,\beta^\mu)$. Choose arbitrarily N scalar fields $\alpha_A(x)$ and a timelike vector field $\beta^\lambda(x)$ on spacetime. This defines a local equilibrium field $\{J^{\lambda}_{A(o)}(x),\ T^{\lambda\mu}_{(o)}(x),\ S^{\lambda}_{(o)}(x)\}$ through the Gibbs relations (21) and (22) applied independently at each point x. These tensors have the usual ideal-fluid structure at each point, and gradients are related by

$$\nabla_{\mu}(P\beta^{\mu}) = \sum_{A} J^{\mu}_{A(o)}\partial_{\mu}\alpha_{A} + T^{\lambda\mu}_{(o)}\nabla_{\mu}\beta_{\lambda} \tag{32}$$

by virtue of (21). However, in general $J^{\lambda}_{A(o)}$ and $T^{\lambda\mu}_{(o)}$ are not conserved, and gradients of α_A, β_λ are not constrained by the conditions (23) for global thermal equilibrium.

To the field of states $\{T^{\lambda\mu}(x),\ J^{\lambda}_{A}(x),\ S^{\lambda}(x),\ldots\}$ that describes the history of a fluid close to equilibrium we can associate a corresponding field of local equilibrium states $\{\alpha_A(x),\ \beta^{\lambda}(x)\}$ (actually a class of such fields - we shall not need to nail down their 4-velocities) by the following prescription. The proximity of the actual state to equilibrium implies that the flux vectors $J^{\lambda}_{A}(x)$ and the timelike eigenvector of $T^{\lambda\mu}(x)$ will all lie within a narrow "ϵ-cone" whose opening angle ϵ may be taken as a measure of the "deviation from equilibrium" (though it is unnecessary and has no intrinsic meaning to single out a particular equilibrium state from which it "deviates").

(a) Choose $u^{\lambda}(x)$ to be an arbitrary unit timelike vector field somewhere within the ϵ-cone.

(b) Fix (N+1) scalar fields $\alpha_A(x)$, $\beta(x)$ by the (N+1) conditions:

$$\begin{aligned} \rho_{(o)}(x) &\equiv T^{\lambda\mu}_{(o)}u_{\lambda}u_{\mu} = \rho(x;u) \equiv T^{\lambda\mu}u_{\lambda}u_{\mu} \\ n_{A(o)}(x) &\equiv -\ J^{\mu}_{A(o)}u_{\mu} = n_A(x;u) \equiv -\ J^{\mu}_{A}u_{\mu} \quad ; \end{aligned} \tag{33}$$

in other words,

$$\delta T^{\lambda\mu}u_{\lambda}u_{\mu} = \delta J^{\mu}_{A}u_{\mu} = 0 \tag{34}$$

From (25) and (34),

$$(\delta S^{\mu} + Q^{\mu})u_{\mu} = 0\ , \tag{35}$$

so that

$$S_{(o)}(x) = S(x;u) - u_{\mu}Q^{\mu}\quad . \tag{36}$$

Thus, with the matching conditions (33), and for arbitrary choice of u^{μ} within the ϵ-cone, the actual and local-equilibrium entropy densities agree to first order in ϵ. Further, it will be true that the entropy density $S(x;u) = -u_{\mu}S^{\mu}(x)$ is highest for the equilibrium state among all states with the same $\{\rho(x;u), n_A(x;u)\}$ if and only if

$$-\ u_{\mu}Q^{\mu} > 0\quad [\text{for all } \delta T^{\lambda\mu},\ \delta J^{\lambda}_{A} \text{ satisfying (34)}]\ , \tag{37}$$

which will hold if Q^{μ} is a future timelike vector.

The first-order agreement between $S(x;u)$ and $S_{(o)}(x)$ enables us easily to disentangle, correct to first order, the thermodynamical pressure $P(x;u)$ from the bulk stress $\pi(x;u)$ in the trace of the stress tensor

$$\frac{1}{3}\Delta_{\lambda\mu}T^{\lambda\mu} = P(x;u) + \pi(x;u) \; . \qquad (38)$$

We have

$$P(x;u) = \frac{\text{Work done in isentropic expansion}}{\text{Change in specific volume } V}$$

$$= - \left[\frac{\partial(\rho(x;u)V)}{\partial V}\right]_{S(x;u)V, n_A(x;u)V}$$

$$= - \left[\frac{\partial(\rho_o(x)V)}{\partial V}\right]_{[S_o + u_\mu Q^\mu]V,\; n_{A(o)}V}$$

by (33) and (36), so that

$$P(x;u) = P_{(o)}(x) + O(\epsilon^2) = P(\alpha_A, \beta) + O(\epsilon^2) \; . \qquad (39)$$

The pressure $P_{(o)}$ of the local equilibrium state matched to the actual state by conditions (33) thus agrees to first order with the actual thermodynamical pressure $P(x;u)$.

Finally, we come to the entropy inequality. Take the divergence of (26), eliminate the divergence of $P\beta^\lambda$ by deployment of (32) and invoke the conservation laws (19) to simplify the other divergences. It is one of the incidental facilities of the covariant formulation that the result is obvious upon inspection:

$$\boxed{0 < \nabla_\mu S^\mu = - \sum_A (\delta J^\mu_A)\partial_\mu \alpha_A - \delta T^{\lambda\mu}\nabla_\mu \beta_\lambda - \nabla_\mu Q^\mu} \; . \qquad (40)$$

In (40), the deviations from local equilibrium,

$$\delta T^{\lambda\mu} = (T - T_{(o)})^{\lambda\mu} \; , \quad \delta J^\lambda_A = J^\lambda_A - J^\lambda_{A(o)} \qquad (41)$$

have time-components constrained by the fitting conditions (33). They contain all information about the viscous stresses, heat flux and diffusion in the off-equilibrium state. We recall that the choice of 4-velocity u^λ within the ϵ-cone of the equilibrium state $(\alpha_A, \beta^\lambda)$ is still arbitrary.

Once a detailed form of Q^λ is specified, linear relations between irreversible fluxes $\delta T^{\lambda\mu}$, δJ^λ_A and gradients $\nabla_{(\mu}\beta_{\lambda)}$, $\partial_\lambda \alpha_A$ follow by familiar arguments. At this point it is important to note that <u>each</u> term on the right-hand side of (40) - not only $\nabla_\mu Q^\mu$ - is of second order in the deviations from local equilibrium. It is therefore a priori unjustified to neglect the second-order contribution Q^λ to the entropy expression (26), even in a <u>linear</u> theory. The key to a complete phenomenological theory thus lies in the specification of Q^λ.

§4. The Traditional (Eckart, Landau-Lifshitz) Theories

The theories of Eckart [36] and Landau-Lifshitz [37] both follow nonrelativistic tradition in making the ansatz

$$Q^\mu = 0 \; . \qquad (42)$$

By (27) and (30), this implies that the spatial entropy flux $\vec{\sigma}$ is a strictly linear function of heat flux $\vec{q}$ and diffusive fluxes $\vec{j}_A$, and depends on no other variables; also that the off-equilibrium entropy density (see (36)) depends only on the densities ρ, n_A and is given precisely by the equilibrium equation of state, $S=S_{(o)}(\rho,n_A)$.

We then have

$$0 < \nabla_\mu S^\mu = -\sum_A (\delta J^\mu_A)\partial_\mu \alpha_A - \delta T^{\lambda\mu}\nabla_\mu \beta_\lambda \tag{43}$$

with

$$\delta J^\mu_A u_\mu = \delta T^{\lambda\mu} u_\lambda u_\mu = 0 \tag{44}$$

and the choice of u^λ is still arbitrary to first order.

The Landau-Lifshitz option - choosing u^λ as the timelike eigenvector of $T^{\lambda\mu}$ - is, by virtue of (44), equivalent to the condition

$$\delta T^{\lambda\mu} u_\mu = 0 \ . \tag{45}$$

In this frame we can therefore identify the shear and bulk stresses $\pi^{\lambda\mu}$, π from the decomposition

$$\delta T^{\lambda\mu} = \pi^{\lambda\mu} + \pi\Delta^{\lambda\mu} \quad , \quad (\pi^{\lambda\mu}u_\mu = \pi^\lambda_\lambda = 0) \tag{46}$$

where (39) has been invoked. The expression (43) then splits into three independent, irreducible pieces:

$$-\sum_A j^\mu_A \partial_\mu \alpha_A - \beta\pi^{\lambda\mu}\langle\nabla_\mu u_\lambda\rangle - \beta\pi\nabla_\mu u^\mu > 0 \ , \tag{47}$$

where

$$j^\lambda_A = \delta J^\lambda_A \quad , \quad \langle X_{\lambda\mu}\rangle \equiv (\Delta^\alpha_\lambda \Delta^\beta_\mu - \tfrac{1}{3}\Delta_{\lambda\mu}\Delta^{\alpha\beta})X_{\alpha\beta} \ , \tag{48}$$

i.e. the angular brackets extract the purely spatial, trace-free part of any tensor.

The simplest way to ensure positivity of the bilinear expression (47) is to assume that (j^λ_A, $\pi^{\lambda\mu}$, π) are linear and purely local function of the gradients. We then obtain uniquely, if the equilibrium state is isotropic ("Curie's principle"),

$$j^\lambda_A = -\sum_B \kappa_{AB}\Delta^{\lambda\mu}\partial_\mu \alpha_B \ , \tag{49}$$

which is an amalgam of Fourier's and Fick's laws (κ_{AB} is a positive matrix), and

$$\pi_{\lambda\mu} = -2\zeta_S\langle\nabla_\mu u_\lambda\rangle \quad , \quad \pi = -\tfrac{1}{3}\zeta_V \nabla_\mu u^\mu \tag{50}$$

the standard Navier-Stokes equations, with ζ_S, ζ_V as shear and bulk viscosities.

The Eckart option (u^λ chosen parallel to J^λ) is mainly of interest if the fluid is simple, so we specialize to this case. Condition (45) is replaced by

$$\delta J^\mu = 0 \ , \tag{51}$$

and the heat flux now appears in the decomposition of $\delta T^{\lambda\mu}$:

$$\delta T^{\lambda\mu} = q^\lambda u^\mu + u^\lambda q^\mu + \pi^{\lambda\mu} + \pi\Delta^{\lambda\mu} \ . \tag{52}$$

Inserting (51), (52) into (43) gives the entropy production in the form

$$q^{\mu}(\partial_{\mu}\beta - \beta\dot{u}_{\mu}) - \beta(\pi^{\lambda\mu}\nabla_{\mu}u_{\lambda} + \pi\nabla_{\lambda}u^{\lambda}) \geq 0 \,. \tag{53}$$

Again making the simplest (linearity and locality) assumption, we obtain Fourier's law

$$q^{\lambda} = -\kappa\Delta^{\lambda\mu}(\partial_{\mu}T + T\dot{u}_{\mu}) \,, \tag{54}$$

(where the term depending on acceleration $\dot{u}_{\mu} \equiv u^{\lambda}\nabla_{\lambda}u_{\mu}$ is sometimes referred to as an effect of the "inertia of heat") and the Navier-Stokes equations (50) as before.

Of course the Navier-Stokes equations in their Landau-Lifshitz and Eckart incarnations are not strictly equivalent, since they involve gradients of two distinct velocities u_J^{λ}, u_E^{λ}, and therefore differ by gradients of the drift $\vec{v}_D = \vec{q}/nmc^2$. Further gradients - spatial gradients of the viscous stresses and the time-derivative of the heat flux - appear (linearly) when we translate Fourier's law from the Landau-Lifshitz version (49) (in its single-component form) to the Eckart form (54) with the aid of the thermodynamical identity

$$n d\alpha = (\rho + P)d\beta + \beta dP \tag{55}$$

and the momentum conservation law $\Delta_{\kappa}^{\lambda}\nabla_{\lambda}T^{\lambda\mu}=0$. Equivalence between the Landau-Lifshitz and Eckart formulations holds if and only if (a) one neglects terms of higher than linear order in the deviations from equilibrium and (b) gradients of $\vec{q}$, π and $\pi^{\lambda\mu}$ are small compared to thermal and velocity gradients.

For a simple fluid, Fourier's law, in the form (54) or (49), and the Navier-Stokes equations (50) comprise a set of 9 equations; together with the 5 conservation laws $\nabla_{\mu}T^{\lambda\mu}=\nabla_{\mu}J^{\mu}=0$, they should suffice, on the basis of naive counting, to determine the evolution of the 14 variables $T^{\lambda\mu}$, J^{λ} from suitable initial data. Unfortunately, this system of equations is of mixed parabolic-hyperbolic-elliptic type. Its pathologies, the most serious of which have only come to light recently [39], probably rule the traditional formulations out of court as viable relativistic theories. We must now go on to consider these problems.

§5. Difficulties with the Traditional Theories: Acausality and Instability

The traditional theories are closely modelled on classical Fourier-Navier-Stokes theory, and it has long been accepted (though not formally proved until 1985 in the case of Eckart's theory!) that they share the classical idiosyncrasy of predicting infinite propagation speeds for thermal and viscous disturbances. Already at the classical level, the parabolic character of the equations has been a source of concern [40,41]. One might reasonably expect signal velocities to be bounded by something like the mean molecular speed, though within the nonrelativistic realm there is no objection in principle to wave front velocities which are infinite, since molecules in the tail of Maxwell's distribution have arbitrarily large velocities. However, a relativistic theory which predicts infinite speeds for causal propagation contradicts the premises on which it is suppositiously based, and must stir uneasiness in even the most pragmatically minded astrophysicist who has to use it. The situation is actually graver than this. Perhaps the most dramatic result of a remarkable series of investigations by Hiscock and Lindblom is the recognition [39] that all "first-order" ($Q^{\lambda}=0$) theories, including the Eckart and Landau-Lifshitz theories, have previously unsuspected pathologies not shared by their classical counterparts: they possess instabilities which grow on a characteristic time-scale

$$\tau = \kappa T/\rho c^4 \sim (\text{sound speed}/c)^4 \times (\text{mean collision time}) \, .$$

It is not my purpose to go into the details of this work here, but it may be useful to outline its general significance and to sketch the methods used, especially as these have also been applied [7], with equally interesting results, to the "extended", "second-order" theories to be discussed in the following section.

In all of these cases, the simple-fluid theory involves 14 independent variables, which we may take to be $T^{\lambda\mu}$, J^{λ} or their equivalent disjunct form:

$$Y_A = (n,T,\pi,u^{\lambda},q^{\lambda},\pi^{\lambda\mu}) \qquad (A=1,\ldots,14) \, . \tag{56}$$

The evolution of these variables is determined by the 5 conservation identities $\nabla_{\mu}T^{\lambda\mu}=\nabla_{\mu}J^{\mu}=0$ and a set of 9 phenomenological laws which, in all the cases we shall consider, are linear in the deviations from equilibrium. The complete system comprises 14 quasilinear equations of the general form

$$C^{AB\mu}(n,T,u^{\mu})\nabla_{\mu}Y_B = D^A(Y) \tag{57}$$

where $C^{AB\cdot}$ is a real, symmetric 14×14 matrix.

A <u>characteristic surface</u> or wave-front history is a hypersurface Σ across which the Y_A are continuous but their first derivatives have jump discontinuities (denoted by square brackets). Since tangential derivatives must be continuous, the jumps must have the form

$$[Y_A]_{\Sigma} = 0 \quad , \qquad [\nabla_{\mu}Y_A]_{\Sigma} = y_A\partial_{\mu}\varphi \tag{58}$$

where $\varphi(x^{\mu})=0$ in the equation of Σ. The jump of (57) across Σ yields

$$C^{AB\mu}(\partial_{\mu}\varphi)y_B = 0 \, . \tag{59}$$

For nontrivial normal jumps y_A, we obtain the characteristic equation that determines Σ:

$$\det(C^{AB\mu}\partial_{\mu}\varphi) = 0 \, . \tag{60}$$

We study systems whose spatial isotropy is not appreciably affected by rotation or gravity, so that gradients of n, T and u^{λ} make no appearance in the coefficients $C^{AB\mu}$. Then the gradient ∇_{μ} is the only coupling that (57) provides between orthogonal spatial components of the tensors $\pi^{\lambda\mu}$ and q^{λ} in the local rest-frame. If, in this frame, we orient the x-axis along the direction in which the wave-front is propagating, we can set (locally)

$$\begin{aligned} &\varphi = x - vt \quad , \qquad u^{\lambda}\partial_{\lambda} = \partial_t \\ &C^{AB\mu}\partial_{\mu}\varphi = C^{AB\mu} - v\, C^{ABt} \quad , \end{aligned} \tag{61}$$

and (60) is an equation of 14th degree in v which is fortunately, reducible. The decouping of longitudinal and transverse modes means that we can write the 14th order system $C\cdot\nabla Y=D$ in the block-diagonal form

$$\begin{bmatrix} \underset{6\times 6}{C_L^\mu} & & & \\ & \underset{3\times 3}{C_{LT_1}^\mu} & & \\ & & \underset{3\times 3}{C_{LT_2}^\mu} & \\ & & & \underset{2\times 2}{cIu^\mu} \end{bmatrix} \nabla_\mu \begin{bmatrix} Y_L \\ Y_{LT_1} \\ Y_{LT_2} \\ Y_T \end{bmatrix} = \begin{bmatrix} D_L \\ D_{LT_1} \\ D_{LT_2} \\ D_T \end{bmatrix}$$

in which the 14-component vector Y has been split into a scalar-longitudinal 6-vector $Y_L=[n,T,\pi,u_x,q_x,\pi_{xx}]$; two longitudinal-transverse 3-vectors (corresponding to the two transverse directions of polarization) $Y_{LT_1}=[u_y,q_y,\pi_{xy}]$ and $Y_{LT_2}=[u_z,q_z,\pi_{xz}]$; and a purely transverse 2-vector $Y_T=[\pi_{yz},\pi_{yy}-\pi_{zz}]$. Equation (60) for v accordingly splits into one 6th-degree and two identical 3rd-degree equations. (The purely transverse modes do not propagate.)

Hiscock and Lindblom proved - as anticipated - that this general scheme, when applied to first-order theories - i.e. theories which adopt $Q^\lambda=0$ and various choices of rest-frame - always yields wave-front speeds v that are superluminal. In the Landau-Lifshitz case, this is evident at once from the simple form (49) to which Fourier's law reduces in that theory:

$$j^\lambda = -\kappa_{11}\Delta^{\lambda\mu}(\partial_\mu\alpha) \;.$$

Its jump across Σ yields the condition

$$\Delta^{\lambda\mu}(\partial_\mu\varphi) = 0 \;,$$

showing that $(\partial_\mu\varphi)$ is timelike, i.e. that the wavefront has a spacelike history Σ. In Eckart's form (54) of Fourier's law, the coupling to acceleration complicates the analysis somewhat; Hiscock and Lindblom showed that the 3-component longitudinal-transverse mode satisfies the equations

$$[\kappa T\partial_t^2 + \zeta_S\partial_x^2 - (\rho + P)\partial_t]Y_{LT}(x,t) = 0$$

which is elliptic!

Acausality was an expected feature of the Eckart-Landau Lifshitz class of theories, but instability was not. The classical theories betray no hint of it. Yet now, with the benefit of hindsight, one can see how it might have easily been anticipated. The crucial point is that, in a Lorentz-invariant theory, superluminal acausality (faster-than-light signal propagation) is not distinguishable from chronological acausality (precedence of response to stimulus). The further link of chronological acausality to instability has long been known; a familiar example is the connection between pre-acceleration and self-acceleration in Dirac's third-order equation of motion for the classical electron. The response is pre-excitable only if the system can pass at least temporarily into a state of negative potential energy, and this implies instability.

Formally, the connection between chronological acausality and anti-damping follows at once from the dispersion relation. Consider the general linear system

$$M(\partial_{\underset{\sim}{x}},\partial_t)Y(\underset{\sim}{x},t) = S(\underset{\sim}{x},t) \quad . \tag{63}$$

If the stimulus S vanishes for $t<t_o$, the solution is expressible as an inverse Fourier-Laplace transform

$$Y(\underset{\sim}{x},t) = \int_{-\infty}^{\infty} d\underset{\sim}{k}\, e^{i\underset{\sim}{k}\cdot\underset{\sim}{x}} \int_{c-i\infty}^{c+i\infty} e^{\Gamma(t-t_o)} M^{-1}(i\underset{\sim}{k},\Gamma)\tilde{S}(\underset{\sim}{k},\Gamma) \ . \tag{64}$$

Placement of the vertical contour of integration L : Im Γ = c depends on the initial conditions. The poles of the integrand are the zeros of the dispersion relation

$$\Delta(\underset{\sim}{k},\Gamma) = \det M(i\underset{\sim}{k},\Gamma) \ .$$

Problems arise with chronological causality, stability or both if $\Delta(k,\Gamma)=0$ has roots $\Gamma_+(\underset{\sim}{k})$ with Re $\Gamma_+>0$ for some real range of $\underset{\sim}{k}$. In this case, for any choice of contour L, poles Γ_+ to the left of L contribute growing modes $e^{\Gamma_+(t-t_o)}$ for $t>t_o$; and poles to the right contribute their residues to $Y(x,t<t_o)$. Thus, no choice of L can preclude both instability and acausality for arbitrary sources and initial conditions.

To derive dispersion relations for the hydrodynamical system (57), Hiscock and Lindblom [39] reduced the system to a linear, homogeneous form by considering small perturbations, $Y=Y_o+\delta Y$, from a homogeneous global equilibrium characterized by constant n_o, T_o, $u^\lambda_{(o)}$, then making a plane-wave ansatz $dY \propto e^{ikx+\Gamma t}$. The wave number k is assumed real, and since this condition is not Lorentz-invariant, the fluid velocity in the direction of propagation is a relevant parameter.

I shall give just a sampling to convey the flavour of the results. The Eckart and Landau-Lifshitz theories may be subsumed into the general class of first-order theories by formally writing a linear composite of the two forms of Fourier's law:

$$q^\lambda = q^\lambda_{Eck} - [(\rho + P)/n]j^\lambda \quad ; \qquad \kappa = \kappa_{Eck} + \kappa_{LL}$$

$$q^\lambda_{Eck} = -\kappa_{Eck}\Delta^{\lambda\mu}(\partial_\mu T + T\dot{u}_\mu) \ , \qquad j^\lambda = -\kappa_{LL}T^2\Delta^{\lambda\mu}\partial_\mu\alpha \quad .$$

Of course, only the "total" heat flux and thermal conductivity κ have physical significance. The simplest nontrivial mode - the 3-component mode $\delta Y_{LT}=(u_y,q_y,\pi_{xy})$ - is governed by the quadratic dispersion relation

$$\kappa_{Eck}T\Gamma^2 - (\rho + P)\Gamma - \zeta_S k^2 = 0$$

for the case in which the fluid is at rest. If κ_{Eck} is nonnegative (which we have no need to assume!) there is one growing mode with (enormous) growth rate

$$\Gamma_+ > c^2 \ \frac{(\rho c^2 + P)}{\kappa_{Eck}T} \quad .$$

In the Landau-Lifshitz limit ($\kappa_{Eck}\to 0$), the growing mode disappears from (65), for the singular reason that its growth rate has become infinite! In any case, a growing mode resurfaces in Landau-Lifshitz theory for waves in a <u>moving</u> fluid; again, the limit $v_{fluid} \to 0$ is singular.

In the explicit dependence of the dispersion relations on κ_{Eck} and the fluid velocity one sees a manifestation of the nontrivial effect on a first-order theory of a change of rest-frame and the resulting adjustment to gradients of deviations from equilibrium that should be made, but cannot be, without sacrificing the first-order character of the theory. It is worth emphasizing here that these difficulties are eliminated in the "second-order" theories to be discussed in the next section.

§6. Alternative Theories: Nonlocal and Extended Thermodynamics

Their serious causality and stability problems point to an unfavourable prognosis on first-order relativistic theories of the Eckart and Landau-Lifshitz type. A post-mortem inquiry is not unduly anticipative; it may actually be past due. In this section we take a critical retrospective look at the foundation of these theories, and then we look at the proposals for new, improved designs now under consideration.

The general framework developed in §2 for covariant, off-equilibrium thermodynamics was built on a minimum of assumptions. In brief, these were: (a) that the theory reduce to the well-established equilibrium theory in that limit; (b) that the concept of entropy retains meaning for at least a small range of deviations from equilibrium; and (c) that the covariant relations between S^λ, $T^{\lambda\mu}$ and J^λ_A established for equilibrium remain valid <u>to first order</u> in the deviations. The last condition just states that the standard and universally accepted linear relations between the fluxes of entropy, heat and diffusion are valid to first order. This led to the general expressions

$$S^\lambda = P(\alpha,\beta)\beta^\lambda - \alpha J^\lambda - \beta_\mu T^{\lambda\mu} - Q^\lambda(\delta J^\mu, \delta T^{\lambda\mu},\ldots) \,, \tag{65}$$

$$\nabla_\mu S^\mu = - \delta J^\mu \partial_\mu \alpha - \delta T^{\lambda\mu}\nabla_\mu \beta_\lambda - \nabla_\mu Q^\mu \tag{66}$$

for the entropy flux and entropy production in a simple fluid. Here, Q^λ denotes an unspecified quantity of at least second order in the deviations δJ^λ, $\delta T^{\lambda\mu}$,... from a nearby equilibrium state (α,β^λ), which can be chosen arbitrarily.

The traditional, first-order theories impose two additional assumptions on this general scheme. They assume <u>entropic linearity</u>, (i.e. linearity of the S-J-T relation) by setting $Q^\lambda=0$ in (65); and, in inferring linear phenomenological laws from (66), they assume <u>locality</u>: the linear dependence of $\delta J^\mu(x)$, $\delta T^{\lambda\mu}(x)$, on the gradients $\partial_\mu\alpha(x)$, $\nabla_\mu\beta_\lambda(x)$ is supposed to be purely local.

One or both of these assumptions must be relaxed if the cul-de-sac reached in §5 is to be avoided. There are strong independent grounds for jettisoning $Q^\lambda=0$ in particular. In the case of a gas, Q^λ can be obtained from kinetic theory. We shall verify in the next Chapter that

$$Q^\lambda(x) = -\tfrac{1}{2}\int N_o f^2 p^\lambda d\omega \qquad (m^3 d\omega = d^3p/|p_o|) \tag{67}$$

up to second order in the deviation $(N-N_o)=N_o f$ of the distribution function $N(x^\alpha, p_\beta)$ from equilibrium. The nonrelativistic equivalent of (67) was noted by Enskog in 1929, and Joseph Meixner [42] recognized and emphasized nearly 50 years ago how neglect of the Q-contribution in conventional theories limits their applicability to small gradients and quasistationary processes.

Since no-one has yet mustered the courage to contemplate extensions of the conventional phenomenological laws that are both nonlocal and nonlinear, current attempts to formulate an improved theory are divided into two broad camps, under the banners "extended thermodynamics" (ET) and "nonlocal thermodynamics" (NLT), according to whether they give up one or other of the linearity and locality postulates. A fully covariant NLT is formidable, and current relativistic NLT [43] compromises the spirit, if not the letter, of covariance by offering a rheomorphic, rather than a causal description of the phenomenological laws: transport coefficients at an event x are taken to depend, not on the entire causal past of x, but (following classical precedent) only on the past history of a "comoving local fluid element".

As an inherently linear theory, NLT is restricted to small deviations from equilibrium. Within this limitation, it is in its broadest (causal) form, the most general theory possible for a prescribed set of thermodynamical variables. It should be able in principle to give a complete account of correlations and fluctuations in media where the range of interparticle forces extends up to or beyond a mean free path. Against this must be set the unwieldiness of nonlocal equations and the enormous input from microphysics required to make the equations operational. In many astrophysical situations - particularly in the relativisitc domain - correlation and memory effects are not of primary interest, and the real need is for a tractable and consistent transport theory coextensive at the macroscopic level with Boltzmann's equation. In such cases, there are no grounds for going beyond a purely local thermodynamical description, as the kinetic-theoretical expression (67) confirms. ET is the simplest, and at present appears the most promising candidate for a theory tailored to meet this specific need. Moreover, it holds out the prospect of straightforward extendability into the nonlinear regime.

I pass on to a rapid sketch of the broad features of NLT and ET, considered as thermodynamical theories.

Nonlocal Thermodynamics

Nonrelativistic NLT goes back 35 years to work by Takizawa [44] and Meixner [45]. The relativistic theory in its rheomorphic version has the same appearance as the classical theory. Schematically, let us write $j(x) \sim \{\delta T^{\lambda}, \delta T^{\lambda\mu}\}$, $E(x) \sim \{-\partial_{\mu}\alpha, -\nabla_{(\mu}\beta_{\lambda)}\}$, where the handwaving symbol "~" indicates that the usual relativistic baggage (metric and Lorentz factors) is appropriately taken care of throughout. The basic idea then is that the local phenomenological laws $j(x)=\sigma(n,T)E(x)$ of conventional theory are generalized to

$$j(\vec{x},t) = \int_{-\infty}^{\infty} \sigma(t-t')E(\vec{x},t')dt' \tag{68}$$

in co-ordinates comoving with an element of fluid. The Fourier transform

$$\sigma(t) = \int_{-\infty}^{\infty} \tilde{\sigma}(\omega)\, e^{-i\omega t}\, d\omega \tag{69}$$

leads to the concept of frequency-dependent transport coefficients:

$$\tilde{j}(x,\omega) = \tilde{\sigma}(\omega)\, \tilde{E}(\vec{x},\omega) \quad . \tag{70}$$

Reality and causality [$\sigma(t-t')=0$ for $t'>t$] require

$$\tilde{\sigma}(\omega) = \tilde{\sigma}^*(-\omega) = \text{regular for } \mathrm{Im}\,\omega > 0 \, . \tag{71}$$

so that Re $\tilde{\sigma}(\omega)$ is an even function.

The total entropy generated in a transition between two equilibrium states is given by the spacetime integral of (66):

$$\Delta S = \int(\nabla_\mu S^\mu)\, d^4x = \int(j(x)E(x) - \nabla_\mu Q^\mu)\, d^4x \, . \tag{72}$$

The integral of the last term vanishes, since $Q^\mu=0$ in the initial and final equilibrium states. We obtain

$$\begin{aligned} \Delta S &= \int d^3\vec{x} \iint \sigma(t - t')\, E(\vec{x},t')\, E(\vec{x},t)\, dt\, dt' \\ &= \int d^3\vec{x} \int_{-\infty}^{\infty} \tilde{\sigma}(\omega)\, \tilde{E}^*(\vec{x},\omega)\, \tilde{E}(\vec{x},\omega)\, d\omega \\ &= 2\int d^3\vec{x} \int_0^{\infty} \mathrm{Re}\, \tilde{\sigma}(\omega) \cdot |E(\vec{x},\omega)|^2\, d\omega \quad . \end{aligned} \tag{73}$$

The entropy produced is positive for all processes $E(\vec{x},\omega)$ if and only if transport coefficients are constrained by

$$\mathrm{Re}\, \tilde{\sigma}(\omega) \geq 0 \, . \tag{74}$$

However, $\nabla_\mu S^\mu \geq 0$ no longer holds as a local condition. In fact, since the form of Q^λ in (65) is left arbitrary, no assumption is made or required about the form or locality of a Gibbs-Duhem relation in the phenomenological theory.

The fully covariant, causal extension of this rheomorphic theory is (in principle) straightforward. The phenomenological laws are now of the form

$$j(x) = \int\sigma(x,x')E(x')d^4x' \tag{75}$$

and the Fourier-transformed transport coefficients $\tilde{\sigma}(\omega,k;n,T;$ cross-sections) now depend also on wave-number, and are constrained by the condition that $\sigma(x,x')$ vanish outside the past null cone of event x.

For materials with short memory, the general causal form of NLT reduces to the linearized version of ET. For example, if we set

$$\sigma(t - t') = \sigma_o \tau^{-1} e^{-(t-t')/\tau} \theta(t-t') \qquad (\tau|\dot{E}/E| \ll 1) \tag{76}$$

in the rheomorphic equation (68), the effect is to introduce a relaxation term into the conventional equations in the manner suggested by Cattaneo [40]:

$$j(x,t) = \int_{-\infty}^{t} \sigma(t - t')E(\vec{x},t')dt' = \sigma_o\{E(\vec{x},t) - \tau\dot{E} + \ldots\} \, . \tag{77}$$

The rheomorphic formulations of relativistic NLT developed so far [43] are derived from the linearized Boltzmann equation by projection-operator techniques. Since the content of these theories is fully equivalent to that of the linearized relativistic Boltzmann equation, whose causal nature has been proved [46], it follows that they must be causal. This makes it plausible that they are also stable.

Extended Thermodynamics and Second-Order Theories

The second-order kinetic formula (67) was the starting point for good deal of work (particularly in Meixner's group [47]) during the 1950's on extending the domain of validity of conventional (nonrelativistic) thermodynamics to shorter scales. At first the thrust of this work was aimed at bringing phenomenological theory into line with the Chapman-Enskog expansion and the phenomenological form of Q^{μ} was accordingly related to gradients of temperature and velocity, with results that were not very perspicuous. The turning point was Müller's 1967 paper [48] which for the first time expressed Q^{α} in terms of the off-equilibrium "forces" $(\vec{q},\pi_{ab},\pi)$ and thus linked phenomenology to the Grad expansion. This marked the birth of what is now known as extended thermodynamics.

As noted in §3, the second-order term Q^{λ} $(\delta J^{\alpha}, \delta T^{\alpha\beta}, \ldots)$ generally depends on an infinite set of "auxiliary" variables $X^{\alpha\beta\gamma\cdots}$ that vanish in equilibrium. In order to arrive at a tractable system of equations for the variables of primary interest - J^{λ} and $T^{\lambda\mu}$ - we shall pragmatically ignore the auxiliary variables. This ansatz is the phenomenological equivalent of Grad's 14-moment approximation in kinetic theory.

For small deviations, it will suffice to retain only the lowest-order (quadratic) terms in the Taylor expansion of Q^{λ} (leading to linear phenomenological laws). The most general form then involves just 5 undetermined coefficients

$$Q^{\lambda} = \frac{1}{2} u^{\lambda}(\beta_o \pi^2 + \beta_1 q^{\mu} q_{\mu} + \beta_2 \pi^{\mu\nu}\pi_{\mu\nu}) - \alpha_o \pi q^{\lambda} - \alpha_1 \pi^{\lambda}_{\mu} q^{\mu} \ . \qquad (78)$$

The condition $- u_{\lambda}Q^{\lambda} \geq 0$ [see (37)], which guarantees that entropy is maximized by equilibrium, requires that the β's be nonnegative; we shall see in a moment that they have the physical significance of relaxation times.

A first-order change of rest-frame (i.e. of the arbitrary neighbouring equilibrium α, β_{μ} introduced in (65)) produces a second-order change in Q^{λ} and thus changes the coefficients α_i, β_i in a well-defined manner. It can be shown, for example, that their values in the Landau-Lifshitz and Eckart frames (the former indicated by a bar) are related by

$$\alpha_i - \bar{\alpha}_i = \bar{\beta}_1 - \beta_1 = [(\rho+P)T]^{-1}, \qquad \beta_o = \bar{\beta}_o \ , \quad \beta_2 = \bar{\beta}_2 \ .$$

With these adjustments, a change of frame leaves the phenomenological laws invariant to first order; no higher derivatives are introduced, contrary to what happens with the "first-order" theories of §4.

The phenomenological laws can be read off from (78) and the expression for entropy production (66). In the Eckart frame, for example, they take the form

$$\begin{aligned} q^{\lambda} &= -\kappa T\Delta^{\lambda\mu}(T^{-1}\partial_{\mu}T + \dot{u}_{\mu} + \beta_1\dot{q}_{\mu} - \alpha_o\partial_{\mu}\pi - \alpha_1\nabla_{\nu}\pi^{\nu}_{\mu}) \\ \pi_{\lambda\mu} &= -2\zeta_S(\nabla_{<\mu}u_{\lambda>} + \beta_2\dot{\pi}_{\lambda\mu} - \alpha\,\nabla_{<\mu}q_{\lambda>}) \\ \pi &= -\frac{1}{3}\zeta_V(\nabla_{\mu}u^{\mu} + \beta_o\dot{\pi} - \alpha_o\nabla_{\mu}q^{\mu}) \end{aligned} \tag{79}$$

which of course reduce to equations (50), (54) of the "standard" Eckart theory when the 5 relaxation and coupling coefficients α_i, β_j are set to zero.

For appropriate values of the coefficients, the system (79) is hyperbolic. In the case of a gas, the new coefficients can be found explicitly [49], and they are purely thermodynamic functions (independent of cross-sections). It turns out that wave-front speeds are finite and comparable with the speed of sound. The fastest mode - one of the two 6-component longitudinal modes Y_L (§5) - approaches a speed of $\sqrt{3/5}$ c in the ultrarelativistic limit $T\to\infty$.

The general class of theories deriving from the quadratic ansatz (78) have been called "second-order theories" by Hiscock and Lindblom [7]. They have demonstrated the striking result that causality and stability are virtually equivalent properties within this class. More precisely, if the coefficients (α_i, β_j) are constrained so that the theory possesses stable equilibrium states, then linear perturbations propagate causally. Conversely, if (α_i, β_j) are such that the theory is causal and (in a suitably defined sense) hyperbolic, then it is also stable.

These are welcome signs of the internal consistency of second-order theories, and schemes of this type are beginning to find astrophysical application (e.g. to radiative transfer [50] and cosmology [51]). But their utility and even soundness remain a subject of debate. One school of thought [43] holds it to be unnecessary or regressive to enlarge the set of thermodynamical variables beyond the standard set (n,ρ,u^{λ}) which describe equilibrium states. These are the variables which - in the case of a gas - determine the so-called normal solutions of Chapman-Enskog theory. Van Kampen [38] argues that his systematic scheme for eliminating "fast" variables (varying on the scale of a mean collision time) from the Boltzmann equation "leads uniquely to $[n,\rho,u^{\lambda}]$ as the only quantities whose rate of variation is determined by the gradients. It does not lead to 'extended non-equilibrium thermodynamics'." However, van Kampen's expansion appears to hinge on the implicit assumption that fast and slow variables have comparable amplitudes. While it is true that large-amplitude fluctuations typically decay on the scale of a mean collision time, it does not follow that <u>all</u> fluctuations decaying on this timescale fall outside the purview of a linearized description of small-amplitude fluctuations - precisely the circumstances for which ET was designed. Processes operating on scales comparable to a mean free path or collision time, which yet involve small deviations from local thermal equilibrium, are comparatively unfamiliar in terrestial conditions, but commonly encountered in high-energy astrophysics - for example, in flickering, marginally thick X-ray, γ-ray and radio sources, or in situations (e.g. neutron star cooling) where transport processes ride on the Fermi surface of a large store of zero-point or rest-mass energy.

In conclusion, a brief reference to two specific versions of extended thermodynamics which are treated in much more detail by their authors elsewhere in this volume. The accession of 5 new phenomenological coefficients heralds an unwelcome inflation both in complexity and indeterminacy of the general second-order schemes vis-á-vis their first-order predecessors. The theories of Carter and of Liu, Müller and Ruggeri [35] are attempts to overcome this problem in different ways.

In the latter theory, the extra indeterminacy is largely pinned down by annexing to the usual conservation and entropy laws (19) a new phenomenological assumption:

$$\nabla_\gamma A^{\alpha\beta\gamma} = I^{\alpha\beta} \tag{80}$$

in which $A^{\alpha\beta\gamma}$, $I^{\alpha\beta}$ are symmetric tensors with traces given by

$$A^{\alpha\beta}{}_{\beta} = - J^{\alpha} , \qquad I^{\alpha}_{\alpha} = 0 \quad . \tag{81}$$

Conditions (80), (81) are modelled on kinetic theory, in which $A^{\alpha\beta\gamma}$ represents the third moment of the distribution function in momentum space, and $I^{\alpha\beta}$ the second moment of the collision term in Boltzmann's equation. (Equations (80), (81) are central to the determination of the distribution function in Grad's 14-moment approximation.) The phenomenological theory is completed by the postulate that the state variables S^λ, $A^{\alpha\beta\gamma}$, $I^{\alpha\beta}$ are invariant functions of $T^{\lambda\mu}$, J^λ only, thus pressing to its logical conclusion the assumption, introduced at the beginning of this section for pragmatic reasons, that a nonequilibrium state may be adequately describable by the hydrodynamic variables $T^{\lambda\mu}$, J^λ alone. The result is a highly nonlinear theory which, for small deviations, reduces to (79) (very nearly - there is some extra indeterminacy) with coefficients (α_i, β_j) in agreement with those obtained from the Grad approximation. The theory is thus an almost exact phenomenological counterpart of the Grad approximation, with the added appeal for continuum mechanists that it can be built up rigorously from a few basic postulates. For nongaseus materials, of course, the physical meaning and status of the postulates (80), (81) are not entirely clear.

Carter's theory is a very elegant derivation of extended thermodynamics from a variational principle. Although in its present form the theory is confined to nonviscous, heat-conducting materials, there would appear to be no obstacle in principle to the inclusion of viscous effects in this formalism.

§7. Internal Structure of Strong Shocks: A Crisis for Extended Thermodynamics?

The close link between "extended" thermodynamical theories and the Grad approximation in kinetic theory has already been remarked. In 1952, examining the interior structure of nonrelativistic shocks on the basis of this approximation, Grad [52] noted a peculiarity that is quite absent from Navier-Stokes theory.

Similar results have been obtained by Anile and his group for other hyperbolic "extended" theories, both relativistic and nonrelativistic [53]. The general result appears to be that hyperbolic theories do not admit a regular shock structure once the speed of the shock front exceeds the highest characteristic velocity. Just as a forced compressive disturbance in an ideal fluid builds up to a shock if its speed exceeds the speed at which sound waves can disperse it, so a "subshock" can be expected to form within a shock layer for speeds exceeding the wave-front velocities of thermo-viscous effects.

The Grad-Müller theory predicts a breakdown of regular shock structure for an upstream Mach number of 1·65. Experimentally, shocks with an apparently smooth internal structure have been observed in Argon for Mach numbers up to 10. This situation has evoked a variety of reactions. My own tentative feeling is that it may well be a genuine crisis for the theory in its present form. It is certainly the most intriguing and fundamental problem now confronting this area. Until it is resolved one can repose little confidence in the viability of "extended thermodynamics" or theories of shock structure in either the relativistic or nonrelativistic regimes. The issue is of practical importance: high-energy shock waves and their structure [54] have far-flung applications to high-energy nucleon collisions [55] as well as numerous astrophysical situations [56].

To understand better the nature of the difficulties it will be useful to recall the hydrodynamical equations for the structure of plane, steady shock layers. Since it is in the nonrelativistic regime that direct conflict with experience arises, and relativity introduces nothing new in principle, we shall not need to go beyond the familiar classical theory.

The general balance laws for particles, momentum and internal energy (ϵ per unit mass) read

$$D_t(nm) + nm\,\mathrm{div}\,\vec{u} = 0 \quad , \qquad nm\,D_t u_a = -\,\partial_b P_{ab}$$
$$nm\,D_t\epsilon + \mathrm{div}\,\vec{q} + P_{ab}\partial_a u_b = 0 \,, \tag{82}$$

in terms of the convective derivative $D_t=\partial_t+\vec{u}\cdot\nabla$. I restate these laws, and their successive first integrals, for steady one-dimensional flow along the x-direction:

$$\partial_x(nmu) = 0 \qquad : \quad nmu = J$$
$$nmu\partial_x u + \partial_x P_{xx} = 0 \qquad : \quad J_u + P_{xx} = aJ \tag{83}$$
$$nmu\partial_x\epsilon + \partial_x q + \partial_x(uP_{xx}) - u\partial_x P_{xx} = 0 \; : \quad J\epsilon + q + P_{xx}u + \frac{1}{2}Ju^2 = (b + \frac{1}{2}a^2)J$$

(J,a,b constant).

For a gas,

$$\epsilon = c_V T \quad , \quad P_{xx} = nkT + \pi_{xx} \quad , \tag{84}$$

and (83) may be written formally as equations for the shear stress and heat flux:

$$\pi_{xx} = J(a - u - kT/mu) \equiv Jf_1(u,T)$$
$$q \;= J[\frac{1}{2}(u - a)^2 + b - c_V T] \equiv Jf_2(u,T) \;. \tag{85}$$

The conservation identities must now be supplemented by a set of phenomenological laws. We adopt Müller's [41] nonrelativistic limit of the "extended" thermodynamical equations (79),

$$q_a = -\kappa(\partial_a T + \beta_1 \dot{q}_a - \alpha_o \partial_a \pi - \alpha_1 \partial_b \pi_{ab})$$

$$\pi_{ab} = -2\zeta_S[\partial_{(a}u_{b)} - \frac{1}{3}\delta_{ab}\mathrm{div}\,\vec{u} + \beta_2\dot{\pi}_{ab} - \alpha_1(\partial_{(a}q_{b)} - \frac{1}{3}\delta_{ab}\mathrm{div}\,\vec{q})] \tag{86}$$

$$\pi = -\frac{1}{3}\zeta_V(\mathrm{div}\,\vec{u} + \beta_o\dot{\pi} - \alpha_o \mathrm{div}\,\vec{q})$$

with $\dot{\Phi} \equiv D_t\Phi$. For steady one-dimensional flow and vanishing bulk viscosity, these reduce to

$$\pi_{xx} = -\frac{4}{3}\zeta_S(\partial_x u - \alpha_1\partial_x q) - 2\zeta_S\beta_2 u\partial_x\pi_{xx}$$

$$q = -\kappa(\partial_x T + \beta_1 u\partial_x q - \alpha_1\partial_x\pi_{xx}) \,. \tag{87}$$

Eliminating π_{xx} and q from (87) by substituting from (85) yields a pair of coupled first-order quasilinear equations for the variables $Z \equiv \binom{u}{T}$:

$$C(Z)\,\partial_x Z = f(Z) \,. \tag{88}$$

In the special case of Navier-Stokes theory the first derivatives of u and T decouple (as can be seen directly from (87)) and C(Z) reduces to a diagonal, nondegenerate matrix:

$$-J\,C = \mathrm{diag}(\frac{4}{3}\zeta_S,\ \kappa) \qquad (\text{Navier-Stokes: } \alpha_1=\beta_1=\beta_2=0) \,. \tag{89}$$

In general, equations (88) are to be solved subject to the asymptotic boundary conditions

$$Z(x=\pm\infty) = Z_\pm \quad , \qquad \partial_x Z(x=\pm\infty) = 0 \tag{90}$$

expressing the requirement that the fluid on either side of the shock layer is in thermal equilibrium. The constants Z_- and Z_+ are the upstream and downstream equilibrium fluid parameters in the (inertial) rest-frame of the shock. They are subject to four compatibility conditions

$$f(Z_\pm) = 0 \,. \tag{91}$$

After elimination of a,b by use of (85), equations (91) reduce to two conditions, which can be regarded, e.g., as fixing the downstream state Z_+ when the upstream parameters Z_- are given arbitrarily. They are, of course, just the familiar Rankine-Hugoniot relations for shock transitions in an ideal fluid.

We first review the Navier-Stokes case. Richard Becker noted in 1922 that the equations admit an elementary and instructive solution for the special value 3/4 of the Prandtl number $c_P\zeta_S/\kappa$. To recover this result, we combine equations (85) and also (for the NS case) (87) to form $\pi_{xx}u+q$:

$$J[\frac{1}{2}(a-u)^2 + b - c_P T] = \pi_{xx}u + q = \frac{4}{3}\zeta_S[\partial_x(-\frac{1}{2}u^2) - \frac{3}{4}\frac{\kappa}{\zeta_S}\partial_x T] \,. \tag{92}$$

Thus, if

$$c_P = \frac{3}{4}\kappa/\zeta_S \qquad (\text{"Becker gas"}) \,, \tag{93}$$

(92) can be integrated at once, with

$$c_P T = \frac{1}{2}(a^2 - u^2) + b \tag{94}$$

as the unique solution that allows $\pi_{xx}u+q$ to vanish for $x=\pm\infty$. (The ansatz (93) is not unrealistic: kinetic theory gives for the Prandtl number of a monatomic gas

$$c_P \frac{\zeta_S}{\kappa} = \frac{5}{2}\frac{k}{m} \cdot \frac{4}{15}\frac{m}{k} = \frac{2}{3} . \tag{95}$$

For air, the value is 0.73.) Substituting (94) into the first of (85) yields

$$\pi_{xx} = J(\text{quadratic function of } u)/u = J(u_- - u)(u - u_+)/u \tag{96}$$

by the boundary conditions. The velocity profile u(x) is then the integral of

$$-\frac{4}{3}\zeta_S \partial_x u = J(u_- - u)(u - u_+)/u \quad . \tag{97}$$

A solution u(x) clearly exists for any assigned pair of boundary values $u_- > u_+$, and decreases monotonically from the upstream to the downstream value.

The proof that a solution of the Navier-Stokes system (88), (90) exists in general has been given in a classic paper by Gilbarg and Paolucci [57]. The underlying idea, in brief, is this. Since the matrix C(Z) is nonsingular in the Navier-Stokes case, the only critical points of (88) arise from f(Z). One can eliminate x and obtain the first-order equation

$$\frac{dT}{du} = \frac{4}{3}\frac{\zeta_S}{\kappa}\frac{f_2(u,T)}{f_1(u,T)} , \tag{98}$$

with f_1, f_2 given by (85). The critical parabolas $f_1=0$ and $f_2=0$ intersect in two points Z_-, Z_+ which are the equilibrium endstates, by (91), corresponding to given values of a,b in (85).

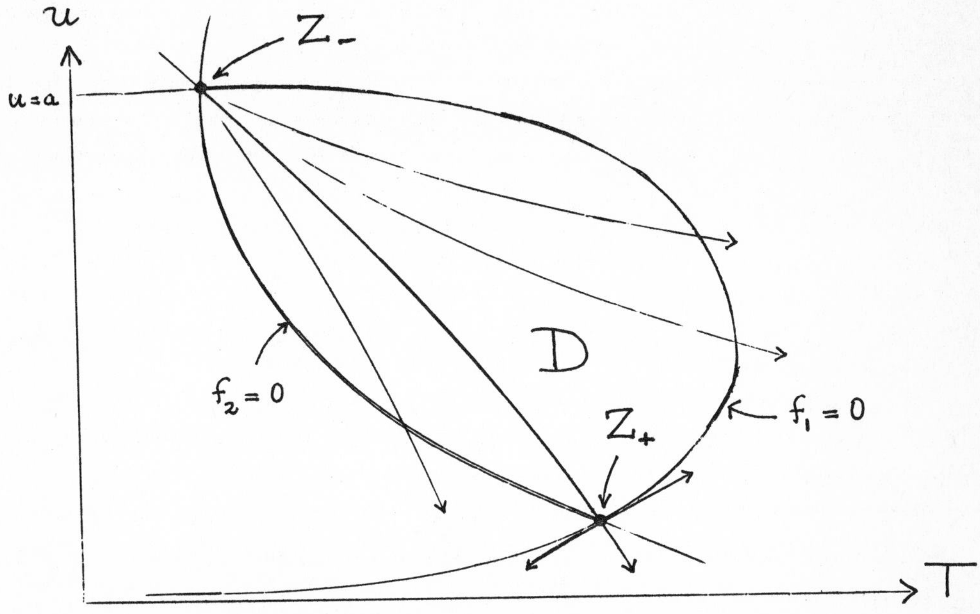

Shock solution for a Navier-Stokes fluid.

Let D be the region $f_1 f_2 < 0$ enclosed between the critical parabolas. Construction of the direction field for (98) shows that integral curves are directed outwards (for increasing x) everywhere on the boundary of D except at the node Z_-. Of the two integral curves through the saddle point Z_+, the one emerging from D must therefore enter Z_- when followed backwards. Thus a unique solution of (98), (90) exists for arbitrary values of a,b - i.e., for arbitrary equilibrium endstates Z_+, Z_- satisfying the Rankine-Hugoniot conditions (91).

We finally turn to the system (88) in the general case when α, β_1, β_2 do not vanish. The possibility that det C = 0 now admits new critical curves into the simple picture shown in the Figure. If these curves segregate Z_- and Z_+ into regions where det C has opposite signs, they will generally obstruct or deflect passage of the integral curve from Z_+ to Z_- and a regular structure for the shock layer will no longer exist. The new curves $u = u_{char}(T)$ are clearly <u>characteristics</u> of the system (88) - loci in the (u,T) plane of stationary jump discontinuities of $\partial_x Z$ - so that $v = -u_{char}(T)$ is the wave-front velocity relative to the fluid of a longitudinal thermo-viscous disturbance that is stationary in the rest-frame of the shock. Thus, as soon as the upstream fluid velocity overshoots a characteristic velocity of the longitudinal mode, one may expect obstruction and breakdown of the shock solution.

The detailed investigation [53] fully confirms these expectations. For the Müller theory (86), the characteristic curve det C = 0 is a fixed oval* intersecting the positive u and T-axes. For sufficiently weak shocks, both Z_- and Z_+ lie within the oval and there is a regular shock structure. When the upstream Mach number $M_- \equiv u_-/c_{s-}$ reaches 1.65, Z_- crosses the oval and the solution ceases to exist.

The collapse of the extended theory for $M_- > 1.65$ contrasts sharply with laboratory experiments which have now established the existence and regular structure of shock waves in Argon for Mach numbers up to at least 10. Moreover, Navier-Stokes theory not only permits regular shocks of arbitrary strength, but provides a tolerable description of their structure up to the highest Mach numbers observed [58].

One can plead extenuating circumstances for the extended theories on two possible grounds:

(a) Shock thicknesses are of the order of a mean free path and thus beyond the scope of any continuum theory.

(b) The extended theories (like Navier-Stokes) are <u>linear</u> theories for small deviations from local thermal equilibrium. They are not applicable inside strong shocks, where deviations are expected to become large.

Neither of these arguments entirely carries conviction, although the issues are admittedly far from clearcut. Measurements for Argon give thicknesses of 7 and 4

---

*Actually an ellipse in the (u,P) plane.

mean free paths respectively for Mach numbers of 1.5 and 2.4 [58]. Numerical simulations by G.A. Bird [59] for a "gas" of 1,000 rigid sphere molecules suggest a thickness of about 4 upstream mean free paths for Mach numbers as high as 10. These numbers are in the borderline range where one might have expected the extended theories to prove their worth.

Nor is there conclusive evidence for very gross departure from local thermal equilibrium even in the strongest shocks. Bird's kinetic model indicates that $1 < T_{long}/T_{lat} \lesssim 2$ for Mach numbers between 3 and 10, where T_{long} and T_{lat} are "temperatures" associated with longitudinal and lateral components of the molecular velocities. Another measure is the ratio of shear stress to thermodynamical pressure, which can be estimated from (85):

$$\frac{\pi_{xx}}{P} = \frac{mu}{kT}(a - u) - 1 \ .$$

We can set $u = Mc_S$, $c_S^2 = \gamma kT/m$ in terms of the local Mach number and sound velocity:

$$\pi_{xx}/P = \gamma Ma/c_S - (\gamma M^2 + 1) \ .$$

The value of a can be found by applying this relation at the downstream end:

$$a = (\gamma M_+^2 + 1)c_S^+/\gamma M_+ \ .$$

In the limit of very strong shocks ($M_- \to \infty$) the Rankine-Hugoniot conditions show that the downstream Mach number decreases to a finite limit, $M_+^2 \to (\gamma-1)/2\gamma$, so that

$$a = c_S^+(\gamma + 1)/\sqrt{2\gamma(\gamma - 1)} \qquad (M_- \to \infty) \ .$$

We estimate the ratio (99) at the sonic point M=1 as a "typical" internal point. Then

$$\left(\frac{\pi_{xx}}{P}\right)_{M=1} = (\gamma + 1)\left\{\frac{c_S^+}{c_S}\sqrt{\frac{\gamma}{2(\gamma-1)}} - 1\right\} \qquad (M_- \to \infty) \ . \tag{99}$$

We assume that $c_S^+ > c_S$, which would be true, for instance, if temperature increases monotonically through the layer. (This is the only assumption made about internal structure, apart from the basic conservation laws.) This finally yields the estimate (for a monatomic gas)

$$(\pi_{xx}/P)_{M=1} > \frac{8}{3}\left(\frac{1}{2}\sqrt{5} - 1\right) = 0.315 \qquad (\gamma = \frac{5}{3},\ M_- \to \infty) \ . \tag{100}$$

Although this is a lower bound, it hardly offers convincing cause to expect a total breakdown of linear phenomenological laws for moderately strong shocks.

While I have no basis or inclination for a dogmatic opinion, to my mind the failure of the present versions of extended thermodynamics to deal satisfactorily with shock structure underlines the distinct possibility that the rather curious wave-front speeds predicted by these theories are chimerical artefacts. Certainly no trace of them is found in the Boltzmann equation, which gives a wave-front speed equal to the largest molecular speed for which the distribution function is non-zero.

However, there is a great diversity of opinion on this issue:

"We are not permitted to indulge our curiosity and discuss [the 13-moment shock structure] solution beyond its proper scope. This is to be compared with the Navier-Stokes equations for which the shock solution exists formally for even greater than "infinite" shock strength; (the differential equations do not recognize zero or even negative temperatures and pressures as being extraordinary). The Navier-Stokes solution begins to differ from the thirteen-moment result at about M=1.2, which point therefore marks the limit of its applicability."

Grad (1952) [52].

"Therefore it appears that theories of the Müller type might be inconsistent with experimental data. However we believe that this could be due to the neglect of non-linear terms in the evolution equations for the heat flux vector and the viscous stress tensor, and not to any inconsistency in the basic postulates of Müller's thermodynamic approach." **Anile and Majorana (1981) [53].**

"The problem of the structure of strong shock fronts must be treated on the basis of the kinetic theory of gases, and hence the many numerical studies concerned with the improvement of the simple [Navier-Stokes] theory ... do not contribute anything new in principle ... and at best are of interest for the case of weak waves only." **Zeldovich and Raizer (1967) [60].**

"Gilbarg and Paolucci ... have shown that ... the Navier-Stokes equation provides at least as good values for shock thickness as does kinetic theory, values, moreover, which are in acceptable agreement with recent experiments..... For these reasons we definitely do not believe it outmoded to use continuum methods in studying the shock layer." **Serrin (1958) [61].**

"Equations have often been successful beyond the limits of their original derivation, and indeed this type of success is one of the hallmarks of a great theory." **David Gilbarg (quoted by Serrin [61]).**

CHAPTER III. ELEMENTS OF COVARIANT KINETIC THEORY

§1. Basic concepts. Synge's Invariant Distribution Function. Transport and Balance Laws.

We have repreatedly mentioned the links between the phenomenology of the previous Chapter and kinetic theory in the case of a gas. It may therefore be of some interest to devote these final pages to a brief sketch of the kinetic theory, developed just far enough to bring out these connections. My hope is that this rudimentary account may serve as an introduction to the extensive systematic treatment of the book by de Groot et al [63].

We consider a simple gas whose particles interact directly only by collisions, and whose motion in gravitational and possibly other background fields is accordingly described by a one-particle Hamiltonian $H(x^{\alpha}, p_{\beta})$.

The momentum space at event x is co-ordinatized by the covariant components p_{α} with metric $g^{\alpha\beta}(x)p_{\alpha}p_{\beta}$ (flat in the (co-) tangent plane at x): (In dignified language, momentum space is the fibre over x of the cotangent bundle.) The mass shell at x is the future sheet of the 3-dimensional hypersurface $H(x^{\alpha}, p_{\beta}) = -m$ (rest-mass) in momentum space. The particle 4-velocity

$$dx^{\alpha}/d\tau = v^{\alpha} = \partial H/\partial p_{\alpha} \qquad (v^{o} > 0) \qquad (1)$$

is also the unit normal to the mass-shell. (In general, v^{α} and p^{α} are not parallel. The simple formulas $H = -(-g^{\alpha\beta}(x)p_{\alpha}p_{\beta})^{1/2}$, $v^{\alpha} = p^{\alpha}/m$ hold only for structureless particles responding to no other forces than gravity.) The invariant element of 3-area on the mass-shell is

$$d\omega = (-g)^{-1/2}\, dp_1 dp_2 dp_3 \int \delta(H + m) dp_o = (-g)^{-1/2}\, d^3p/v^{o}\ . \qquad (2)$$

Synge's (1934) invariant distribution function $N(x^{\alpha}, p_{\beta})$ for the assembly is defined by the statement: the number of particle world-lines with momenta in the range $(p_{\mu}, d\omega)$ that cross a target 3-area $d\Sigma_{\mu}$ in spacetime in the direction of its normal is given by

$$dN = N(x,p)d\omega \cdot \epsilon\, v^{\mu}\, d\Sigma_{\mu}\ . \qquad (3)$$

(The sign factor ϵ was introduced in the first equation of Chapter 1.) For any given set of curvilinear spacetime co-ordinates, if we choose $d\Sigma_{\mu}$ as an element of the 3-space x^{o}= const., we find

$$d\Sigma_{\mu} = -(-g)^{1/2}\, \epsilon_{\mu 123} dx^1 dx^2 dx^3 = -(-g)^{1/2}\, \delta^{o}_{\mu} d^3x\ ,$$

$$dN = N\, d\omega \cdot (-g)^{1/2}\, v^{o} d^3x = N\, d^3x\, d^3p \qquad (4)$$

where (2) has been used. This identifies the invariant N(x,p) as the conventional distribution function in phase space and, incidentally, proves the invariance of the 6-volume $d^3x\, d^3p$ under arbitrary co-ordinate transformations.

We can now proceed to obtain a general transport equation for N(x,p). Given an arbitrary scalar field $F(x^\kappa, p_\lambda)$ defined over phase space, we define its configuration space (i.e. spacetime) gradient $\nabla_\mu F$ by

$$\nabla_\mu F^\mu(x^\kappa, p_\lambda) \equiv \frac{\partial F}{\partial x^\mu} + \frac{\partial F}{\partial p_\lambda} \Gamma^\kappa_{\lambda\mu} p_\kappa \; . \tag{5}$$

What does this mean? It is the <u>covariant</u> gradient holding the <u>vector</u> p_λ (not its components) fixed; in that case the components must change in accordance with $dp_\lambda - \Gamma^\kappa_{\lambda\mu} p_\kappa dx^\mu = 0$ in an infintesimal displacement dx^μ. The extension of this definition to tensorial functions over phase space should be obvious. In particular, ∇_μ reduces to its ordinary meaning for tensor fields not depending on momentum and

$$\nabla_\mu p_\lambda = 0 \quad , \quad \nabla_\mu \Phi^{\cdot\cdot}_{\cdot\cdot}\,(g_{\alpha\beta}(x), p_\lambda) = 0 \; . \tag{6}$$

It follows that for particles subject to no fields other than gravity,

$$\nabla_\mu H(x,p) = 0 \longrightarrow \nabla_\mu v^\lambda(x,p) = 0 \quad , \quad \nabla_\mu d\omega = 0 \; . \tag{7}$$

Evolution of N along phase trajectories is given by

$$\frac{dN(x,p)}{d\tau} = v^\mu \partial_\mu N + \frac{dp_\mu}{d\tau} \frac{\partial N}{\partial p_\mu} \quad . \tag{8}$$

Individual terms on the right-hand side are not separately invariant under co-ordinate transformations; but (8) can be re-expressed in the manifestly invariant form

$$\frac{dN}{d\tau} = v^\mu \nabla_\mu N + \frac{\delta p_\mu}{\delta\tau} \frac{\partial N}{\partial p_\mu} \quad . \tag{9}$$

Assume for definiteness and simplicity that there are no external fields other than gravity. Then from $\delta p_\mu / \delta\tau = 0$ and (7)

$$dN/d\tau = \nabla_\mu (N v^\mu) \; . \tag{10}$$

If there were no collisions, this expression would, of course, be zero.

When a collision occurs, the smooth evolution $dN/d\tau = 0$ is interrupted by a discontinuous change in N. We may think of a collision as a point in spacetime where a smooth particle world-line is created or destroyed. Statistically, the effect of such changes can be ascribed to a "collision term" C[N]. For the nett number of particle world-lines with 4-momenta $(p_\mu, d\omega)$ created by collisions in a small volume dV_4 we can write $C[N(x,p)]dV_4 d\omega$, so that, by (3),

$$\int_{\partial V_4} N(x,p) \epsilon v^\mu d\Sigma_\mu = \int_{V_4} C[N](-g)^{1/2}\, d^4x \quad . \tag{11}$$

This leads to the general <u>transport equation</u>

$$\boxed{dN/d\tau = \nabla_\mu (N v^\mu) = C[N]} \tag{12}$$

Multiplication of (12) by an arbitrary tensorial function $\Phi^{\cdot\cdot}_{\cdot\cdot}(x,p)$ and integration over momentum space, recalling (7), leads to the <u>balance law</u>

$$\boxed{\nabla_\mu(\int N\,\Phi\, v^\mu d\omega) = \int \Phi\, C[N]\, d\omega} \quad . \tag{13}$$

The right-hand side represents the rate of production per unit 4-volume of the property $\Phi^{\cdot\cdot}_{\cdot\cdot}(x,p)$ by collisions. If collisions preserve the total 4-momentum and number of the colliding particles then C[N] must have the properties:

$$\int \Phi\, C[N]\, d\omega = 0 \qquad \text{if} \qquad \Phi = 1,\ p_\lambda\ ; \tag{14}$$

and (13) gives the <u>conservation laws</u>

$$\nabla_\mu J^\mu = \nabla_\mu T^{\lambda\mu} = 0 \tag{15}$$

where

$$J^\mu = \int N\, v^\mu d\omega \quad , \qquad T_\lambda{}^\mu(x) = \int N(x,p)\, p_\lambda v^\mu d\omega \quad . \tag{16}$$

§2. Boltzmann Equation. H-theorem. Thermal Equilibrium

In a dilute simple gas, elastic binary collisions give the dominant contribution to C[N]. We have to consider the probability that two incoming particles, with momenta in the ranges $(p_\mu, d\omega)$, $(p'_\mu, d\omega')$, will collide to produce outgoing particles with momenta $p_{*\mu}$, $p'_{*\mu}$. Boltzmann's ansatz assumes the incoming particles uncorrelated and sets this probability proportional to the product $N(x,p)d\omega \cdot N(x',p')d\omega'$. We shall follow this prescription here, which will lead us to classical (Boltzmann)* statistics.

Taking account of inverse collisions, Boltzmann's form of the collision term is then given by

$$\boxed{C[N(x,p)] = \int W(N_* N'_* - NN')\, d\omega'\, d\omega_*\, d\omega'_*} \tag{17}$$

in which we abbreviate $N_* \equiv N(x,p_*)$ etc., and the transition probability $W(p,p';p_*,p'_*)$ is trivially symmetric in p and p', and in p_* and p'_*, and is assumed to satisfy "microscopic reversibility":

$$W(p,p';\ p_*,p'_*) = W(p_*,p'_*;\ p,p') \ . \tag{18}$$

---

*Fermi or Bose statistics could easily be incorporated by introducing (dis-) occupation probability factors $\Delta(x,p_*)\Delta(x,p'_*)$ for the <u>final</u> states, to allow for Pauli exclusion or Bose-Einstein effects, so that, e.g., $\Delta(x,p_*)=0$ for fermions if the phase cell at (x,p_*) is fully occupied. The formal definition is

$$\Delta(x,p) = 1 + \epsilon h^3\, g^{-1}\, N(x,p)$$

$$\epsilon = \begin{cases} +1 \text{ (bosons)} \\ -1 \text{ (fermions)} \\ \ \ 0 \text{ (classical)} \end{cases} ; \qquad g = \text{spin weight} = \begin{cases} 2 \text{ (electrons, photons)} \\ 1 \text{ (neutrinos)} \end{cases}$$

These symmetries imply the identity

$$\int \Phi(x,p) C[N] d\omega = \frac{1}{4} \int \delta[\Phi] W(N_* N'_* - NN') d^4\omega \tag{19}$$

for any (tensor) function $\Phi(x,p)$, where

$$\delta[\Phi] \equiv \Phi + \Phi' - \Phi_* - \Phi'_* \ . \tag{20}$$

It follows that the integral (18) vanishes if $\delta[\Phi]=0$, for example if $\Phi=1$ or p_λ, which confirms that the ansatz (17) yields the conservation laws (15).

Next, we multiply the Boltzmann equation (12) by the derivative $\phi'(N)$ of an arbitrary scalar function $\phi(N)$, integrate over momentum space and apply (19). The result is

$$\nabla_\mu \left(\int \phi(N) v^\mu d\omega \right) = \frac{1}{4} \int \delta[\phi'(N)] C[N] d\omega \ . \tag{21}$$

This leads us at once to <u>Boltzmann's H-theorem</u>. Defining the entropy flux*

$$S^\mu(x) = - \int \{ N \, \ell n(Nh^3) - N \} \, v^\mu \, d\omega \tag{22}$$

and using (21) gives

$$\nabla_\mu S^\mu = \frac{1}{4} \int W \{ \ell n(N_* N'_*) - \ell n(NN') \} (N_* N'_* - NN') \, d^4\omega \geq 0 \ . \tag{23}$$

For local thermal <u>equilibrium</u> the entropy production vanishes, and (23) implies

$$\ell n \, N + \ell n \, N' = \ell n \, N_* + \ell n \, N'_*$$

which requires ℓn N to be an additive collision invariant, necessarily of the form

$$\ell n(Nh^3) = \alpha(x) + \beta_\lambda(x) p^\lambda \ . \tag{24}$$

The local equilibrium distribution for Boltzmann statistics thus takes the form

$$N_o(x,p) = h^{-3} \exp[\alpha(x) + \beta_\lambda(x) p^\lambda] \tag{25}$$

where β_λ is necessarily future timelike to ensure convergence of the moments (16), (22) of the distribution function. These are readily evaluated in the rest-frame with 4-velocity $u^\lambda \equiv \beta^\lambda/\beta \overset{*}{=} \delta^\lambda_o$ in local Lorentz co-ordinates:

$$p^\lambda \overset{*}{=} m(\sinh \chi \cdot \hat{e}, \cosh \chi) \ ; \quad d\omega = \frac{d^3p}{v^o} \overset{*}{=} \sinh^2\chi \, \sin\theta \, d\chi \, d\theta \, d\varphi$$

where $\hat{e}$ is a unit spatial vector. The results are

$$J^\lambda_{(o)} = \int N_o v^\lambda d\omega = nu^\lambda \quad , \quad T^{\lambda\mu}_{(o)} = \rho u^\lambda u^\mu + P\Delta^{\lambda\mu}$$

$$S^\lambda_{(o)} = P\beta^\lambda - \alpha J^\lambda_{(o)} - \beta_\mu T^{\lambda\mu}_{(o)} \ , \tag{26}$$

$$\rho + P = 4\pi(m/h)^3 \, e^\alpha \, K_3(\beta)/\beta, \quad P = n/\beta = nkT = 4\pi(m/h)^3 \, e^\alpha \, K_2(\beta)/\beta$$

---

*For quantum gases, one takes account of the "entropy of the holes" and the integrand of (22) is generalized to $N \, \ell n(Nh^3) - \epsilon^{-1} h^{-3} g \, \Delta \, \ell n \, \Delta$, from which the classical result is recoverable as the limit $\epsilon \to 0$. The equilibrium distribution function which replaces (25) is then

$$N_o(x,p) = gh^{-3} [\exp(-\alpha - \beta_\lambda p^\lambda) - \epsilon]^{-1} \ .$$

where K_n are modified Bessel functions; their detailed properties are not of interest to us here.

The local equilibrium form (25) makes the collision integral vanish: $C[N_o]=0$. If thermal equilibrium holds globally in some region, it then follows from the Boltzmann equation (12) that

$$v^\mu \nabla_\mu N_o = N_o v^\mu(\partial_\mu \alpha + p^\lambda \nabla_\mu \beta_\lambda) = 0$$

must hold for all p^λ on the mass-shell. We thus derive

$$\partial_\mu \alpha = \nabla_{(\mu}\beta_{\lambda)} = 0 \tag{27}$$

as necessary conditions for global equilibrium, in accordance with the phenomenological result (25) in Chapter I.

§3. Small Deviations from Local Thermal Equilibrium

For a gas that departs slightly from local thermal equilibrium, we may choose (independently at each point x) an arbitrary local equilibrium distribution $N_o(x,p)$ of the form (25) that is close to the actual distribution $N(x,p)$ and set

$$N(x,p) = N_o(x,p)[1 + f(x,p)] \, . \tag{28}$$

(Bear in mind that the local equilibrium parameters $\alpha(x)$, $\beta_\lambda(x)$ that appear in N_o are unconstrained, and do not satisfy (27).)

Substitution in (16), (22) yields

$$J^\lambda(x) = J^\lambda_{(o)}(x) + \delta J^\lambda(x) \, , \qquad T^{\lambda\mu}(x) = T^{\lambda\mu}_{(o)}(x) + \delta T^{\lambda\mu}(x)$$

where $J^\lambda_{(o)}$, $T^{\lambda\mu}_{(o)}$ have the ideal fluid form (26), and

$$\delta J^\lambda(x) = \int N_o f v^\lambda d\omega \, , \qquad \delta T^{\lambda\mu}(x) = \int N_o f p^\lambda v^\mu d\omega \, . \tag{29}$$

Further, since

$$\delta(N \, \ell n(Nh^3) - N) = \ell n(N_o h^3)\delta N + \frac{1}{2} N_o^{-1}(\delta N)^2 + \ldots$$

up to terms of second order, we obtain

$$\delta S^\lambda = S^\lambda - S^\lambda_{(o)} = - \int(\alpha + \beta_\mu p^\mu)N_o f v^\lambda d\omega + \frac{1}{2}\int N_o f^2 v^\lambda d\omega + \ldots$$

which corroborates two relations - (25) and (67) - that played a key role in the phenomenological considerations of Chapter II:

$$\boxed{\delta S^\lambda = - \alpha \delta J^\lambda + \beta_\mu \delta T^{\mu\lambda} - Q^\lambda} \tag{30}$$

$$\boxed{Q^\lambda = - \frac{1}{2}\int N_o f^2 v^\lambda d\omega + \ldots} \tag{31}$$

The neighbouring local equilibrium distribution N_o contains 5 free functions $\alpha(x)$, $\beta^\lambda(x) = \beta(x)u^\lambda(x)$ which are largely at our disposal. Exactly as in phenomenological theory, we choose the hydrodynamical velocity $u^\lambda(x)$ arbitrarily, and then fix $\alpha(x)$, $\beta(x)$ by the conditions

$$u_\mu \delta J^\mu = u_\lambda u_\mu \delta T^{\lambda\mu} = 0 \quad . \tag{32}$$

It then follows from (30) and (31) that

$$- u_\mu \delta S^\mu = u_\mu Q^\mu > 0 \ , \tag{33}$$

confirming that equilibrium maximizes the entropy density under the constraints (32) on particle and energy densities (compare (37) of the preceding Chapter), and that $S^\lambda_{(o)}$, $T^{\lambda\mu}_{(o)}$ give the entropy and thermodynamical pressure correctly to first order in deviations.

§4. Linearized Boltzmann Equation

A kinetic approach to nonequilibrium processes has the formidable task of solving Boltzmann's integro-differential equation

$$v^\mu \nabla_\mu N(x,p) = C[N(x,p)]$$

$$C[N(x,p)] = \int W(p,p'; p_*,p_*')(N_* N_*' - NN')d\omega' d\omega_* d\omega_*' \tag{34}$$

for the distribution function

$$N(x,p) = N_o(x,p)(1 + f(x,p)) \ . \tag{35}$$

Once f is known, it is straightforward to relate the deviations δJ^λ, $\delta T^{\lambda\mu}$ given by (29), which encapsulate heat flux and viscous stresses, to gradients $\partial_\lambda \alpha$, $\nabla_{(\mu}\beta_{\lambda)}$ and thus arrive at phenomenological laws.

For small deviations, the collision term may be linearized in f to obtain the linearized Boltzmann equation

$$\mathcal{L}[f(x,p)] = - v^\mu[\nabla_\mu f(x,p) + \partial_\mu \alpha + p^\lambda \nabla_{(\mu}\beta_{\lambda)}] = - N_o^{-1} v^\mu \nabla_\mu N \tag{36}$$

where the linear integral operator $\mathcal{L}$ is defined by (recalling the notation (20))

$$\mathcal{L}[f(x,p)] \equiv \int W(p,p'; p_*,p_*')N_o' \delta[f] d\omega' d\omega_* d\omega_*' \quad . \tag{37}$$

It is convenient to set up a Hilbert space for the operator $\mathcal{L}$. For any pair of functions F(x,p), G(x,p) on phase space, define the inner product

$$(F,G) \equiv \int N_o(x,p)F(x,p)\mathcal{L}[G(x,p)]d\omega = \frac{1}{4}\int W \, N_o N_o' \delta[F]\delta[G]d^4\omega \tag{38}$$

where we have deployed the general identity (20). It follows at once that

$$(F,G) = (G,F) \ , \qquad (F,F) > 0 \ . \tag{39}$$

Equation (23) for the entropy production may now be re-expressed as

$$\nabla_\mu S^\mu = (f,f) > 0 \ . \tag{40}$$

The inhomogeneous equation (36) is to be solved for f(x,p) with $\partial_\mu \alpha$, $\nabla_{(\mu}\beta_{\lambda)}$ assumed given. Associated with the fact that f=1, p^λ are 5 solutions of the associated homogeneous equation, the left-hand side satisfies 5 identities (orthogonality relations)

$$\int N_o \mathcal{L}[f]d\omega = (1,f) = 0 \quad ; \qquad \int N_o p^\lambda \mathcal{L}[f]d\omega = [p^\lambda, f] = 0$$

for arbitrary functions f(x,p). This imposes 5 compatibility conditions on the inhomogeneous part of the right-hand side of (36). These are automatically satisfied in consequence of the conservation identities $\nabla_\mu J^\mu = \nabla_\mu T^{\lambda\mu} = 0$.

§5. Structure of the Solution. Chapman-Enskog and Grad approximations.

To discern the general form of the solution of (36), we note that $\beta^\mu = \beta u^\mu$ is the only vector involved in the integral operator $\mathcal{L}$ as a parameter. Hence the effect of $\mathcal{L}$ on various functions of p^λ must be as follows:

$$\mathcal{L}[F(\gamma)] = F_1 \quad ; \qquad \mathcal{L}[F(\gamma)p^\lambda] = F_2 p^\lambda + F_3 u^\lambda$$

$$\mathcal{L}[F(\gamma)p^\kappa p^\lambda] = F_4 p^\kappa p^\lambda + 2F_5 u^{(\kappa}p^{\lambda)} + F_6 u^\kappa u^\lambda + F_7 \Delta^{\kappa\lambda} \quad . \tag{41}$$

Here $\gamma = -v_\mu u^\mu = (1-\vec{v}^2)^{-1/2}$ is the Lorentz dilation factor associated with momentum $p_\lambda = mv_\lambda$ in the fluid's rest-frame, $F(\gamma)$ is an arbitrary function and $F_i(\beta,\gamma)$ are seven undetermined scalar functions. From the linearity of $\mathcal{L}$ it then follows that

$$\mathcal{L}[F(\gamma)\Delta^\lambda_\mu p^\mu] = F_2(\beta,\gamma)\Delta^\lambda_\mu p^\mu \quad , \qquad \mathcal{L}[F(\gamma)\langle p^\kappa p^\lambda\rangle = F_4(\beta,\gamma)\langle p^\kappa p^\lambda\rangle \quad . \tag{42}$$

(The angular brackets, as in (48) of Chapter II, denote the purely spatial, trace-free part.)

We attempt to resolve the right-hand side of (36) into pieces resembling the right-hand side of (42) by splitting v^λ into its spatial and time-components:

$$v^\lambda = \vec{v}^\lambda + \gamma(\vec{v})u^\lambda \quad , \qquad \vec{v}^\lambda \equiv \Delta^\lambda_\mu v^\mu \tag{43}$$

The result is

$$\mathcal{L}[f] + v^\mu \nabla_\mu f(x,p) = \gamma\dot{\alpha} - m\gamma^2\dot{\beta} + \frac{1}{3}m\beta(\gamma^2 - 1)\nabla_\mu u^\mu$$

$$+ \vec{v}^\mu[\partial_\mu\alpha - m\gamma(\partial_\mu\beta - \beta\dot{u}_\mu)] + \langle v^\lambda v^\mu\rangle m\beta\langle\nabla_\mu u_\lambda\rangle \tag{44}$$

where $\dot{\Phi} \equiv u^\mu \partial_\mu \Phi$.

The easiest way to proceed now is to follow Chapman and Enskog: assume that deviations from local equilibrium are <u>quasistationary</u>, so that in lowest approximation we can neglect $\nabla_\mu f$. To the same approximation, $\dot{\alpha}$ and $\dot{\beta}$ are proportional to $\nabla_\mu u^\mu$, and the vectors $\partial_\mu\alpha$ and $(\partial_\mu\beta - \beta\dot{u}_\mu)$ are parallel, as one readily verifies from the conservation laws and the thermodynamical identity (55) of Chapter II. The general form of the solution is now apparent from (42):

$$f(x,p) = A_o(\beta,\gamma)(\nabla_\mu u^\mu) + A_1(\beta,\gamma)\vec{v}^\lambda(\partial_\lambda\alpha) + A_2(\beta,\gamma)\langle v^\lambda v^\mu\rangle\langle\nabla_\mu u_\lambda\rangle + f_{hom}(x,p) \tag{45}$$

Here, $f_{hom}(x,p) = C_\mu(x)p^\mu + C_5(x)$ is the general solution of $\mathcal{L}[f]=0$; its arbitrariness merely reflects the arbitrariness of the way that we choose to fit N_o to N in (35).

Introducing (45) into (29), we obtain the desired relation between the thermoviscous forces $\delta T^{\lambda\mu}$, δJ^{λ} and thermodynamical gradients. Their irreducible pieces π, q^{λ}, $\pi_{\lambda\mu}$ come out, unsurprisingly, proportional to $\nabla_{\mu}u^{\mu}$, $\Delta^{\lambda\mu}\partial_{\mu}\alpha$, $\langle\nabla_{\mu}u_{\lambda}\rangle$ with coefficients ζ_V, κ, ζ_S proportional to (A_o,A_o), (A_1, A_1), (A_2, A_2) (as can be seen immediately from (40)) and thus are manifestly positive. Since $\nabla_{\mu}f$ was discarded, gradients of heat flux and viscous stress make no appearance in these equations: they are the "standard" Landau-Lifshitz-Eckart phenomenological laws.

The alternative procedure, following Grad, is to enter (36) with the ansatz

$$f(x,p) = b_{\lambda\mu}(x)p^{\lambda}p^{\mu} + f_{hom}(x,p) \tag{46}$$

where we may assume $g^{\lambda\mu}b_{\lambda\mu}=0$, since any trace could be absorbed into f_{hom}. Thus, there are 9 nontrivial undetermined spacetime fields $b_{\lambda\mu}(x)$. These can be placed into linear correspondence, on the one hand, with the 9 thermoviscous stresses $\pi_{\lambda\mu}$, π and q^{λ} by way of (29); and, on the other hand, with $\partial_{\mu}\alpha$ and $\nabla_{\mu}\beta_{\lambda}$ by (36) with deployment of (41).* The result is again a set of phenomenological laws with the difference that $\nabla_{\mu}f$ is not neglected and hence gradients of $\pi_{\lambda\mu}$, π and q^{λ} now enter the equations. We have arrived at the kinetic equivalent of Müller's equations, (79) of Chapter II.

This is about as far as we can pursue the story without entering into a lot of elementary but slightly sordid detail.

ACKNOWLEDGEMENTS

I am indebted to fellow participants of the Noto C.I.M.E. School for interactions and comments which significantly influenced the final form of these lectures; I should like to mention particularly A.M. Anile, B. Carter, M.L. Ekiel-Jezewska, L. Lindblom, A. Majorana, S. Motta, I. Müller, T. Ruggeri and C.G. van Weert.

The considerate arrangements and warm hospitality of the organizers - Professors Roberto Conti, Marcello Anile, Yvonne Choquet-Bruhat and Pietro Zecca and their colleagues - was deeply appreciated by all of us, as was the apt choice of site. During the proceedings, it was a stirring thought that Syracuse, the birthplace and home of the illustrious founder of our subject, Archimedes, was less than 30 km away.

---

*To obtain a definite set of equations for $b_{\lambda\mu}(x)$, one forms second moments of the Boltzmann equation by setting $\Phi=p^{\kappa}p^{\lambda}$ in (13).

REFERENCES

[1] C. Cattaneo (editor), Relativistic Fluid Dynamics. Edizioni Cremonese, Rome 1971.
[2] S. Chandrasekhar, Phys. Rev. Letters 24, 611 (1970).
[3] J.L. Friedman and B.F. Schutz, Astrophys. J. 222, 281 (1978).
[4] C. Cutler and L. Lindblom, Astrophys. J. 314, 234 (1987).
[5] J.L. Friedman, J.R. Ipser and L. Parker, Astrophys. J. 304, 115 (1986).
[6] L.L. Lindblom and W.A. Hiscock, Astrophys. J. 267, 384 (1983).
[7] W.A. Hiscock and L. Lindblom, Ann. Physics 151, 466 (1983). See also their contribution to this volume.
[8] C. Cattaneo, Compt. Rend. Acad. Sci. (Paris) 247, 431 (1958).
[9] J.L. Synge, Relativity: The Special Theory. North-Holland, Amsterdam 1956.
[10] W. Israel, Lettere al Nuovo Cimento 7, 860 (1973).
I. Bailey, Ann. Phys. 119, 76 (1979).
I. Bailey and W. Israel, Ann. Physics 130, 188 (1980).
[11] M.M. Som, M.L. Bedran and E.P. Vasconcellos-Vaidya, Physics Letts. A117, 169 (1986).
[12] V.H. Hamity, Phys. Rev. 187, 1745 (1969).
[13] D. van Dantzig, Nederl. Akad. Wetensch. Proc. 42, 608 (1939).
[14] O. Klein, Rev. Mod. Phys. 21, 531 (1949).
[15] N.G. van Kampen, Phys. Rev. 173, 295 (1968).
[16] W. Israel and J.M. Stewart, in General Relativity and Gravitation (ed. A. Held), Vol. 2, p. 491. Plenum, New York 1980.
[17] W. Israel, in Proceedings of the Third Moscow Seminar on Quantum Gravity (ed. M.A. Markov, V.A. Berezin and V.P. Frolov), p. 265. World Scientific, Singapore 1985.
[18] C.W. Bernard, Phys. Rev. D9, 3312 (1974).
[19] K. Chou, Z. Su, B. Hao and L. Yu, Physics Reports 118, 1 (1985).
A.J. Niemi and G.W. Semenoff, Ann. Physics 152, 105 (1984).
E. Calzetta and B.L. Hu, Phys. Rev. D35, 495 (1987).
R.D. Jordan, Phys. Rev. D33, 444 (1986).
A.J. Niemi, Physics Letts. 203B, 425 (1988).
[20] T. Arimitsu, M. Guida and H. Umezawa, Physica 148A, 1 (1988).
M. Ban and T. Arimitsu, Physica 146A, 89 (1987).
H. Matsumoto, Z. Physik C34, 335 (1987).
[21] O. Eboli, R. Jackiw and S.-Y. Pi, Phys. Rev. D (in press, 1988).
[22] L. van Hove, Physics Reports 137, 11 (1986).
[23] N.P. Landsman and Ch. G. van Weert, Physics Reports 145, 141 (1987).
[24] W.G. Dixon, Arch. Ration. Mech. Anal. 80, 159 (1982).
W. Israel, Physics Letts. 86A, 79 (1981).
[25] W. Israel, J. Non-Equilib. Thermodyn. 11, 295 (1986).
[26] T.L. Ho and N.D. Mermin, Phys. Rev. B21, 5190 (1980).
[27] D.A. Kirzhnitz, Soviet Physics Uspekhi 9, 692 (1967).
[28] M. Ichiyanagi, Progr. Theor. Phys. 66, 1 (1981).
T. Toyoda, Ann. Physics 141, 154 (1982).
[29] M.A.B. Beg and R.C. Furlong, Phys. Rev. D31, 1370 (1985).
[30] S. Coleman and E. Weinberg, Phys. Rev. D7, 1888 (1973).
[31] A.B. Migdal, Rev. Mod. Phys. 50, 107 (1978).
[32] H.E. Haber and H.A. Weldon, Phys. Rev. D25, 502 (1982).
J.I. Kapusta, Phys. Rev. D24, 426 (1981).
E.J. Ferrer, V. de la Incera and A.E. Shabad, Nuovo Cimento 98A, 245 (1987).
[33] D. Pines and M.A. Alpar, Nature 316, 27 (1985).
G.E. Volovik, Soviet Physics Uspekhi 27, 363 (1984).
E.B. Sonin, Rev. Mod. Phys. 59, 87 (1987).
D.M. Sedrakyan, K.M. Shakhabasyan and A.G. Movsisyan, Astrofizika 19, 175 (1983).

[34] J. Casas-Vásquez, D. Jou and G. Lebon (editors), Recent Developments in Non-equilibrium Thermodynamics. Springer Verlag, Berlin 1984.
[35] I.-S. Liu, I. Müller and T. Ruggeri, Ann. Physics 169, 191 (1986).
[36] C. Eckart, Phys. Rev. 58, 919 (1940).
[37] L.D. Landau and E.M. Lifshitz, Fluid Mechanics, §127. Pergamon Press, London 1958.
[38] N.G. van Kampen, J. Stat. Phys. 46, 709 (1987).
[39] W.A. Hiscock and L. Lindblom, Phys. Rev. D31, 725 (1985).
[40] C. Cattaneo, Sulla Conduzione del Calore. Atti del Seminario Matematico e Fisico della Universitá di Modena 3 (1948).
[41] I. Müller, Zeitschr. f. Physik 198, 329 (1967).
[42] J. Meixner, Ann. Physik 39, 333 (1941).
I. Prigogine, Physica 15, 272 (1949).
[43] C.G. van Weert, Physica 111A, 537 (1982); Ann. Physics 149, 133 (1982).
M. Ekiel-Jezewska and L.A. Turski, Preprint, Institute for Theoretical Physics, Polish Academy of Sciences 1982; also ref. [34], p. 414.
[44] E.I. Takizawa, Mem. Fac. Engng, Nagoya 5, 1 (1953).
[45] J. Meixner, Zeitschr. f. Physik 139, 30 (1954).
[46] C. Cercignani, Phys. Rev. Letters 50, 1122 (1983).
M. Dudyński and M.L. Ekiel-Jezewska, Phys. Rev. Letters 55, 2831 (1985).
[47] E.g., H.G. Reik, Zeitschr. f. Physik 148, 156 (1957).
[48] I. Müller, Zeitschr. f. Physik 198, 329 (1967).
[49] W. Israel and J.M. Stewart, Ann. Physics 118, 341 (1979).
[50] M.A. Schweizer, Astrophys. Journ. 258, 798 (1982).
D. Jou and D. Pavon, Astrophys. Journ. 291, 447 (1985).
[51] V.A. Belinskii, E.J. Nikomarov and I.M. Khalatnikov, Soviet Physics JETP 50, 213 (1979).
[52] H. Grad, Commun. Pure Appl. Math. 5, 257 (1952).
[53] A.M. Anile and A. Majorana, Meccanica 16, 149 (1981).
A. Majorana and S. Motta, J. Non-Equilib. Thermodyn. 10, 1 (1985).
[54] K.S. Thorne, Astrophys. J. 179, 897 (1973).
P.A. Koch, Phys. Rev. 140A, 1161 (1965).
G.F. Chapline and T.A. Weaver, Phys. Fluids 33, 1884 (1979).
A.M. Anile, J.C. Miller and S. Motta, Phys. Fluids 26, 1450 (1983).
[55] M.I. Sobel, P.J. Siemens, J.P. Bondorf and H.A. Bethe, Nucl. Phys. A251, 502 (1975).
[56] R.D. Blandford and C.F. McKee, Mon. Not. Roy. Astron. Soc. 180, 343 (1977).
[57] D. Gilbarg, Amer. J. Math. 73, 256 (1951).
D. Gilbarg and D. Paolucci, Journ. Rat. Mech. Anal. 2, 617 (1953).
[58] F.S. Sherman and L. Talbot, in Rarefied Gas Dynamics (ed. F.M. Devienne), p. 161. Academic Press, New York 1960.
K. Hansen and D.F. Hornig, Journ. Chem. Phys. 33, 913 (1960).
M.N. Kogan, Rarefied Gas Dynamics, p. 335. Plenum Press, New York 1969.
[59] G.A. Bird, J. Fluid Mech. 30, 479 (1967).
[60] Ya. B. Zel'dovich and Yu. P. Raizer, Physics of Shock Waves and High Temperature Hydrodynamic Phenomena, Vol. 2, p. 476. Academic Press, New York 1967.
[61] J. Serrin, Handbuch der Physik, Vol. 8, Pt. 1, p. 227. Springer, Berlin 1958.
[62] S.R. de Groot, W.A. van Leeuwen and C.G. van Weert, Relativistic Kinetic Theory. North-Holland, Amsterdam 1980.

RELATIVISTIC PLASMAS

Harold Weitzner
Courant Institute of Mathematical Sciences
New York University
251 Mercer Street, New York, New York 10012 USA

I. INTRODUCTION

We shall examine several models that describe relativistic plasmas of some interest for laboratory applications. Our intent is to apply relatively standard techniques of applied mathematics to these different models to show what types of results are more or less readily available. For the most part our focus will be on problem formulation and solution rather than deep analysis of problems of model formulation. The laboratory plasmas most closely related to the problems we treat are charged particle beams and microwave power generators, including free electron lasers. In a brief series of lectures we cannot present detailed studies of any of these complicated devices, but we can give an elementary exploration of models and types of analysis which are possible. Most of these models are characterized by being a representation of a non-charge neutral plasma, while much of the more standard plasma physics relates to charge neutral plasmas. The interested reader should study R.C. Davidson, Theory of Nonneutral Plasmas (Benjamin, Reading, MA, 1974).

We start with the motion of a single particle in given electromagnetic fields and we generate an adiabatic invariant for motion in fields of appropriate structure. We use this invariant in later discussion of the relativistic Vlasov model. We then turn to a trivial extension of single particle motion. We examine a zero temperature, but relativistic plasma. In this model all particles have identical trajectories, but still we find it convenient to view the system as a fluid rather than a collection of single particles. We examine a simple steady beam flow and we contrast the results with the well-known Bennett pinch flow. We then examine the more complex problem of steady helically symmetric steady flows. Such flows are of interest in some free electron lasers.

We next turn to a kinetic model and we present the relativistic Vlasov equation. We characterize one particularly simple plasma beam steady flow which depends on one coordinate only. When we turn to a plasma beam steady state which is only axisymmetric we find that we can generate beam steady flows by the use of our single particle adiabatic invariant. We then use the relativistic Vlasov equation to generate a cool fluid model of a plasma. We examine simple beam steady flows based on this model and we compare them with the cold plasma model, the Bennett pinch model, and the Vlasov-adiabatic model. We discuss very briefly variants of this cool plasma fluid model. Finally we comment on ideal and non-ideal relativistic magnetohydrodynamics.

The notation for relativistic problems is unfortunately far from standard and we apologize for employing our own version. We take $c = 1$ and a rationalized system of units so that in usual coordinates $(\underset{\sim}{x},t) = (x_i,t) = (x^\mu)$ Maxwell's equations are

$$\underset{\sim}{\nabla}\times\underset{\sim}{E} = -\underset{\sim}{B}_{,t} \tag{1}$$

$$\underset{\sim}{\nabla}\times\underset{\sim}{B} = \underset{\sim}{J} + \underset{\sim}{E}_{,t} , \tag{2}$$

together with the subsidiary relations

$$\underset{\sim}{\nabla}\cdot\underset{\sim}{E} = Q \tag{3}$$

$$\nabla\cdot B = 0 , \tag{4}$$

where Q is the charge density, $\underset{\sim}{J}$ is the current density, and

$$Q_{,t} + \underset{\sim}{\nabla}\cdot\underset{\sim}{J} = 0 \tag{5}$$

must hold. We introduce the usual metric $g_{\mu\nu}$ such that the nonzero elements of $g_{\mu\nu}$ are $g_{11} = g_{22} = g_{33} = -g_{44} = 1$, and $g^{\mu\nu}$ is defined such that $g_{\mu\nu}g^{\nu\lambda} = \delta^\lambda_\mu$. The electromagnetic field tensor $F_{\mu\nu}$ is antisymmetric and $E_i = E_{i4} = -F_{4i}$, $i=1,2,3$ and $B_1 = F_{23} = -F_{32}$, $B_2 = F_{31} = -F_{13}$, and $B_3 = F_{12} = -F_{21}$. For an ordinary velocity $\underset{\sim}{v} = (v_i)$, we define $\gamma^{-2} = 1 - v_i^2$ and the associated four velocity is $u^\mu = (\gamma v_i,\gamma)$ and $u^\mu u_\mu = -1$. Although we have not written Maxwell's equations in an explicity covariant form, it is well-known that they are covariant and we do not use the explicity covariant forms. The Lorentz force on a particle of charge e is $e(\underset{\sim}{E} + \underset{\sim}{v}\times\underset{\sim}{B})_i = eF^{i\mu}u_\mu/\gamma$. Finally, we denote the rest mass, the only mass we introduce, by m.

We may always introduce a vector potential $\underset{\sim}{A}$ such that

$$\underset{\sim}{B} = \underset{\sim}{\nabla}\times\underset{\sim}{A} , \tag{6}$$

and then

$$\underset{\sim}{E} = -\underset{\sim}{\nabla}\phi + \underset{\sim}{A}_{,t} . \tag{7}$$

The potentials may be modified by the introduction of a gauge transformation

$\underset{\sim}{A} \rightarrow \underset{\sim}{A} + \underset{\sim}{\nabla}\lambda$, $\phi \rightarrow \phi + \lambda_{,t}$, although we can usually identify easily what gauge we desire.

II. SINGLE PARTICLE MOTION

We may write the relativistic extension of Newton's second law as

$$(\gamma \underset{\sim}{v})_{,t} = \underset{\sim}{u}_{,t} = (e/m)(\underset{\sim}{E} + \underset{\sim}{v} \times \underset{\sim}{B}) \ . \tag{8}$$

Again, although the form is not explicitly covariant, the equation is well-known to be covariant. When we dot $\underset{\sim}{v}$ into (8) we find conservation of energy in the usual form

$$\gamma_{,t} = (e/m)\underset{\sim}{E} \cdot \underset{\sim}{v} \ , \tag{9}$$

and we could introduce a momentum energy four vector (u^i, γ), just as one expects. We should add that we do not emphasize covariant forms since in the solution of specific problems, it is highly desirable to reduce the results to the more explicit non-covariant variables. We may recast the single particle motion in Hamiltonian formalism with the definition

$$H(p^i, x^i, t) = [(p^i - eA^i)^2 + m^2]^{1/2} + e\phi \tag{10}$$

and Hamilton's equations

$$\dot{x}^i = H_{,p^i}$$

$$\dot{p}^i = -H_{,x^i}$$

are easily shown to be equivalent to (8) with the identification

$$\underset{\sim}{u} = (\underset{\sim}{p} - e\underset{\sim}{A})/m \ . \tag{11}$$

We may readily generate canonical transformations from the variational form

$$p^i dx^i - H\,dt = \overline{p}^i d\overline{q}^i - \overline{H}\,dt + dW \ , \tag{12}$$

where W is the generator of the transformation. We can transform easily from x,y,z coordinates to cylindrical coordinates $r, \theta, \overline{z}$ and canonical momenta $p_r, p_\theta, \overline{p}_z$

with the generator $W = -rp_r - \theta p_\theta - \bar{z}\,\bar{p}_z + (x^2+y^2)^{1/2}p_r + p_\theta \tan^{-1}(y/z) + z\bar{p}_z$ so that $r = (x^2+y^2)^{1/2}$, $\theta = \tan^{-1} y/x$, $\bar{z} = z$, $p_x = p_r \cos\theta - (p_\theta/r)\sin\theta$, $p_y = p_r \sin\theta + (p_\theta/r)\cos\theta$, $\bar{p}_z = p_z$, and

$$H(p_r,p_\theta,p_z,r,\theta,z,t) = \{[(p_r-eA_r]^2 + (\frac{p_\theta}{r} - eA_\theta)^2 + (p_z-eA_z)^2] + m^2\}^{1/2} + e\phi \quad (13)$$

We can now construct a relativistic adiabatic invariant. We consider an electromagnetic field with axial symmetry, so that the potentials are independent of θ. We then infer, since $H,_\theta = 0$ that p_θ is a constant and the Hamiltonian is a function of p_r, p_z, r, z, and t only. In many cases we are interested in fields and in motions that vary slowly with the coordinate z and with the time t. Specifically, we assume $\underset{\sim}{A} = \underset{\sim}{A}(r,\varepsilon z,\varepsilon t)$ and $\phi = \phi(r,\varepsilon z,\varepsilon t)$. For such fields $B_r = -\varepsilon A_{\theta,\varepsilon z}$, $B_\theta = \varepsilon A_{r,\varepsilon z} - A_{z,r}$, $B_z = (rA_\theta),_r/r$. Thus, the magnetic field is mainly in the θ and z directions with a small r component. In axisymmetric problems we may always make a gauge transformation and assume $A_r \equiv 0$. We choose this gauge throughout these notes.

C.S. Gardner (Phys. Rev. 115, 791, (1959)) has laid out a general theory of adiabatic invariants of Hamiltonian systems based on the idea of repeated canonical transformations. We have cast the motion of a relativistic particle in Hamiltonian form in order to take advantage of this analysis. We have extended this type of analysis in Phys. Fluids 24, 2280 (1981) and we may readily conclude that the quantity

$$I_r^o = \oint p_r dr \quad (14)$$

is the first term in the asymptotic expansion of an adiabatic invariant. In (14) the integration is carried out with H, p_θ, p_z, εz, and εt all held constant, and thus $I_r^o = I_r^o(H,p_\theta,p_z,\varepsilon z,\varepsilon t)$. Consistent with the usual theory of adiabatic invariants I_r^o is the first term in a series I_r^n such that

$$\frac{d}{dt}\sum_{n=0}^{N} I_r^n \varepsilon^n = O(\varepsilon^{N+1}) \ . \quad (15)$$

We do not carry out all the details of the proof, but we sketch the argument. We work first in the r,p_r phase plane and we hold εz, p_z, and εt fixed. We assume that the curves H = const. are closed and we introduce an area preserving transformation such that $dr dp_r = dI_r^o d\chi^o$. Such a transformation always exists and has an explicit representation. Since the transformation is area preserving, there is a generator such that

$$p_r dr = I_r^o d\chi^o + dW(\chi^o, r) .$$

Since, we have held p_z, εz, and εt fixed we conclude that $W = W(\chi^o, r, p_z, \varepsilon z, \varepsilon t)$. We may now introduce a canonical transformation in the full space

$$p_r dr + p_z dz - H\, at = I_r^o d\chi^o + \overline{p}_z d\overline{z} - \overline{H}\, at + d\left(W^o(\chi^o, r, \overline{p}_z, \varepsilon z, \varepsilon t) + \overline{p}_z(z-\overline{z})\right) .$$

It is easy to see that

$$p_r = \frac{\partial W^o}{\partial r}$$

$$I_r^o = -\frac{\partial W^o}{\partial \chi^o}$$

$$p_z = \overline{p}_z = \varepsilon \frac{\partial W^o}{\partial \varepsilon z}$$

$$\varepsilon\overline{z} = \varepsilon z + \varepsilon \frac{\partial W^o}{\partial \overline{p}_z}$$

$$\overline{H} = H + \varepsilon \frac{\partial W}{\partial \varepsilon t} .$$

Since $\overline{H} = H(I_r^o, \overline{p}_z, \varepsilon\overline{z}, \varepsilon t) + O(\varepsilon)$, we see that

$$\dot{I}_r^o = O(\varepsilon)$$

$$\dot{\overline{p}}_z = O(\varepsilon)$$

$$\varepsilon\dot{\overline{z}} = O(\varepsilon)$$

If we repeat the transformation on I_r^o and χ^o we find

$$(\dot{I}_r^o + \varepsilon\dot{I}_r') = O(\varepsilon^2)$$

$$\dot{\bar{\bar{p}}}z = O(\varepsilon)$$

$$\dot{\bar{\bar{\varepsilon\zeta}}} = O(\varepsilon) \ .$$

After N stops we recover the adiabatic invariant series in (15).

Instead of this invariant one could also derive the extension of the non-relativistic magnetic moment adiabatic invariant to the relativistic case. One must require that relativistic larmor radius is much smaller than the scale length of the electromagnetic field variations and also that in appropriate units $\underset{\sim}{E}\cdot\underset{\sim}{B}$ must be small. Other derivations of the relativistic magnetic moment have been given, but Gardner's procedure would apply directly.

III. COLD RELATIVISTIC PLASMA

We now consider a relativistic fluid with zero pressure. We introduce a localized invariant particle number density n, flow velocity $\underset{\sim}{v}$, and corresponding flow reduced momentum $\underset{\sim}{u}$. Conservation of mass becomes

$$(n\gamma)_{,t} + \nabla\cdot n\underset{\sim}{u} = 0 \ , \tag{16}$$

or

$$\partial_\mu P^\mu = 0 \tag{17}$$

where $P^\mu = nu^\mu$. For a cold fluid $\gamma^2 = 1 + \underset{\sim}{u}\cdot\underset{\sim}{u} = 1/(1-\underset{\sim}{v}\cdot\underset{\sim}{v})$ and hence

$$P^\mu P_\mu = -n^2 \tag{18}$$

Conservation of momentum is equivalent to the motion of a single particle, or

$$\underset{\sim}{u}_{,t} + (\underset{\sim}{v}\cdot\underset{\sim}{\nabla})\underset{\sim}{u} = (e/m)(\underset{\sim}{E} + \underset{\sim}{v}\times\underset{\sim}{B}) \tag{19}$$

In a more symmetrical form

$$n\gamma\underset{\sim}{u}_{,t} + n(\underset{\sim}{u}\cdot\underset{\sim}{\nabla})\underset{\sim}{u} = (ne/m)(\underset{\sim}{E}\gamma + \underset{\sim}{u}\times\underset{\sim}{B}) \ , \tag{20}$$

or in conservation form

$$\partial_\mu(P^\mu\underset{\sim}{u}) = (ne/m)(\underset{\sim}{E}\gamma + \underset{\sim}{u}\times\underset{\sim}{B}) \ . \tag{21}$$

Conservation of energy is a consequence of (19), and is obtained by dotting u^i into (19) and one finds

$$\gamma_{,t} + (\underset{\sim}{v}\cdot\underset{\sim}{\nabla})\gamma = (e/m)\underset{\sim}{E}\cdot\underset{\sim}{v} \,, \tag{22}$$

or in conservation form

$$\partial_\mu(P^\mu\gamma) = (e/m)n\underset{\sim}{E}\cdot\underset{\sim}{u} \,. \tag{23}$$

Clearly (21) and (23) can be combined in a manifestly covariant form although (20) or (21) is adequate in a cold fluid without the additional equation of energy conservation. The cold fluid model is coupled to Maxwell's equations through the current relations

$$Q = \sum_s e_s n_s \gamma_s \tag{24}$$

$$\underset{\sim}{J} = \sum_s e_s n_s \gamma_s \underset{\sim}{v}_s \,, \tag{25}$$

where the subscript s is an index of the species, electron, or various ions. The system of Maxwell's equations, and fluid equations for each species of conservation of mass (16) or (17), conservation of momentum (19), (20), or (21), and the relations (24), (25) constitutes the cold relativistic plasma model. The model uses both fluid and single particle properties to characterize the system.

A particularly simple example of a cold plasma steady flow is a flow with all quantities, when represented in cylindrical coordinates, functions of r only. For each species we take $\underset{\sim}{u} = (0, v(r), w(r))$, $n = n(r)$, $\gamma^2(r) = 1 + [v(r)]^2 + [w(r)]^2$, while for the electromagnetic fields, $\underset{\sim}{B} = (0, B_\theta(r), B_z(r))$ and $\underset{\sim}{E} = (E_r(r), 0, 0)$. It is trivial to obtain that for each species

$$-v^2/r = (e/m)(E_r\gamma + vB_z - wB_\theta) \tag{26}$$

and for the electromagnetic fields

$$(rEr)_{,r}/r = \sum_s (n_s\gamma_s e_s) \tag{27}$$

$$B_{z,r} = -\sum_s (n_s v_s e_s) \tag{28}$$

$$(rB_\theta),r/r = \sum_s (n_s w_s e_s) \,. \tag{29}$$

Clearly many steady solutions are possible as each species of particle has three unknowns n(r), v(r), and w(r) constrained by one relation (26), while the electromagnetic fields are then specified by (27), (28), and (29) and the value of B_z at r = 0.

Despite the richness of the solution space one particular example has been studied repeatedly, and we include it here. The Bennett pinch consists of the beam flow of electrons and ions in which the ions are assumed at rest and momentum balance for the ions is ignored. The electron flow is in the z direction only, so that $v_e = 0$, and the ion number density n_i is set equal to $\gamma_e n_e f(r)$, where f(r) is the so called neutralization fraction. In this approximation

$$w_e B_\theta = E_r \gamma_e \tag{30}$$

$$(rEr)_{,r}/r = en_e\gamma_e(f(r) - 1) \tag{31}$$

$$(rB_\theta)_{,r}/r = -en_e w_e \ . \tag{32}$$

A direct consequence of this system is the relation

$$\frac{d \log \gamma_e}{d \log (rB_\theta)} = (1 - f\gamma_e^2) \ , \tag{33}$$

a functional relationship between γ_e, f, and (rB_θ) which is independent of r. For the even more special case in which w_e and γ_e are constant we conclude

$$f = \gamma_e^{-2} \tag{34}$$

and $n_e(r)$ is entirely arbitrary. Thus, given any constant values of f, w_e, and γ_e satisfying (34) and any number density $n_e(r)$, a steady flow exists.

Although it is somewhat out of logical order it is reasonable to comment here on the warm version of the Bennett pinch, see W.H. Bennett, Phys. Rev. 45, 890 (1934) and R.C. Davidson, Theory of Nonneutral Plasmas (Benjamin, Reading, MA, 1974). We do not modify the ions but we assume that there is an isothermal pressure force $[\nabla(n_e(r)\kappa T_e)]/n_e(r)$ added to the left hand side of (26). It is straightforward to verify that provided

$$f > \gamma_e^{-2} \tag{35}$$

there is a steady flow with γ_e and f constant but $n_e(r)$ is no longer arbitrary and

$$n(r) = n(0)/(1 + r^2/r_B^2)^2 \tag{36}$$

where r_B is the Bennett radius, $r_B^2 = 8kT_e/[n_e e^2(f-\gamma_e^{-2})]$. The addition of a simple ad hoc thermal force appears to modify the beam flow dramatically as a profile is uniquely specified by (36) and the equality (34) is replaced by the inequality. If the square of the beam radius is to be much larger than $8kTe\gamma_e^2/(n_e e^2)$, the electron Debeye distance squared, then one must again be close to satisfying (34) rather than merely (35). We shall return to examine the relevance of the two models when we consider the Vlasov model and a cool plasma model.

We now return to our cold relativistic plasma model (16) and (20) together with the definitions (24), (25) and Maxwell's equations (1)-(4). One type of free electron laser (FEL) of some interest, see T.C. Marshall, Free Electron Lasers (Macmillan, New York, 1985), consists of a steady helically symmetric flow of electrons. This flow is often unstable and so a small electromagnetic signal applied at one point of the system will travel down the laser growing as it travels. Thus, one may amplify electromagnetic waves or use such a device to generate microwaves. Many types of FEL's are possible and we concentrate on FEL's with relatively high density and large radius. Such devices are usually described as non-tenuous and thick. We apply standard fluid dynamic techniques to characterize steady helically symmetric electron flows.

In helical symmetry all unknowns, when written in cylindrical coordinates, are functions of r and $\phi = \theta - kz$ only. The quantity $2\pi/k$ is the helical wavelength. We introduce the notation $\underset{\sim}{u} = (u,v,w)$. We may satisfy two of the four Maxwell equations in steady state by the introduction of an electrostatic potential

$$-(e/m)\underset{\sim}{E} = \underset{\sim}{\nabla}\Phi \ , \tag{37}$$

and a magnetic flux function $\psi(r,\phi)$ such that

$$-(e/m)B_r = -\psi,_\phi/r \tag{38}$$

$$-(e/m)(B_\theta - krB_z) = \psi,_r \ . \tag{39}$$

We have introduced the factor $-(e/m)$ to simplify the notation. We must finally satisfy (2) and (3) of the Maxwell system. Conservation of mass allows us to introduce a stream function $\chi(r,\phi)$ such that

$$(e^2/m)nu = -\chi,_\phi/r \tag{40}$$

$$(e^2/m)(nv - nkrw) = \chi,_r \ . \tag{41}$$

The $\hat{r}$ and $\hat{\theta}-kr\hat{z}$ components of (2) imply

$$-(e/m)(B_z + krB_\theta) = \chi \tag{42}$$

so that

$$-(e/m)B_\theta = (kr\chi + \psi_{,r})/(1 + k^2r^2) \tag{43}$$

$$-(e/m)B_z = (\chi - kr\psi_{,r})/(1 + k^2r^2) , \tag{44}$$

and (38), (43), (44) give $\underset{\sim}{B}$ in terms of χ and ψ. Only one component of (2) remains, and

$$\Delta^*\psi \equiv [r\psi_{,r}/(1+k^2r^2)]_{,r}/r + \psi_{,\phi\phi}/r^2$$

$$= -(e^2/m)nw - kr\chi_{,r}(1+k^2r^2) - 2k\chi/(1+k^2r^2)^2 . \tag{45}$$

Finally, the remaining equation of the Maxwell system is

$$(r\Phi_{,r})_{,r}/r + \Phi_{,\phi\phi}(1+k^2r^2)/r^2 = (e^2/m)n\Upsilon . \tag{46}$$

We now turn to the analysis of momentum conservation (19). Conservation of energy, (22), is one component of (19) and we infer that there is a function $E(\chi)$ such that

$$\Upsilon - \Phi = E(\chi) . \tag{47}$$

Further, if we take the component of (19) in the direction $kr\hat{\theta} + \hat{z}$ we find easily that there is a function $F(\chi)$ such that

$$krv + w + \psi = F(\chi) . \tag{48}$$

Thus, we may complete the representations (40), (41) with

$$(e^2/m)nv = [\chi_{,r} + (e^2/m)nkr(F-\psi)]/(1+k^2r^2) \tag{49}$$

$$(e^2/m)nw = [(e^2/m)n(F-\psi) - kr\chi_{,r}]/(1+k^2r^2) \tag{50}$$

Finally from (40), (47), (49) and (50)

$$(ne^2/m)^2 = |\underset{\sim}{\nabla}\chi|^2/\{(1+k^2r^2)[(E+\Phi)^2 - 1] - (F-\psi)^2\} \tag{51}$$

where

$$|\underset{\sim}{\nabla}\lambda|^2 = \lambda_{,r}^2 + \lambda_{,\phi}^2(1+k^2r^2)/r^2 \tag{52}$$

and (45), (46) become

$$\Delta^*\psi = (ne^2/m)[\psi - F(\chi)]/(1+k^2r^2) - 2k\chi/(1+k^2r^2)^2 \tag{53}$$

$$(r\Phi_{,r})_{,r}/r + \Phi_{,\phi\phi}(1+k^2r^2)/r^2 = (ne^2/m)(E(\chi) + \Phi) \ . \tag{54}$$

When we obtain an equation which characterizes χ (51), (53) and (54) will constitute a closed system. In view of the identity

$$0 = (n\underset{\sim}{u}\cdot\underset{\sim}{\nabla})(\underset{\sim}{u}\cdot\underset{\sim}{\nabla}\chi) = \underset{\sim}{u}\cdot(n\underset{\sim}{u}\cdot\underset{\sim}{\nabla})\underset{\sim}{\nabla}\chi + \underset{\sim}{\nabla}\chi\cdot[(n\underset{\sim}{u}\cdot\nabla)\underset{\sim}{u}]$$

so that in terms of the nonlinear operator L we obtain from (19)

$$L\chi \equiv \chi_{,r}^2\chi_{,\phi\phi} - 2\chi_{,r}\chi_{,\phi}\chi_{,r\phi} + \chi_{,\phi}^2\chi_{,rr} \tag{55}$$

$$\begin{aligned} L\chi = &- \{\chi_{,r}/[r(1+k^2r^2)]\}[\chi_{,\phi}^2 + k^2r^4(e^2n/m)^2(\psi-F)^2/(1+k^2r^2)] \\ &- (ne^2/m)^2[\phi+E(\chi)]r^2\underset{\sim}{\nabla}\chi\cdot\underset{\sim}{\nabla}\Phi + (rne^2/m)^2[\psi-F(\chi)]\underset{\sim}{\nabla}\chi\cdot\underset{\sim}{\nabla}\psi/(1+k^2r^2) \\ &+ (\underset{\sim}{\nabla}\chi)^2((ne^2/m)\{\chi + 2k[\psi-F(\chi)]/(1-k^2r^2)\}/(1+k^2r^2) - r\chi_{,r}/(1+k^2r^2)^2) \ . \end{aligned} \tag{56}$$

The system (53)-(56) together with the definition (51) characterizes cold relativistic helically symmetric flows.

In a paper with A. Fruchtman and P. Amendt, Phys. Fluids 30, 539 (1987) we have presented two types of perturbation expansions of this system. Here, we will consider another type of expansion. The system (53)-(56) consists of two quasilinear elliptic equations, plus one quasilinear hyperbolic equation. The real characteristics are the streamlines χ = const. and in many problems of interest these streamlines close on themselves. Periodicity on the closed characteristics raises fundamental questions of mathematical interest in the appropriate data to specify in order guarantee the existence and uniqueness of a solution. Resonances on the closed characteristics also are of particular physical interest as they are associated with instabilities of flows and wave amplification, see "A Thick Beam Free Electron Laser", by A. Fruchtman, to be published in Phyics of Fluids, 1987. The resonances also affect the steady flow perturbation expansions contained in the first reference of this paragraph and in our subsequent analysis.

In many cases one is interested in flows in which the variation in the radial direction is much faster than the variation in the z direction. We may easily obtain such flows by an expansion in the small parameter k. If we require that the flow be largely in the z direction, then we introduce scaled variables such that

$$\chi = \chi_o + k\chi_1(r) + k^2\chi_2(r,\phi) + O(k^3) \tag{57a}$$

$$\psi = k\psi_1 + O(k^2) \tag{57b}$$

$$\Phi = k\Phi_1 + O(k^2) \tag{57c}$$

$$ne^2/m = k\nu_1(r) + k^2\nu_2(r,\phi) + O(k^3) \tag{57d}$$

$$E(\chi) = E_o(\chi_1) + k\big(E_o'(\chi_1)\chi_2 + E_1(\chi_1)\big) + O(k^2) \tag{57e}$$

$$F(\chi) = F_o(\chi_1) + k\big(F_o'(\chi_1)\chi_2 + F_1(\chi_1)\big) + O(k^2) \tag{57f}$$

In lowest order (56) reduces to

$$\chi_{1,r} = r\nu_1(r)\chi_o \tag{58}$$

so that from (51)

$$\chi_o^2 r^2 = E_o(\chi_1)^2 - 1 - F_o^2(\chi_1) \tag{59}$$

In next order the electromagnetic fields are specified by the relations

$$\psi_1 = -\frac{1}{2}r^2 - \int_o^r dr' \, \tilde{F}(\chi_1(r'))/r' + \tilde{\psi}_1(r,\phi) , \tag{60}$$

where

$$\tilde{F}_{,\chi_1} = F_o(\chi_1) , \quad \tilde{F}(0) = 0 , \quad \text{and} \quad \Delta\tilde{\psi}_1 = 0$$

while

$$\Phi_1 = -\int_o^r dr' \, \tilde{E}(\chi_1(r'))/r' + \tilde{\Phi}_1(r,\phi) \tag{61}$$

where

$$\tilde{E}_{,\chi_1} = E_o(\chi_1) , \quad \tilde{F}(0) = 0 , \quad \text{and} \quad \Delta\tilde{\Phi}_1 = 0 .$$

The stream function χ_2 is now given by

$$\chi_{2,\phi\phi} + \chi_2 = - r\chi_o\nu_1(E_o(\chi_1)\Phi_{1,r} + F_o(\chi_1)\psi_{1,r})$$
$$+ r^2\nu_1(\chi_1 - 2F_o(\chi_1)) + \nu_1[F_o(F_1-\psi_1) - E_o(E_1+\Phi_1)]/\chi_o \, . \qquad (62)$$

If any of the terms on the right hand side of (62) are proportional to cos ϕ or sin ϕ then (62) has no periodic solutions. Such terms may come from the homogeneous solutions of Laplace's equation added to (60) and (61). If any terms with explicit ϕ dependence occur in ψ_1 or Φ_1, then the functions $E_o(\chi_1)$ and $F_o(\chi_1)$ must also be chosen so that no terms proportional to cos ϕ or sin ϕ appear on the right hand side of (62). The amplitude of the terms in $\chi_2(r,\phi)$ which are proportional to cos ϕ or sin ϕ is determined by the consistency of the next order expansion of (56). In next order one must guarantee that the analog of (62) contain no secular terms, i.e., terms proportional to cos ϕ, sin ϕ. This condition will generate a differential equation relating to unknown amplitude of $\chi_2(r,\phi)$ and amplitudes of $\psi_2(r,\phi)$ and $\Phi_2(r,\phi)$. This problem is quite similar to the self-excited helical flow studied in our original FEL reference. The resonance is quite clear and explicit here and unlike the previous expansions which admitted higher harmonic resonances. This variant scaling generates steady flows rather different from those previously studied and it indicates the wealth of phenomena hidden in the system (53)-(56). We shall not explore the questions further here. We remind the reader that the stability of these flows, and hence the utility of these flows as a microwave source or amplifier is studied by a consistent, linearized stability analysis of the system (16), (19), and Maxwell's equations.

IV. THE RELATIVISTIC VLASOV EQUATION

We introduce a particle number density distribution function $f(x^\mu,u^i)$. In an early paper Clemmow and Willson, Proc. Cambridge Philos. Soc. 53, 222 (1957) described the model. It is a simple exercise to verify that the phase space measure $du^1du^2du^3/\gamma(\underset{\sim}{u}) \equiv d\omega$ is invariant under a Lorentz transformation. Further, the Liouville equation consistent with the relativistic form of Newton's laws (8) is

$$f_{,t} + \underset{\sim}{v}\cdot\underset{\sim}{\nabla}\underset{\sim}{f} + (e/m)(\underset{\sim}{E} + \underset{\sim}{v}\times\underset{\sim}{B})\cdot\underset{\sim}{\nabla}_{\underset{\sim}{u}}f = 0 \, . \qquad (63)$$

We may rewrite (63) in the covariant form by multiplying by $\gamma(\underset{\sim}{u})$

$$u^\mu\partial_\mu f + (e/m)F^{i\mu}u_\mu\partial_{u^i}f = 0 \, , \qquad (64)$$

and we then assume that f is a relativistic invariant. The invariance of the form (64) is not immediately obvious as the index i is only summed one to three, but a simple calculation verifies easily its invariance. It is also useful to introduce moments of $f(x^\mu, u^i)$, and we use

$$h = \int f d\omega \tag{65}$$

$$h\langle u^\mu\rangle = \int f u^\mu d\omega \tag{66}$$

$$h\langle u^\mu u^\nu\rangle = \int f\, u^\mu u^\nu d\omega\ , \tag{67}$$

etc., and we also introduce the variances

$$\Theta^{\mu\nu} = \int f(u^\mu - \langle u^\mu\rangle)(u^\nu - \langle u^\nu\rangle)d\omega \tag{68}$$

$$S^{\mu\nu\lambda} = \int f(u^\mu - \langle u^\mu\rangle)(u^\nu - \langle u^\nu\rangle)(u^\lambda - \langle u^\lambda\rangle)d\omega \tag{69}$$

and

$$T^{\mu\nu\lambda\sigma} = \int f(u^\mu - \langle u^\mu\rangle)(u^\nu - \langle u^\nu\rangle)(u^\lambda - \langle u^\lambda\rangle)(u^\sigma - \langle u^\sigma\rangle)d\omega\ . \tag{70}$$

The aim of these notes has been to treat relativistic problems as minor variants of non-relativistic problems. When we come to relativistic kinetic theory, we find a fundamental difference from non-relativistic theory. We are used to considering the moments or variances of a distribution function as essentially arbitrary, subject only to the constraint that there is a non-negative distribution function of which they are moments. However, when we examine (65)-(70) we see that integration is over three reduced momenta but moments corresponding to four reduced momenta appear. Since $u^\mu u^\nu g_{\mu\nu} = -1$ we readily obtain the identities

$$\Theta^{\mu\nu}g_{\mu\nu} = -h(1 + \langle u^\mu\rangle\langle u_\mu\rangle) \tag{71}$$

$$S^{\mu\nu\lambda}g_{\nu\lambda} = -2\Theta^{\mu\nu}\langle u_\nu\rangle \tag{72}$$

$$T^{\mu\nu\lambda\sigma}g_{\lambda\sigma} = -h(1 + \langle u^\lambda\rangle\langle u_\lambda\rangle)\Theta^{\mu\nu} - 2S^{\mu\nu\lambda}\langle u_\lambda\rangle\ . \tag{73}$$

Now (71)-(73) and obvious extensions make clear that the moments of a relativistic kinetic theory are not arbitrary, but they are constrained by specific identities.

Before we discuss (63) further we observe that another series of identities, obtained by multiplication of (63), by 1, u^μ, $u^\mu u^\nu$, $u^\mu u^\nu u^\lambda$, etc., and integration over all reduced velocities, are

$$\partial_\mu(h\langle u^\mu\rangle) = 0 \tag{74}$$

$$\partial_\mu\{h\langle u^\mu\rangle\langle u^\nu\rangle + \Theta^{\mu\nu}\} = (e/m)F^{\nu\mu}h\langle u_\mu\rangle \tag{75}$$

$$\partial_\mu\{h\langle u^\mu\rangle\langle u^\nu\rangle\langle u^\lambda\rangle + \langle u^\mu\rangle\Theta^{\nu\lambda} + \langle u^\nu\rangle\Theta^{\lambda\mu} + \langle u^\lambda\rangle\Theta^{\mu\nu} + S^{\mu\nu\lambda}\}$$

$$= (e/m)\{F^{\nu\mu}\Theta^{\lambda\rho}g_{\mu\rho} + F^{\lambda\mu}\Theta^{\nu\rho}g_{\mu\rho} + F^{\nu\mu}\langle u_\mu\rangle h\langle u^\lambda\rangle + F^{\lambda\mu}\langle u_\mu\rangle h\langle u^\nu\rangle\} . \tag{76}$$

We shall discuss this system much further shortly, but we first return to (63) and (64).

One of the other consequences of (64), of which (74) is a special case, is that for any smooth function F(f)

$$\left(\int \gamma\, F(f)d\omega\right)_{,t} + \nabla\cdot\left(\int \underset{\sim}{v}F(f)d\omega\right) = 0 . \tag{77}$$

If one selects $F(f) = f \log f - f$ then the first term in (77) is usually identified with the entropy density and one says that entropy of the system is conserved. Equally one observes that (63) and (64) are reversible in time. Nevertheless, it is well-known by now that the non-relativistic Vlasov equation admits of irreversible phenomena, most explicitly collisionless shocks. Thus, we expect this system also to admit irreversible phenomena. Physically, when one couples the relativistic Vlasov equation to Maxwell's equations, the electric and magnetic fields in (63) and (64) contain an extremely long range collisional interaction between particles as well as propagating and dynamically independent radiation fields. If there were a satisfactory relativistic version of a Fokker-Planck collision term that could be added to the right hand side of (63) and (64) to represent the shorter range collisions, then one would have a system that manifestly caused entropy to increase. The difficulty with the addition of a collision term is that there are not forms that are generally accepted to be correct and generally relevant. It is likely that a satisfactory kinetic theory with a collision term would also require that the radiation field be treated as an independent ensemble with its own distribution function. The interested reader should turn to the plasma physics literature of the 1960's where a multitude of papers on the topic exist.

We now turn to the solution of a simple problem with this model. The first, and the simplest, cold plasma problem we treated was the characterization of a steady beam flow which depends on r only. It is of some interest to return to this problem here. We consider steady solutions of (63), and solutions which depend on r only. Clearly the characteristics of (63) are the single particle trajectories of Section II. We may then represent a solution of (63) as a function of the constants of motion of the Hamiltonian motion given by (10). If the flows and fields depend on r only, then the Hamiltonian is independent of θ and z, so that p_θ and p_z are constants of the motion. Thus, one might take as a solution of (63) $f(H,p_\theta,p_z)$, where $f(H,p_\theta,p_z)$ is integrable but essentially arbitrary. We may then easily calculate the contribution of a given particle species to the charge or current density

$$Q_s = e_s \int F \, dp_r dp_\theta dp_z / r \tag{78}$$

$$\underset{\sim}{J}_s = e_s \int [(\underset{\sim}{p} - e_s \underset{\sim}{A})/m_s] dp_r dp_\theta dp_z / r \; , \tag{79}$$

so that $J_{sr} \equiv 0$ and

$$Q_s = e_s \int F\big(H(\alpha,\beta,\gamma),\; r\beta + re_s A_\theta,\; \gamma + e_s A_z\big) d\alpha d\beta d\gamma \tag{80}$$

$$J_{s_\theta} = e_s \int (\beta + e_s A_\theta) F d\alpha d\beta d\gamma \tag{81}$$

$$J_{sz} = e_s \int (\gamma + e_s A_\theta) F d\alpha d\beta d\gamma \tag{82}$$

where

$$H = (\alpha^2 + \beta^2 + \gamma^2 + m^2)^{1/2} + e\phi \; . \tag{83}$$

When (80)-(83) is coupled with the steady state version of Maxwell's equations we find three nonlinear ordinary differential equations for ϕ, A_θ, and A_z. This system must automatically satisfy the moment relations (71)-(76). Thus, at very low temperatures the Vlasov model must match the cold plasma model. If we were to give a particular, simple form for $f(H,p_\theta,p_z)$ we might be able to say more. Since we lack a physically convincing relativistic Boltzmann equation for a plasma with only coulomb collisions, it is hard to decide on a physically relevant analog of thirteen moments or Hermite expansion of a distribution function. Thus, it is not at all clear how to pick F. We do not pursue this matter further.

We next turn to a slightly more complicated steady flow. We consider an axisymmetric steady flow. Without loss of generality we may select a gauge in which $A_r \equiv 0$. We can give an explicit, algebraic expression for only two constants of the motion, namely H and p_θ. If we were to take a distribution function of these two constants of the motion only, $f = f(H,p_\theta)$ we would readily conclude that

$$J_{sz} = e_s \int (p_z - e_s A_z) f(H,p_\theta) dp_r dp_\theta dp_z / r \equiv 0 \; . \tag{84}$$

Thus, we cannot set up the simplest case with a flow in the z direction. It is the great difficulty in the concrete representation of such a steady flow that leads one to consider almost immediately numerical computation. We certainly expect that simple steady flows exist which vary with r and z and which have substantial flow velocities in the z direction. We cannot address this problem in all generality, but we can examine steady axisymmetric flows which vary slowly in the z direction. That is, all physical variables are functions of r and εz. Again, without loss of generality $A_r \equiv 0$. We can now apply the analysis of the second section, and although we do not have an additional exact constant of the motion, we do have an adiabatic invariant to some order in ε. Specifically, $I_r = \oint p_r dr$, see (14), is a constant of the motion, where $H = [p_r^2 + (p_\theta/r - e_s A_\theta)^2 + (p_z - e_s A_z)^2]^{1/2} + e\phi$ and H, p_θ, p_z, and εz are held constant in the integration. Formally $H = H(I_r, p_\theta, p_z, \varepsilon z)$ and now

$$J_{sz} = e_s \int (p_z - e_s A_z) f(H, p_\theta I_r) dI_r dp_\theta dp_z / cr \frac{\partial I_r}{\partial p_r} \neq 0 , \tag{85}$$

where $\partial I_r/\partial p_r$ is evaluated at a fixed value of r and $p_r = p_r(I_r, r, H, p_\theta, p_z, \varepsilon z)$. Although the formalism is highly implicit and not at all obviously simply computable, we can at last lay out in more or less explicit terms the conditions for a long, thin steady axisymmetric flow. For a low temperature flow, meaning that the distribution function is highly peaked about one point or surface, one may need additional terms in the adiabatic invarient series. This flow again satisfies the moment series (71)-(76), so that in the cold limit one should approximate the cold beam steady flow.

This simple example of the difficulties in the representation of axisymmetric flows should indicate to the reader the great practical desirability and utility of continuum or fluid models, as opposed to kinetic models. We shall now use our Vlasov model as a basis for the construction of a cool fluid model. With P. Amendt in a paper, Phys. Fluids <u>28</u>, 949 (1985) we have modified the approach of W.A. Newcomb, Phys. Fluids <u>25</u>, 846 (1982) to this problem. Before we advance the model, we obtain a few additional consequences of our system. We return to the system of constraints (71)-(73) and the system of differential equations. We may write (75) and (76) in non-conservation form

$$h\langle u^\mu\rangle \partial_\mu \langle u^\nu\rangle + \partial_\mu \Theta^{\mu\nu} = (e/m) F^{\nu\mu} h \langle u_\mu \rangle \tag{86}$$

and

$$h\langle u^\mu\rangle\partial_\mu(\Theta^{\nu\lambda}/h) + \partial_\mu S^{\mu\nu\lambda} + \Theta^{\mu\lambda}\partial_\mu\langle u^\nu\rangle + \Theta^{\mu\nu}\partial_\mu\langle u^\lambda\rangle$$

$$= (e/m)(F^{\nu\mu}\Theta^{\rho\lambda}g_{\mu\rho} + F^{\lambda\mu}\Theta^{\rho\nu}g_{\mu\rho}) \ . \tag{87}$$

Clearly (74) represents conservation of mass, (75) and (86) conservation of momentum, and (76) and (87) conservation of energy or a generalized equation of state. We also need certain consequences of (74)-(76) and (86), (87). From (86) we find easily

$$h\langle u^\mu\rangle\partial_\mu(\langle u^\nu\rangle\langle u_\nu\rangle) + \partial\langle u_\nu\rangle\partial_\mu\Theta^{\mu\nu} = 0 \ , \tag{88}$$

and then

$$h\langle u^\mu\rangle\partial_\mu(\Theta^{\nu\lambda}g_{\nu\lambda}/h + \langle u^\nu\rangle\langle u_\nu\rangle) + \partial_\mu(S^{\mu\nu\lambda}g_{\nu\lambda} + 2\Theta^{\mu\nu}\langle u_\nu\rangle) = 0 \tag{89}$$

In equation (89) we have an explicit verification of the consistency of the constraints (71) and (72).

V. A COOL RELATIVISTIC FLUID MODEL

We will next take the system of moment equations of the relativistic Vlasov equation (74)-(76) and the equivalent forms (86)-(89) and we will hypothesize a cool fluid model. We know that in given electric and magnetic fields the solution of the relativistic extension of Newton's laws vary continuously with the data. Thus, if the support in reduced momentum space of the initial data for the relativistic Vlasov equation lies in a small domain, then for some later time interval, the support will also lie in a small domain. Hence it is reasonable to look for solutions of the relativistic Vlasov equation with small variances. Specifically, it is reasonable to look for solutions with $\Theta^{\mu\nu} = O(\varepsilon^2)$, $S^{\mu\nu\lambda} = O(\varepsilon^3)$ and $T^{\mu\nu\lambda\kappa} = O(\varepsilon^4)$, etc. Amendt, Phys. Fluids 29, 1458 (1986), has shown how to introduce the cold scaling invariantly. We assume

$$1 + \langle u^\mu\rangle\langle u_\mu\rangle = O(\varepsilon^2) \ , \tag{90}$$

and this defines the parameter ε. We may then require that

$$\Theta^{\mu\nu} = O(\varepsilon^2)h \tag{91}$$

$$S^{\mu\nu\lambda} = O(\varepsilon^3)h \ . \tag{92}$$

If we carry our system through $O(\varepsilon^2)$, then we may drop $S^{\mu\nu\lambda}$ in (76), (87), and (89). We may no longer require that the constraints (71) and (72) hold exactly; we may only demand

$$\Theta^{\mu\nu}g_{\mu\nu} = -h(1 + \langle u^\mu\rangle\langle u_\mu\rangle) + O(\varepsilon^3) \tag{93}$$

$$\Theta^{\mu\nu}\langle u_\nu\rangle = O(\varepsilon^3) \ . \tag{94}$$

It is easy to show, that if the system with $S^{\mu\nu\lambda}$ set to zero is mathematically well-posed and if (90), (91), (93), and (94) hold initially then they hold for some time interval. Thus, we may consider as a cool plasma model the system

$$\partial_\mu(hw^\mu) = 0 \tag{95}$$

$$\partial_\mu(hw^\mu w^\nu + \Theta^{\mu\nu}) = (e/m)F^{\nu\mu}hw_\mu \tag{96}$$

$$hw^\mu\partial_\mu(\Theta^{\nu\lambda}/h) + \Theta^{\mu\nu}\partial_\mu w^\lambda + \Theta^{\mu\lambda}\partial_\mu w^\nu = (e/m)(F^{\nu\mu}\Theta^{\rho\lambda}g_{\mu\rho} + F^{\lambda\mu}\Theta^{\rho\nu}g_{\mu\rho}) \ , \tag{97}$$

together with the constraints on the initial data

$$\Theta^{\mu\nu} = O(\varepsilon^2) \tag{98}$$

$$\Theta^{\mu\nu}g_{\mu\nu} = -h(1 + w^\mu w_\mu) + O(\varepsilon^3) \tag{99}$$

$$\Theta^{\mu\nu}w_\nu = O(\varepsilon^3) \ . \tag{100}$$

We shall examine some simple properties of this model.

A direct calculation shows that the model proposed is indeed a hyperbolic system of partial differential equations provided that in the Lorentz frame in which w^μ has a only a nonzero four component, Θ^{ij} is positive definite. Recently A.M. Anile and co-workers have started consideration of a more sophisticated version of the truncation scheme in which they can guarantee that the constraints (71), (72) hold exactly and that the overall system is symmetric hyperbolic. Such improvements are clearly a major step forward, but we do not explore them further here. We continue with an examination of some properties of our simpler system (95)-(100).

It is of interest to decide what type of information we may reasonably expect to be able to obtain from our system. There are essentially two potential types of phenomena contained in a system such as (95)-(97) subject to constraints (98)-(100). First, one may have a flow which is essentially a cold plasma flow with small corrections in the momentum equation coming from the small pressure term in (96). In this case the equations of state characterizes those pressure tensors which are consistent with the given flow. Second, there may be particular flows in which the pressure forces, although small, may have a direct, significant effect on the macroscopic flow variables. Such possibilities may appear when geometrical effects may make the magnitude of $\partial_\mu(hw^\mu w^\nu)$ comparable with $\partial_\mu\Theta^{\mu\nu}$. We shall explore both possibilities in the following simple examples.

We consider first a flow which is a function of r only and for which only w_θ, w_z, and $w^4 = \gamma$ are nonzero. We also define $\Theta^{i4} = P_i$, $\Theta^{44} = U$, and assume $\underset{\sim}{E} = \hat{r}E_r$, $\underset{\sim}{B} = \hat{\theta}B_\theta + \hat{z}B_z$. We find that provided $r = 0$ lies in the flow domain

$$\theta_{r\theta} = \theta_{rz} = P_1 = 0 \; , \tag{101}$$

and

$$[-hw_\theta^2 + (r\Theta_{rr}),_r - \Theta_{\theta\theta}]/r = (e/m)(\gamma E_r + w_\theta B_z - w_z B_\theta)h \tag{102}$$

$$[\Theta_{rr}(rw_\theta),_r - 2w_\theta\Theta_{\theta\theta}]/r = (e/m)\left(B_z(\Theta_{\theta\theta}-\theta_{rr}) - B_\theta\Theta_{\theta z} + E_r P_\theta\right) \tag{103}$$

$$\Theta_{rr}w_{z,r} - 2w_\theta\Theta_{\theta z}/r = (e/m)\left(B_z(\Theta_{\theta z} - B_\theta(\Theta_{zz}-\Theta_{rr})) + E_r P_z\right) \tag{104}$$

$$\Theta_{rr}\gamma,_r - 2w_\theta P_\theta/r = (e/m)\left(B_z P_\theta - B_\theta P_z + E_r(\Theta_{rr}+U)\right) \tag{105}$$

while the constraints become

$$1 + w_\theta^2 + w_z^2 = \gamma^2 + O(\varepsilon^2) \tag{106}$$

and the "solution" of (100) is

$$\gamma P_\theta = \Theta_{\theta\theta}w_\theta + \Theta_{\theta z}w_z + O(\varepsilon^3) \tag{107}$$

$$\gamma P_z = \Theta_{\theta z}w_\theta + \Theta_{zz}w_z + O(\varepsilon^3) \tag{108}$$

$$\gamma^2 U = \Theta_{\theta\theta}w_\theta^2 + 2\Theta_{\theta z}w_\theta w_z + \Theta_{zz}w_z^2 + O(\varepsilon^3) \tag{109}$$

If we adjoin Maxwell's equation

$$(rE_r),_r/r = e\gamma h + Q^{ext} \tag{110}$$

$$-B_{z,r} = ew_\theta h + J_\theta^{ext} \tag{111}$$

$$(rB_\theta),_r/r = ew_z h + J_z^{ext} \tag{112}$$

where the superscript ext refers to external charges or current, then we have a complete system. It is easy to show, however, that (103)-(105) together with

(106)-(109) are a dependent but consistent system, so that, to the accuracy of the model we may omit (105) and obtain γ from the algebraic relation (100). Thus, we may construct beam steady flows if we give $h(r)$, $w_\theta(r)$, $w_z(r)$, $\gamma(r)$, $E_r(r)$, $B_\theta(r)$, and $B_z(r)$ corresponding to a cold beam profile and then the tensor $\Theta^{\mu\nu}$ must satisfy (103), (104), (107), (108), and (109). This is clearly an example of a system in which the beam system is essentially unmodified by thermal corrections, but the pressure tensor is constrained in form. Our analysis clearly shows that the severe constraint of the Bennett profile (36) is not typical of a cool plasma beam, but rather that the cold beam system is much closer to the cool system. The Bennett profile is discussed in greater detail in a paper of Amendt and Weitzner: "Cool Electron Beam Steady Flows," to be published in Physics of Fluids (1987).

We next examine the second case in which thermal effects may indeed modify the steady flow. We consider an axisymmetric steady flow with all flow components nonzero. Furthermore, we assume that the variation of quantities in the z direction is slower than in the r directions; specifically all variables are functions of r and of $\varepsilon^2 z$, where ε is the scaling factor in the pressure tensor, so that $\Theta^{\mu\nu} = hO(\varepsilon^2)$. Conservation of mass becomes

$$hw_r = -\varepsilon^2 \chi,_\zeta/r \tag{113}$$

$$hw_z = \chi,_r/r \ , \tag{114}$$

where $\zeta = \varepsilon^2 z$. We may also introduce a magnetic flux function ψ such that

$$B_r = -\varepsilon^2 \psi,_\zeta/r \tag{115}$$

$$B_z = \psi,_r/r \ , \tag{116}$$

while Maxwell's equations reduce to

$$B_\theta = e\chi/r \tag{117}$$

$$r(\psi,_r/r),_r + \varepsilon^4 \psi,_{\zeta\zeta} = -erhw_\theta \tag{118}$$

$$\underset{\sim}{E} = -\nabla\Phi \tag{119}$$

and

$$(r\Phi,_r),_r/r + \varepsilon^4 \Phi,_{\zeta\zeta} = -eh\gamma(1-f) \ , \tag{120}$$

where f is the ion neutralization fraction.

Conservation of momentum reduces to the three equations

$$-w_\theta^2/r = (e/m)(E_r\gamma + w_\theta B_z - w_z B_\theta) + O(\varepsilon^2) \tag{121}$$

$$r^2h\big(w_r(rw_\theta)_{,r}/r + \varepsilon^2 w_z w_{\theta,\zeta}\big) + (r^2\Theta_{r\theta})_{,r} = (e/m)\varepsilon^2 h(\chi_{,\zeta}\psi_{,r} - \chi_{,r}\psi_{,\zeta}) \tag{122}$$

$$h(w_r w_{z,r} + \varepsilon^2 w_z w_{z,\zeta}) + (r\Theta_{rz})_{,r}/r = (e/m)\varepsilon^2 h\{-\gamma\Phi_{,\zeta} - \chi_{,\zeta}B_\theta/r + w_\theta\psi_{,\zeta}/r\}\ . \tag{123}$$

In (122) and (123) each of the terms is $O(\varepsilon^2)$, and the pressure tensor appears non-trivially.

In order to discover whether $\Theta_{r\theta}$ or Θ_{rz} can be nonzero in a particular flow we must examine the equations of state. We give only the six equations that characterize Θ_{ij}, where i and j are space indices. In these calculations we have eliminated Θ_{i4} by means of (100), which gives the remaining components of $\Theta^{\mu\nu}$. We define

$$\alpha = (e/m)(B_z + E_r w_\theta/\gamma) + 2w_\theta/r \tag{124}$$

and

$$\beta = (e/m)(B_\theta - E_r w_z/\gamma)\ , \tag{125}$$

and after a little calculation we find from (97)

$$\alpha\Theta_{r\theta} - \beta\Theta_{rz} = 0 \tag{126}$$

$$\alpha\Theta_{\theta\theta} - \beta\Theta_{\theta z} = -\Theta_{rr}[(rw_\theta)_{,r}/r + (e/m)B_z] \tag{127}$$

$$\alpha\Theta_{z\theta} - \beta\Theta_{zz} = -\Theta_{\theta\theta}[w_{z,r} - (e/m)B_\theta] \tag{128}$$

$$\Theta_{r\theta}[(rw_\theta)_{,r}/r + (e/m)B_z] = 0 \tag{129}$$

$$\Theta_{rz}[w_{z,r} - (e/m)B_\theta] = 0 \tag{130}$$

$$\Theta_{rz}[(rw_\theta)_{,r}/r + (e/m)B_z] + \Theta_{r\theta}[w_{z,r} - (e/m)B_\theta] = 0\ . \tag{131}$$

Since we wish to explore the possibility that thermal effects modify the momentum balance relations, we assume that at least one of $\Theta_{r\theta}$ and Θ_{rz} is nonzero. It then follows easily from (129)-(131) that

$$(rw_\theta)_{,r}/r + (e/m)B_z = (w_{z,r} - (e/m)B_\theta) = 0\ ,$$

and hence

$$rw_\theta = -(e/m)\psi = 0 \tag{132}$$

$$[\chi,_r/(rh)],_r - (e^2/m)\chi/r \ , \tag{133}$$

while from (126)-(128)

$$\alpha\Theta_{r\theta} - \beta\Theta_{rz} = \alpha\Theta_{\theta\theta} - \beta\Theta_{\theta z} = \alpha\Theta_{z\theta} - \beta\Theta_{zz} = 0 \ . \tag{134}$$

The relations (118), (132), and (133) show that $h(r,z)$ and boundary conditions completely determine the flow and magnetic field variables χ, ψ, and w_θ. The relation (121) determines E_r, and Poisson's equation (120) gives the neutralization fraction $f(r,z)$. Finally $\Theta_{r\theta}$ and Θ_{rz} are given by (122) and (123). Hence, the state is completely specified by $h(r,z)$, but (134) is not yet satisfied. If either α or β is nonzero then it follows that the matrix Θ_{ij} must have a zero eigenvalue, and hence the pressure tensor Θ_{ij} is not positive definite, as is required. Thus, in order to generate an acceptable steady flow state we must adjoin the two relations $\alpha = 0$ and $\beta = 0$, where α and β are given by (124) and (125). Since we have at our disposal only the one free function $h(r,z)$, we conjecture that the system is overdetermined and finally $\Theta_{r\theta} = \Theta_{rz} = 0$.

We suspect that generally the pressure tensor generates only small corrections in the underlying cold relativistic plasma flow. One may then ask what is the role of the cool relativistic fluid model. We see three significant uses of the model. First, it demonstrates quantitatively the connection between solutions of the Vlasov equation at low temperature and the cold relativistic fluid model. Second, it restricts the class of admissible pressure tensors to those satisfying (97), (99), and (100). Third, it functions as a basis for the construction of generalized relativistic fluid plasma models, as is being carried out by Anile and co-workers.

VI. RELATIVISTIC MAGNETOHYDRODYNAMCS

We shall give a very brief discussion of the formulation of ideal and non-ideal relativistic magnetohydrodynamics. Before we do so it is perhaps worthwhile to examine very briefly from a purely physical point of view the nature of the difficulty in the description of a relativistic, single fluid model plasma. In order to do so it is useful to return to Maxwell's equations in dimensional form. We use gaussian units and we set $\varepsilon_o = \mu_o = 1$

$$\nabla \cdot E = 4\pi Q \tag{135}$$

$$\nabla \times \mathbf{E} = -\mathbf{B}_{,t}/c \tag{136}$$

$$\nabla \times B = (4\pi\mathbf{J} + \mathbf{E}_{,t})/c \tag{137}$$

and

$$Q_{,t} + \nabla \cdot \mathbf{J} - 0 \ . \tag{138}$$

We assume that the characteristic distance is L, the characteristic time T, and velocity characteristic of the system is $V = L/T$. We will scale all quantities relative to B, the magnetic field. From (137) we take

$$J \sim cB/L \tag{139}$$

and from (136)

$$\mathbf{E} \sim LB/(Tc) \ , \tag{140}$$

so that

$$Q \sim \nabla \cdot \mathbf{E} \sim B/(Tc) \ , \tag{141}$$

but

$$Q_{,t}/(\nabla \cdot \mathbf{J}) = (L^2/T^2)/c^2 = V^2/c^2 \ . \tag{142}$$

Thus, in a non-relativistic plasma

$$\nabla \cdot \mathbf{J} = 0$$

and the system becomes

$$\nabla \times \mathbf{E} = -\mathbf{B}_{,t}/c \tag{143}$$

$$4\pi\mathbf{J}/c = \nabla \times \mathbf{B} \tag{144}$$

and

$$4\pi Q = \nabla \cdot \mathbf{E} \ , \tag{145}$$

so that Q is determined by (145) rather than Q determining $\nabla \cdot \mathbf{E}$, and similarly $\mathbf{J}$ is determined by $\mathbf{B}$ from (144). On the other hand $\mathbf{B}$ is determined by (143) once a constituative relation is given that determines $\mathbf{E}$ as a function of the other variables. Such a relation is usually called Ohm's law. However, in a relativistic system with $V \sim c$ it is clear that the role of the charge density is crucial to the formulation of a relativistic single fluid plasma model.

We consider a single fluid with the usual thermodynamic variables e, ρ, and p which satisfies the standard thermodynamic relation $TdS = de + pd(1/\rho)$ and which is characterized by a four-velocity w^μ, $w^\mu w_\mu = -1$. Conservation of mass assumes the form

$$\partial_\mu(\rho w^\mu) = 0 \ . \tag{146}$$

The fluid possesses an energy momentum tensor

$$T^{\mu\nu} = (\rho + \rho e + p)w^\mu w^\nu + pg^{\mu\nu} \tag{147}$$

and conservation of energy momentum assumes the form

$$\partial_\mu T^{\mu\nu} = f^\nu_{EM} + f^\nu_{OH} \ , \tag{148}$$

where f^ν_{EM} is the electromagnetic force and f^ν_{OH} is that part of the force coming from the Ohm's law. We give expressions for f^ν_{EM} and f^ν_{OH} shortly. We mention in passing that the interested reader should consult A. Lichnerowicz, Relativistic Hydrodynamics (W.A. Benjamin, New York, 1967), and also K.O. Friedrichs, Comm. Pure Appl. Math. 27, 749 (1974). It is easy to introduce a formal generalization of the non-relativistic form of Ohm's law

$$\underset{\sim}{J} = \sigma(\underset{\sim}{E} + \underset{\sim}{v}\times\underset{\sim}{B}) \ . \tag{149}$$

We introduce the current four-vector J^μ corresponding to $(\underset{\sim}{J},Q)$ and we then hypothesize

$$J^\mu = \varepsilon w^\mu + \sigma F^{\mu\nu}w_\nu \ , \tag{150}$$

where σ is the conductivity of the medium and ε is the invariant rest charge carried by the fluid. With J^μ defined we may then introduce the electromagnetic force

$$f^\mu_{EM} = F^{\mu\nu}J_\nu \ , \tag{151}$$

the usual relativistic Lorentz force. Since Ohmic heating dissipates energy there must be a dissipation term corresponding to f^{μ}_{OH} and

$$f^{\mu}_{OH} = -\sigma(F^{\nu\lambda}w_{\lambda}F_{\nu\kappa}w^{\kappa})w^{\mu} \ . \tag{152}$$

In the rest frame

$$f^{\mu}_{OH} = -(\sigma \underset{\sim}{E}\cdot\underset{\sim}{E})\delta^{\mu}_{4} \ , \tag{153}$$

so that f^{μ}_{OH} is the Ohmic energy loss for the system. We do not give Maxwell's equations explicitly in covariant form. They are, of course, unchanged from (1)-(4).

Two special cases of this system are well understood. If we set $\sigma = 0$ but $\varepsilon \neq 0$, so that we examine a charged fluid of the type studied earlier in these lectures, then Lichnerowicz has shown tha the system is strictly hyperbolic. Thus, compatible initial data of differentiability up to some order will generate a unique solution in an appropriate class. Such a problem is then considered well-posed, and would be quite satisfactory. The other limit which is well understood is the case $\sigma \to \infty$ in which case we would set

$$F^{\mu\nu}w_{\nu} = 0 \ , \tag{154}$$

and

$$f^{\mu}_{OH} = 0 \ . \tag{155}$$

Exactly as in the non-relativistic version one may view (154) as defining $\underset{\sim}{E}$ in terms of $\underset{\sim}{B}$ and $\underset{\sim}{w}$, then $\underset{\sim}{B}$ is determined by

$$\underset{\sim}{B}_{,t} = -\underset{\sim}{\nabla}\times\underset{\sim}{E}$$

and $\underset{\sim}{J}$ is defined as

$$\underset{\sim}{J} = \underset{\sim}{\nabla}\times\underset{\sim}{B} - \underset{\sim}{E}_{,t} \ . \tag{156}$$

Friedrichs has shown that this system may be recast as a symmetric hyperbolic system and thus again the initial value problem is well-posed. If σ is neither infinite nor zero, the problem is far more complex. The system may be cast as a hyperbolic system, but one which is neither strictly hyperbolic nor symmetric hyperbolic. One can prove existence and uniqueness of solutions, but in C^{∞}

function classes they are generally considered too restrictive. Friedrichs has attempted to remedy this situation by modifying the electromagnetic force and the internal energy function. He has succeeded in recasting the system as a symmetric hyperbolic system, but as he notes there is no physical basis for his modifications; their role is to modify and improve the mathematical structure. To this worker's knowledge, there is little advance on these problems in recent years. We believe that much of the underlying problem is a real, physical problem. Since one must have a dynamical relation to advance and determine the electric charge density, it is unlikely that such a simple force as (150) represents the real current density vector generated by a two-charged component fluid. We prefer to consider coupled two-charged fluid models.

Acknowledgment

This work was supported by the U.S. Office of Naval Research.

AN IMPROVED RELATIVISTIC WARM PLASMA MODEL

A.M.Anile and S. Pennisi
Dipartimento di Matematica
Citta' Universitaria
Viale A.Doria 6
95125 Catania (Italy)

Sec.1.-Introduction

Relativistic intense charged particle beams are of great interest in several areas of plasma physics and technology [1].
A fundamental description of these beams is usually based upon the Vlasov or Vlasov-Boltzmann equation. Calculations for specific situations are then performed by using numerical simulation techniques. A kinetic approach, however, has some drawbacks. In particular, within a kinetic framework, the actual calculation of equilibrium configurations in an arbitrary geometry is time consuming and very difficult. Furthermore a stability analysis of such equilibrium configurations is an almost impossible task except in very special cases. These drawbacks could be avoided, at least in part, by adopting a fluid model. Obviously a fluid model cannot provide an accurate microscopic description but could be adequate if one is mainly interested in the gross features of a configuration.
Relativistic fluid models can be constructed by considering the moment equations arising from the relativistic Vlasov equation and adopting a suitable closure approximation. For particle beams the closure approximation must be based on the assumption that the particle distribution function represents a warm fluid, i.e. the dispersion of the velocity about the mean is small. Based on this approximation models have been proposed by Siambis [2], Newcomb [3], Amendt and Weitzner [4] . The latter, according to our opinion, is the most satisfactory because it is fully covariant and complete (i.e. they provide a minimal set of field equations). The present work is based on the Amendt and Weitzner model and we believe that it represents a considerable improvements upon theirs. The Amendt and Weitzner relativistically covariant warm fluid model could be improved significantly in two points . The first point is related to the constraint equations which must be satisfied by the moments and which arise from the fact that the moments must arise from a distribution fucntion. In the Amendt and Weitzner model these constraints are satisfied only

approximately. The second point is that the Amendt and Weitzner model can be shown to lead to a hyperbolic system but it is not known whether such a system is equivalent to a symmetric one (for symmetric hyperbolic systems one has a much more satisfactory mathematical theory [5]).

In this paper we present a fluid model which solves the above inconveniences of the Amendt and Weitzner one. More precisely, for our model the constraint equations are satisfied exactly and furthemore our model leads to a symmetric hyperbolic system. We shall achieve these results by using concepts and methods arising from a seemingly totally unrelated area, viz the theory of relativistic extended thermodynamics of Liu, Muller and Ruggeri [6] . We stress that we are not advocating such a theory but only using it as a mathematical tool.

The plan of the paper is the following. In Sec. 2 we start from the relativistic Vlasov equation and derive the moment equations. In Sec.3 we give an invariant definition of the warm plasma in such a way that the constraints are satisfied exactly, and we give physical examples (a low temperature ideal gas near thermal equilibrium and the case of a monoenergetic distribution function). In Sec.4 we treat the closure problem. We assume the same smallness parameter as the other authors (i.e. the small spread in momentum space) but we propose a different closure procedure which leads to a symmetric hyperbolic system.

Sec.2-The relativistic Vlasov equation.

Let u^μ denote the four velocity of a particle, which, in inertial coordinates (x^μ) in Minkowski coordinates, writes

$$u^\mu=\Gamma(v^i,1)$$

where $\Gamma = (1-v^i v_i)^{-1/2}$ is the Lorentz factor and v^i the three velocity. Let $u^i=\Gamma v^i$ and introduce the one particle invariant distribution function $f(x^\mu, u^i)$. Then the Lorentz covariant Vlasov equation writes [4]

$$u^\mu \frac{\partial f}{\partial x^\mu} + \frac{e}{m} F^{i\mu} u_\mu \frac{\partial f}{\partial u^i} = 0 \qquad (1)$$

with e the electric charge, m the particle mass . The following quantities related to the moments of the distribution function f can be defined

$$h = \int f\omega \tag{2}$$

$$hw^{\mu} = \int f u^{\mu} \omega \tag{3}$$

$$\Theta^{\mu\nu} = \int f(u^{\mu}-w^{\mu})(u^{\nu}-w^{\nu})\omega \tag{4}$$

$$S^{\mu\nu\lambda} = \int f(u^{\mu}-w^{\mu})(u^{\nu}-w^{\nu})(u^{\lambda}-w^{\lambda})\omega \tag{5}$$

where $\omega = \dfrac{du^1 du^2 du^3}{u^0}$ is the invariant measure on the velocity space.

The quantity h can be interpreted as an invariant particle density, w^{μ} as an average four-velocity.

From the definitions (4) and (5) the following relationships are obtained

$$\Theta^{\mu}_{\ \mu} = -\,h\,(1 + w^{\mu} w_{\mu}) \tag{6}$$

$$S^{\mu\nu}_{\ \ \ \nu} = -\,2\,\Theta^{\mu\nu} w_{\nu} \tag{7}$$

By taking the moments od the Vlasov equation (1), assuming that f vanishes sufficiently fast at infinity in the velocity space, we obtain

$$\frac{\partial h w^{\mu}}{\partial x^{\mu}} = 0 \tag{8}$$

$$\frac{\partial}{\partial x^{\mu}}\left(h\, w^{\mu} w^{\nu} + \Theta^{\mu\nu} \right) = \frac{e}{m}\, h\, F^{\mu\nu} w_{\mu} \tag{9}$$

$$\frac{\partial}{\partial x^{\mu}}\left(h\, w^{\mu} w^{\nu} w^{\alpha} + w^{\mu}\Theta^{\nu\alpha} + w^{\nu}\Theta^{\mu\alpha} + w^{\alpha}\Theta^{\mu\nu} + S^{\mu\nu\alpha}\right) =$$

$$= \frac{e}{m}\left(F^{\nu\mu}\Theta^{\alpha}_{\ \mu} + F^{\alpha\mu}\Theta^{\nu}_{\ \mu} + F^{\nu\mu} w_{\mu} h w^{\alpha} + F^{\alpha\mu} w_{\mu} h w^{\nu}\right) \tag{10}$$

The set of equations (8), (9), (10) obviously is not a closed system. An outstanding problem is that of closing the system by making judicious assumptions on $S^{\mu\nu\alpha}$ and determining the range of validity of the approximation.

Sec.3 <u>Invariant definition of warm plasma</u>

We introduce the following quantity

$$\varepsilon^2 = -(1 + w^\mu w_\mu) \tag{11}$$

and define a warm plasma by the requirements :

i) $\varepsilon \ll 1$

ii) in the local rest frame of the plasma, in which

$$w^\mu = (0,0,0\ \bar{w}^4) \quad \text{with} \quad \bar{w}^4 = \sqrt{1+\varepsilon^2},$$

the following representation holds

$$\frac{\bar{\Theta}^{ij}}{h} = [\bar{K}^{ij} + \frac{1}{3}(1 - \bar{K}^l{}_l)\delta^{ij}]\varepsilon^2 + \frac{1}{3}\bar{K}^{44}\delta^{ij}\varepsilon^4$$

$$\frac{\bar{\Theta}^{i4}}{h} = \bar{K}^{i4}\varepsilon^3 , \quad \frac{\bar{\Theta}^{44}}{h} = \bar{K}^{44}\varepsilon^4 \tag{12.1}$$

$$\frac{\bar{S}^{ijk}}{h} = [\bar{K}^{ijk} - \frac{3}{5}\delta^{(ij}\bar{K}^{k)l}{}_l]\varepsilon^3 + \frac{6}{5}\delta^{(ij}\bar{K}^{k)4}\varepsilon^3\sqrt{1+\varepsilon^2} +$$

$$+ \frac{3}{5}\delta^{(ij}\bar{K}^{k)44}\varepsilon^5 \tag{12.2}$$

$$\frac{\bar{S}^{ij4}}{h} = [\bar{K}^{ij4} - \frac{1}{3}\bar{K}^{4l}{}_l\delta^{ij}]\varepsilon^4 + \frac{2}{3}\delta^{ij}\bar{K}^{44}\varepsilon^4\sqrt{1+\varepsilon^2} +$$

$$+ \frac{1}{3}\delta^{ij}\bar{K}^{444}\varepsilon^6 \tag{12.3}$$

$$\frac{\bar{S}^{i44}}{h} = \bar{K}^{i44}\varepsilon^5, \quad \frac{\bar{S}^{444}}{h} = \bar{K}^{444}\varepsilon^6 \tag{12.4}$$

where overbars denote components in the plasma local rest frame and $K^{\mu\nu}$, $K^{\mu\nu\lambda}$ are dimensionless tensors at most of order zero in ε, such that $K^{\mu\nu} = K^{(\mu\nu)}$, $K^{\mu\nu\lambda} = K^{(\mu\nu\lambda)}$. It is easy to check that the above ordering satisfies the constraints (6) and (7) exactly.

It is obvious that the ordering (12) is based upon the physically intuitive idea that, for a sufficiently collimated beam, the higher order off-center moments are of decreasing order of magnitude with respect to the <u>spread parameter</u> ε . This requirement could have been expressed in a much simpler ordering, as in Amendt and Weitzner [4] , viz

$$\frac{\bar{\Theta}^{ij}}{h} = \bar{K}^{ij}\varepsilon^2 , \quad \frac{\bar{\Theta}^{4i}}{h} = \bar{K}^{4i}\varepsilon^3, \quad \frac{\bar{\Theta}^{44}}{h} = \bar{K}^{44}\varepsilon^4 , \quad \frac{\bar{S}^{ijk}}{h} = \bar{K}^{ijk}\varepsilon^3$$

$$\frac{\bar{S}^{ij4}}{h} = \bar{K}^{ij4}\varepsilon^4 , \quad \frac{\bar{S}^{i44}}{h} = \bar{K}^{i44}\varepsilon^5, \quad \frac{S^{444}}{h} = \bar{K}^{444}\varepsilon^6$$

with $K^{\mu\nu}, K^{\mu\nu\lambda}$ dimensionless tensors of order one.

However the above ordering does not statisfy the constraints (6) and (7) exactly. Therefore we think that it is convenient to add higher-order terms in order to make it satisfy the constraints (6) and (7) exactly. This can be achieved as follows. Let us introduce the quantities

$$\Theta'^{\alpha\beta} = \Theta^{\alpha\beta} + \frac{1}{3}\left(\frac{\Theta^{\mu\nu}w_\mu w_\nu}{w^\lambda w_\lambda} + hw^\mu w_\mu + h\right)h^{\alpha\beta} \tag{13}$$

$$S'^{\alpha\beta\gamma} = S^{\alpha\beta\gamma} + \left[\frac{2}{w^\mu w_\mu} w_\lambda w_\delta \Theta^{\lambda\delta} + \frac{S^{\mu\nu\lambda} w_\mu w_\nu w_\lambda}{(w^\delta w_\delta)^2}\right] w^{\langle\alpha} h^{\beta\gamma\rangle} + \tag{14}$$

$$+ \frac{3}{5} h^{\langle\alpha\beta} h^{\gamma\rangle\delta} \left(2\Theta_{\delta\mu} w^\mu + \frac{S_{\delta\mu\nu} w^\mu w^\nu}{w^\lambda w_\lambda}\right)$$

where $h^{\alpha\beta} = g^{\alpha\beta} - \dfrac{w^\alpha w^\beta}{w^\lambda w_\lambda}$ is the projection tensor onto the three space orthogonal to w^μ.

It is easily seen that the constraints (6) and (7) are equivalent to

$$\Theta'^{\alpha\beta} h_{\alpha\beta} = 0 \tag{15.1}$$

$$S'^{\alpha\beta\gamma} h_{\alpha\beta} = 0 \tag{15.2}$$

In the local rest frame of the fluid, in which $\overline{w}^\mu = (0,0,0,\sqrt{1+\varepsilon^2})$, the constraints (15.1,2) set restrictions only on the spatial components Θ'^{ij}, S'^{ijk}. Therefore one could try to impose the intuitive ordering of Amendt and Weitzner preciously discussed to the variables $\Theta'^{\alpha\beta}$, $S'^{\alpha\beta\gamma}$, taking into account the restrictions (15.1,2). Quite simply one assumes

$$\frac{\overline{\Theta}'^{ij}}{h} = (\overline{K}^{ij} - \frac{1}{3}\overline{K}^{l}{}_{l}\delta^{ij})\varepsilon^2, \quad \frac{\overline{\Theta}'^{i4}}{h} = \overline{K}^{i4}\varepsilon^3, \quad \frac{\overline{\Theta}'^{44}}{h} = \overline{K}^{44}\varepsilon^3 \tag{16.1}$$

$$\frac{\overline{S}'^{ijk}}{h} = (\overline{K}^{ijk} - \frac{3}{5}\delta^{(ij}\overline{K}^{k)l}{}_{l})\varepsilon^3 \tag{16.2}$$

$$\frac{\overline{S}'^{ij4}}{h} = (\overline{K}^{ij4} - \frac{1}{3}\overline{K}^{4l}{}_{l}\delta^{ij})\varepsilon^4 \tag{16.3}$$

$$\frac{\overline{S}'^{i44}}{h} = \overline{K}^{i44}\varepsilon^5 \quad , \quad \frac{\overline{S}'^{444}}{h} = \overline{K}^{444}\varepsilon^6 \tag{16.4}$$

It is easy to read off from the above ordering that the ordering (12) is recovered.

In order to give a physical interpretation of the ordering (12) it is convenient to change to another set of variables.

Let $p^{\mu}=mu^{\mu}$ be the four momentum of a particle. Then, following Liu, Muller and Ruggeri [6] we define

$$V^{\mu} = \int fp^{\mu}\,\pi \tag{17}$$

$$T^{\mu\nu}= \int fp^{\mu}p^{\nu}\pi \tag{18}$$

$$A^{\mu\nu\alpha}= \int fp^{\mu}p^{\nu}p^{\alpha}\pi \tag{19}$$

where $\pi= \dfrac{dp^1dp^2dp^3}{p^4}$ and f is the distribution function.

The relationship with the former variables is

$$V^{\mu} = m^3h\ w^{\mu} \tag{20}$$

$$T^{\mu\nu} =m^4\ (\ \Theta^{\mu\nu} + h\ w^{\mu}w^{\nu}\ ) \tag{21}$$

$$A^{\mu\nu\alpha}= m^5\ (\ S^{\mu\nu\alpha} +3\ w^{(\nu}\Theta^{\mu\alpha)} + h\ w^{\mu}w^{\nu}w^{\alpha}\ ) \tag{22}$$

and the constraints (6), (7) read

$$T^{\mu}{}_{\mu} = -\ h\ m^4, \qquad A^{\mu\nu}{}_{\nu} = -m^2\ V^{\mu} \quad .$$

The energy-momentum tensor $T^{\mu\nu}$ can be decomposed in the usual way

$$T^{\mu\nu} = E\ U^{\mu}U^{\nu} + ph^{\mu\nu} + 2U^{(\mu}q^{\nu)} + t^{<\mu\nu>} \tag{23}$$

where

$$U^{\alpha} =V^{\alpha}\ (-V^{\mu}V_{\mu}\ )^{-1/2}$$

$$n = \sqrt{-V^{\mu}V_{\mu}}$$

is the rest frame particle number density,

$$h^{\mu\nu}=g^{\mu\nu}+U^{\mu}U^{\nu}$$

is the projection tensor onto the 3-space orthogonal to U^{α},

$$p=\frac{1}{3}h_{\mu\nu}T^{\mu\nu}$$

is the isotropic pressure,

$$E=T^{\mu\nu}U_{\mu}U_{\nu}$$

is the total rest frame energy density,

$$q^{\alpha}=-h^{\alpha}_{\mu}U_{\nu}T^{\mu\nu}$$

is the heat flux vector,

$$t^{\langle\alpha\beta\rangle}=\left[h^{\alpha}_{\mu}h^{\beta}_{\nu}-\frac{1}{3}h^{\alpha\beta}h_{\mu\nu}\right]T^{\mu\nu}$$

is the stress tensor.

In terms of the former variables we have

$$p=\frac{1}{3}m^{4}h_{\mu\nu}\theta^{\mu\nu}$$

$$E=hm^{4}+m^{4}h_{\mu\nu}\theta^{\mu\nu}$$

$$q^{\alpha}=-\frac{1}{\sqrt{-w^{\mu}w_{\mu}}}m^{4}h^{\alpha}_{\mu}w_{\nu}\theta^{\mu\nu}$$

$$t^{\langle\alpha\beta\rangle}=m^{4}\left[h^{\alpha}_{\mu}h^{\beta}_{\nu}-\frac{1}{3}h^{\alpha\beta}h_{\mu\nu}\right]\theta^{\mu\nu}$$

$$n=m^{3}h\sqrt{-w^{\mu}w_{\mu}}$$

$$U^{\alpha}=\frac{1}{\sqrt{-w^{\mu}w_{\mu}}}w^{\alpha}$$

Now let us consider the case when the distribution function f is the Jüttner distribution function f_{eq} corresponding to thermal equilibrium [6] for a non-degenerate relativistic gas,

$$f_{eq} \propto e^{-\frac{U_\alpha}{K_B T} p^\alpha}$$

where T is the absolute temperature, K_B the Boltzmann constant.

In this case one obtains [6]

$$T^{\mu\nu} = E u^\mu u^\nu + p_{eq} h^{\mu\nu}$$

with

$$E = nm\left(G - \frac{1}{\gamma}\right) ,$$

$$p_{eq} = \frac{nm}{\gamma}$$

where $\gamma = \frac{m}{K_B T}$ and $G = \frac{K_3(\gamma)}{K_2(\gamma)}$ with $K_n(x)$ the Kummer function of order n [7].

Now, in general,

$$n^2 = m^\sigma h^2 (1 + \varepsilon^2)$$

and

$$E = hm^4 + 3p \qquad ,$$

whence

$$1 + \varepsilon^2 = \frac{m^2 n^2}{(E-3p)^2} .$$

It follows that, at thermal equilibrium

$$\varepsilon^2 = \frac{1}{\left(G - \frac{4}{\gamma}\right)^2} - 1 \qquad (24)$$

from which it follows that the limit $\varepsilon \longrightarrow 0$ corresponds to $\gamma \longrightarrow \infty$. Furthermore, from the well known properties of the function G, we have

$$G=1+\frac{5}{2\gamma}+O(\frac{1}{\gamma})$$

whence

$$\varepsilon^2=\frac{3K_B T}{m}+O(\frac{1}{\gamma}) . \tag{25}$$

This shows that $\varepsilon \longrightarrow 0$ corresponds to a low temperature plasma.

It is easy to see that the ordering (12) is trivially satisfied at thermal equilibrium (for a low temperature plasma).

It is reasonable to assume that for a low temperature (warm) plasma close to thermal equilibrium the ordering (12) is also satisfied.

Another example of physical interest is when the distribution function f corresponds to a monoenergetic distribution in a given Lorentz frame (lab frame),

$$f_{me}=\frac{ah}{4\pi sha}\frac{\sqrt{1+E_o}}{E_o^2}e^{\frac{a^i u^i}{E_o}}\delta(|u|-E_o) \tag{26}$$

where E_o is the beam energy, u^i the spatial components of the four-velocity, $|u|=\sqrt{u^i u^i}$, a^i a spatial three-vector characterizing the spread in momentum space, $a=\sqrt{a^i a^i}$.

It is easy to check that in this case, which corresponds to a state very far from thermal equilibrium, one has

$$\varepsilon^2=E_o^2(1-b^2) \tag{27}$$

where $b=\coth a-\frac{1}{a}$ is the ratio of the drift velocity to particle velocity.

Also in this case the ordering (12) is trivially satisfied.

Sec.4-Closure of the hierarchy of moment equations.

The ordering (12) suggests that we close the moment equations (8), (9), (10) by systematically neglecting terms wich are of higher order in ε.

For istance a <u>cold plasma model</u> can be obtained if we neglect terms of order ε^2 and keep only those of order 1. In this case we obtain the following equations

$$\frac{\partial}{\partial x^{\mu}} (hw^{\mu}w^{\nu}) = \frac{e}{m} F^{\nu\mu}w_{\mu} ,$$

while eq. (10) is automatically satisfied.

Consistently with the ordering (12), a <u>warm plasma model</u> can be obtained if we neglect the term $S^{\mu\nu\alpha}$ in eq. (10). In fact, in the model proposed by Amendt & Weitzner [4], one puts

$$S^{\mu\nu\alpha}=0 \; . \tag{28}$$

As recognired by Amendt & Weitzner, this choice although it seems reasonable and leads to a hyperbolic system, however suffers from the difficulty that it does not satisfy the contraint equation (7).

The aim of this section is to remedy this and other difficulties by proposing a more satisfactory choice.

Before we embark on this matter let us comment briefly on the closure problem for the hierarchy of moment equations.

In the so called Grad method of moments for describing the transition to the hydrodinamic limit one from the Boltzmann equation truncates the hierarchy of moment equations by putting equal to zero one of the higher (off-center) moments [7]. In this way one obtains a system of conservation equations which, under suitable condition, is hyperbolic and leads to finite wave speeds (unlike the case of the Chapman-Enskog method [7]). A similar procedure is adopted in the framework of the causal theory of irreversible thermodynamics [8].

However these approaches are quite different from the one which is discussed here.

Our approach is based on a precise ordering of the various components of the off-center moments, depending on a physically well defined smallness parameter ε.

On the contrary, in the framework of the Grad method or of the causal theories of irreversible thermodynamics, apparently there is no justification for putting equal to zero one of the higher off-center moments. In fact, for a system close to thermal equilibrium, the distribution function can be written as

$$f=f_{eq}+\lambda\delta f$$

where λ measures the closeness to equilibrium, and one sees that, for an arbitrary δf, the off-center moments are all of the same order $O(\lambda)$.

Now we return to the problem of improving upon the Amendt & Weitzner model. Our idea is the following. We look for $S^{\mu\nu\alpha}$ as function of the previous moments h, w^{α}, $\theta^{\mu\nu}$ such that it satisfies the contraint (7) exactly and preserves the ordering (12). This idea is somehow terminiscent of one of the basic tenets of extended thermodynamics [6] where, in a completely different context than the one considered here instead of putting one of the off-center moments equal to zero, one sets it equal to a function of the previous moments.

Therefore, we shall look for a "constitutive function"

$$S^{\mu\nu\alpha}=\hat{S}^{\mu\nu\alpha}(h,\ w^{\beta},\ \theta^{\sigma\beta}) \tag{29}$$

such that:

i) the constraint (7) is satisfied exactly,

ii) the ordering (12) is preserved.

Instead of using these variables it is more convenient to use the variables V^{μ}, $T^{\mu\nu}$, $A^{\mu\nu\alpha}$ and seek "constitutive functions"

$$A^{\mu\nu\alpha}=\hat{A}^{\mu\nu\alpha}(V^{\lambda},\ T^{\lambda\rho}) \tag{30}$$

such that the constraint: i) $A^{\mu\nu}{}_{\nu_R}=-m^2V^{\mu}$ is verified and, ii) the corresponding (12) for $S^{\mu\nu\alpha}$ is preserved.

We shall solve this problem in the case where the plasma is near thermal equilibrium. As will be apparent later this solution might be applicable also to cases where the plasma is very far from thermal equilibrium.

Now, at thermal equilibrium, the heat flux q_{α} and the stress tensor $t^{\langle\mu\nu\rangle}$ vanish and the pressure p reduces to the equilibrium pressure p_{eq}.

Therefore near thermal equilibrium, we shall seek constitutive functions $\hat{A}^{\mu\nu\alpha}(V^{\lambda},\ T^{\lambda\rho})$ which are linear in q_{α}, $t^{\langle\mu\nu\rangle}$ and $\pi=p-p_{eq}$. The most general form which satisfies the constraint (7), is isotropic and has the required symmetry properties is easily shown to be

$$A^{\mu\nu\alpha}=(C_1^{\prime\,0}+C_1^{\pi}p)\left[U^{\alpha}U^{\mu}U^{\nu}+\frac{1}{2}\,g^{(\alpha\mu}U^{\nu)}\right]-$$

$$- \frac{nm^2}{2} g^{\langle\alpha\mu}U^{\nu\rangle} - 3C_3[g^{\langle\alpha\mu} + 6U^{\langle\alpha}U^{\mu}]q^{\nu\rangle} + 3C_5 t^{\langle\langle\alpha\mu\rangle}U^{\nu\rangle} \tag{31}$$

where the quantities $C_1'^{0}$, C_1^{π}, C_3, C_5 depend only on the thermal equilibrium variables n, T.

Now we shall determine the coefficients $C_1'^{0}$, C_1^{π}, C_3, C_5 up to an arbitrary function of density by following an approach taylored on the entropy principle of Liu, Müller and Ruggeri [6]. We stress here that we are using the entropy principle only as a heuristic and mathematical method and we are not committed to any deep physical interpretation.

In terms of the variables V^{μ}, $T^{\mu\nu}$, $A^{\mu\nu\alpha}$, equations (8), (9), (10) can be rewritten as

$$\frac{\partial V^{\mu}}{\partial x^{\mu}} = 0 \tag{32}$$

$$\frac{\partial}{\partial x^{\mu}} T^{\mu\nu} = enF^{\nu\mu}u_{\mu} \tag{33}$$

$$\frac{\partial}{\partial x^{\mu}} A^{\mu\nu\alpha} = e(F^{\nu\mu}T^{\alpha}_{\mu} + F^{\alpha\mu}T^{\nu}_{\mu}) + I^{\nu\alpha} \tag{34}$$

where we have added a term $I^{\nu\alpha}$ representing the effect of localized collisions among the electrons.

According to the entropy principle method one assumes that one can introduce a four-vector h^{α} (called the entropy flux) as a constitutive function of V^{μ}, $T^{\mu\nu}$,

$$h^{\alpha} = \hat{h}^{\alpha}(V^{\mu}, T^{\lambda\rho}) \tag{35}$$

such that the entropy production inequality

$$\frac{\partial h^{\alpha}}{\partial x^{\alpha}} \geq 0 \tag{36}$$

holds as a consequence of the field equations (32), (33), (34).

According to Liu's theorem [6] the constraint (36) can be exploited by introducing Lagrange multipliers ξ, Λ, $\Sigma_{\nu\alpha}$ corresponding to eqs. (32), (33), (34) respectively, such that the following inequality

$$\frac{\partial h^{\alpha}}{\partial x^{\alpha}}+\zeta\frac{\partial V^{\alpha}}{\partial x^{\alpha}}+\Lambda_{\beta}\left(\frac{\partial T^{\beta\alpha}}{\partial x^{\alpha}}-eF^{\beta\alpha}V_{\alpha}\right)+\Sigma_{\beta\gamma}\left[\frac{\partial A^{\alpha\beta\gamma}}{\partial x^{\alpha}}-\right.$$

$$\left.-eF^{\alpha\gamma}T^{\beta}_{\alpha}-eF^{\alpha\beta}T^{\gamma}_{\alpha}-I^{\beta\gamma}\right]\geq 0 \qquad (37)$$

holds for all fields V^{μ}, $T^{\mu\nu}$. The quantities ζ, Λ_{β}, $\Sigma_{\beta\gamma}$ are considered as constitutive quantities, functions of V^{μ}, $T^{\lambda\rho}$.

The inequality (37) is slightly different from the one considered by Liu, Müller and Ruggeri, because there is an extra term appearing due to the compling with the electromagnetic field $F^{\mu\nu}$. Due to the arbitrariness of $F^{\mu\nu}$ the following consequences are obtained from eq. (37)

$$\frac{\partial h^{\alpha}}{\partial x^{\alpha}}+\zeta\frac{\partial V^{\alpha}}{\partial x^{\alpha}}+\Lambda_{\beta}\frac{\partial T^{\beta\alpha}}{\partial x^{\alpha}}+\Sigma_{\beta\gamma}\left[\frac{\partial A^{\alpha\beta\gamma}}{\partial x^{\alpha}}-I^{\beta\gamma}\right]\geq 0 \qquad (38)$$

$$\Lambda_{[\alpha}V_{\beta]}+2T^{\gamma}_{\ [\alpha}\Sigma_{\beta]\gamma}=0 \qquad (39)$$

Eq. (38) is exactly the entropy inequality studied by Liu, Müller and Ruggeri [6]. Liu, Müller and Ruggeri [6] show how to exploit eq. (38) in order to obtain restrictions on the coefficients of the representation (31). By assuming an expression for h^{α} which is quadratic in the deviations from thermal equilibrium q_{α}, $t^{\langle\alpha\beta\rangle}$, π, they satisfy eq. (38) up to the third order from thermal equilibrium, to the first order from equilibrium for ζ, Λ_{β}, $\Sigma_{\beta\gamma}$, $A^{\alpha\beta\gamma}$, $I^{\beta\gamma}$.

Before we make use of their results for $A^{\alpha\beta\gamma}$ we must check that eq. (39) holds, at least to first order from thermal equilibrium.

Now, at thermal equilibrium, $T^{\mu\nu}$ has the form

$$T^{\mu\nu}_{eq}=EU^{\mu}U^{\nu}+p_{eq}h^{\mu\nu}$$

with

$$E=nm\left(G-\frac{1}{\gamma}\right), \qquad p_{eq}=\frac{nm}{\gamma}.$$

Furthermore, from Ref. [6] one has the following expressions for Λ_{α}, $\Sigma_{\beta\gamma}$

$$\Lambda^{\alpha}=N_1U^{\alpha}+N_2q^{\alpha} \qquad (40)$$

$$\Sigma^{\alpha\beta}=M_1t^{\langle\alpha\beta\rangle}+M_2h^{\alpha\beta}+M_3U^{(\alpha}q^{\beta)} \qquad (41)$$

with the coefficients complicated expressions functions of γ, n, and satisfying the relationship

$$N_2+\frac{mG}{2}M_3=0 \ .$$

By substituting the expressions (40), (41) into eq. (39) and using the above relationship, eq. (39) is satisfied identically.

Now we can use the expression for $A^{\mu\nu\alpha}$ obtained by Liu, Müller and Ruggeri.

The coefficients in eq. (31) are given by

$$\begin{cases} C_1^{0'}=nm^2\left[1+\frac{6}{\gamma}G\right]-\frac{nm}{\gamma}\cdot C_1^{\pi} \\[2ex] C_1^{\pi}=-6m\dfrac{1}{\frac{3}{\gamma}-\left[2-\frac{20}{\gamma^2}\right]G-\frac{13}{\gamma}G^2+2G^3}\left\{2-\frac{5}{\gamma^2}+\left[\frac{19}{\gamma}-\frac{30}{\gamma^3}\right]G-\right. \\[2ex] \qquad \left. -\left[2-\frac{45}{\gamma^2}\right]G^2-\frac{9}{\gamma}G^3-\frac{m^3}{3}\left[1-\frac{1}{\gamma^2}+\frac{5}{\gamma}G-G^2\right]\frac{K_BA}{n\gamma^5}\right\}, \qquad (42) \\[2ex] C_3=-\dfrac{m}{\gamma}\dfrac{1+\frac{6}{\gamma}G-G^2-\frac{m^3}{15}\frac{A}{n\gamma^5}}{1+\frac{5}{\gamma}G-G^2} \\[2ex] C_5=m\left(\frac{6}{\gamma}+\frac{1}{G}\right)-\frac{m^4}{15G}\frac{K_BA}{n\gamma^5} \end{cases}$$

with A an arbitrary function of the variable $\frac{n\gamma}{K_2(\gamma)}$.

From the corresponding expression for $A^{\mu\nu\alpha}$ we can now obtain $S^{\mu\nu\alpha}$ and finally we can check that the ordering (7) holds.

The identification of the parameter ε^2 is now slightly different than in the pure thermal equilibrium case. From the definition of total energy density $E=hm^4+m^4h_{\mu\nu}\theta^{\mu\nu}$ we obtain

$$G-\frac{1}{\gamma}=\frac{h^{-1}h_{\mu\nu}\theta^{\mu\nu}+1}{\sqrt{-w^{\mu}w_{\mu}}}\,. \tag{43}$$

Now, assuming the ordering of eq. (7) <u>only for</u> $\theta^{\mu\nu}$ we get

$$G-\frac{1}{\gamma}=\left[1+\varepsilon^{2}\right]^{-1/2}(1+\varepsilon^{2}+\varepsilon^{4}K^{44}) \tag{44}$$

whence we see that the limit $\varepsilon\longrightarrow 0$ corresponds again to $\gamma\longrightarrow\infty$.

In the Appendix we prove the following theorem.

<u>Theorem</u>.

Let assume the ordering of eq. (12) only for $\theta^{\mu\nu}$ and that the $\lim_{\gamma\to\infty} A\left[\frac{n\gamma}{K_2(\gamma)}\right]$ exists and is finite. Then the ordering of eq. (12) for $S^{\mu\nu\alpha}$ follows. Also from the appendix we obtain an explicit expression for $S^{\alpha\beta\gamma}$ in terms of the variables h, w^{μ}, $\theta^{\mu\nu}$,

$$S^{\alpha\beta\gamma}=-2(1+w^{\tau}w_{\tau})(-w^{\delta}w_{\delta})^{-1/2}\theta^{\mu\nu}\left[h^{(\alpha}{}_{\mu}h^{\beta}{}_{\nu}-\frac{1}{3}h^{(\alpha\beta}h_{\mu\nu}\right]w^{\gamma)}+$$

$$+4(1+w^{\tau}w_{\tau})(-w^{\delta}w_{\delta})^{-3/2}\left[\frac{1}{5}h^{(\alpha\beta}(-w^{\lambda}w_{\lambda})+w^{(\alpha}w^{\beta}\right]h^{\gamma)}{}_{\mu}\theta^{\mu\nu}w_{\nu}-$$

$$-(1+w^{\tau}w_{\tau})(-w^{\delta}w_{\delta})^{-3/2}\left\{\frac{h}{9}(1+w^{\lambda}w_{\lambda})^{2}+\left[\theta^{\mu\nu}w_{\mu}w_{\nu}(-w^{\lambda}w_{\lambda})^{-1}-\right.\right.$$

$$\left.\left.-\frac{h}{6}(1+w^{\lambda}w_{\lambda})^{2}\right]\left(\frac{2}{15}\alpha_{1}+2\right)\right\}(w^{(\alpha}w^{\beta}-w^{\sigma}w_{\sigma}h^{(\alpha\beta})w^{\gamma)}-$$

$$-2h^{(\alpha\beta}\left(\frac{w^{\gamma)}w_{\delta}}{w^{\lambda}w_{\lambda}}-\frac{3}{5}h^{\gamma)\delta}\right]\theta_{\delta\mu}{}^{\mu}w \tag{45}$$

where

$$\alpha_{1}=\alpha_{1}(h)=\lim_{\varepsilon\to\infty}-\frac{m^{3}k}{3n\gamma^{2}}A\left[\frac{n\gamma}{K_{2}(\gamma)}\right]$$

is an arbitrary function of h.

The above expression for $S^{\alpha\beta\gamma}$ provides the sought for closure for the system of moment equations (8), (9), (10).

Now we ask ourselves what is the mathematical structure of the resulting system. For the simple closure $S^{\alpha\beta\gamma}=0$ Amendt & Weitzner [4] were able to show that the resulting system was, under suitable conditions hyperbolic.

For our choice, eq.(45), the situation looks much more promising. In fact, one of the consequences of the entropy principle approach is that, in terms of the Lagrange multiplier variables ξ, Λ_α, $\Sigma_{\beta\gamma}$, the system (32), (33), (34) turns out to be symmetric. Furthermore, under the condition that the quantity

$$h'^{\alpha}u_{\alpha}$$

(where $h'^{\alpha}=h^{\alpha}+\xi V^{\alpha}+\Lambda_{\beta}T^{\beta\alpha}+\Sigma_{\beta\gamma}A^{\alpha\beta\gamma}$) is a convex function of the variables ξ, Λ_β, $\Sigma_{\beta\gamma}$, the system is symmetric hyperbolic.

Now the above condition certainly holds for the ideal gas state equation at thermal equilibrium and by continuity in a neighbourhood of equilibrium.

Therefore, with the closure given by eq. (41), we also obtain that, near thermal equilibrium, the resulting system is symmetric hyperbolic, which is, from the mathematical view point, a much stronger result than mere hyperbolicity.

Appendix

Theorem.

Let us assume that in the local rest frame of the fluid, $\theta^{\alpha\beta}$ satisfies the ordering (12.1) with $\overline{K}^{\mu\nu}$ quantities of order zero in ε.

Then the expression for $S'^{\mu\nu\alpha}$ that can be obtained from eq. (14) and (31) with the coefficients $C_1'^{0}$, C_1^{π}, C_3, C_5 given by eqs. (42), *to the leading order in* ε, under the assumption that

$$\lim_{\gamma\to\infty} - \frac{m^3 k_B}{3n\gamma^2} A\left(\frac{n\gamma}{K_2(\gamma)} \right) = \alpha_1(h)$$

exists and is finite, satisfies the following ordering $(16)_{2-4}$ with

$$\overline{K}^{ijk} - \frac{3}{5}\delta^{(ij}\overline{K}^{k)l}{}_{l} = 0 \;; \qquad \text{(A1)}$$

$$\overline{K}^{ij4} - \frac{1}{3}\delta^{ij}\overline{K}^{4l}{}_{l} = \frac{2}{3}\left(\overline{K}^{ij} - \frac{1}{3}\delta^{ij}\overline{K}^{l}{}_{l}\right) \;; \qquad \text{(A2)}$$

$$\overline{K}^{i44} = \frac{4}{3}\overline{K}^{i4} \;; \qquad \overline{K}^{444} = \left(\overline{K}^{44} - \frac{1}{6}\right)\left(\frac{2}{15}\alpha_1 + 2\right) + \frac{1}{9} \qquad \text{(A3)}$$

Proof.

From eq. (14) and (31) we obtain, in the local rest frame of the fluid, the expression $(16)_{2-4}$ for $S'^{\mu\nu\alpha}$ with

$$\overline{K}^{ijk} - \frac{3}{5}\delta^{(ij}\overline{K}^{k)l}{}_{l} = 0$$

$$\overline{K}^{ij4} - \frac{1}{3}\delta^{ij}\overline{K}^{4l}{}_{l} = \varepsilon^{-2}\left(\frac{C_5}{m} - \sqrt{1+\varepsilon^2}\right)\left(\overline{K}^{ij} - \frac{1}{3}\overline{K}^{l}{}_{l}\delta^{ij}\right)$$

$$\overline{K}^{i44} = -\varepsilon^{-2}\left(5\frac{C_3}{m} + 2\sqrt{1+\varepsilon^2}\right)\overline{K}^{i4}$$

$$\overline{K}^{444} = \varepsilon^{-6}\left[\sqrt{1+\varepsilon^2}\left(\frac{3}{\gamma}G - \varepsilon^2 - 3\overline{K}^{44}\varepsilon^4\right) + \frac{C_1^{\pi}}{6m}\left(\varepsilon^2 + \overline{K}^{44}\varepsilon^4 - \frac{3}{\gamma}\sqrt{1+\varepsilon^2}\right)\right].$$

Now, from $G = \dfrac{K_3(\gamma)}{K_2(\gamma)}$ and from (44) we obtain

$$G=1+\frac{5}{2}\frac{1}{\gamma}+\frac{15}{8}\frac{1}{\gamma^2}-\frac{15}{8}\frac{1}{\gamma^3}+\frac{135}{2^7}\frac{1}{\gamma^4}+$$

$$+\frac{45}{2^5}\frac{1}{\gamma^5}+O\left[\frac{1}{\gamma^6}\right]$$

$$\frac{1}{\gamma}=\frac{1}{3}\varepsilon^2+\frac{2}{3}\left[K^{44}-\frac{1}{3}\right]\varepsilon^4+\frac{1}{9}\left[-8K^{44}+\frac{59}{24}\right]\varepsilon^6+O(\varepsilon^8)\ .$$

We note that a comparison between (24) and (44) in thermal equilibrium gives $K_{44}=\frac{1}{6}-\frac{1}{12}\varepsilon^2+O(\varepsilon^4)$.

Then the following expansions can be derived

$$G^2=1+5\frac{1}{\gamma}+10\frac{1}{\gamma^2}+\frac{45}{8}\frac{1}{\gamma^3}-\frac{15}{4}\frac{1}{\gamma^4}+\frac{135}{2^7}\frac{1}{\gamma^5}+O\left[\frac{1}{\gamma^6}\right]$$

$$G^3=1+\frac{15}{2}\frac{1}{\gamma}+\frac{195}{8}\frac{1}{\gamma^2}+\frac{305}{8}\frac{1}{\gamma^3}+\frac{45\cdot 59}{2^7}\frac{1}{\gamma^4}-$$

$$-\frac{315}{2^5}\frac{1}{\gamma^5}+O\left[\frac{1}{\gamma^6}\right]$$

$$1-\frac{1}{\gamma^2}+\frac{5}{\gamma}G-G^2=\frac{3}{2}\frac{1}{\gamma^2}+\frac{15}{4}\frac{1}{\gamma^3}+O\left[\frac{1}{\gamma^4}\right]$$

$$2+\frac{17}{2}\frac{1}{\gamma^2}+\left(\frac{10}{\gamma}+\frac{60}{\gamma^3}\right)G-\left[2+\frac{27}{2}\frac{1}{\gamma^2}\right]G^2=$$

$$=\frac{15}{4}\frac{1}{\gamma^4}+45\frac{1}{\gamma^5}+O\left[\frac{1}{\gamma^6}\right]$$

$$\frac{3}{\gamma}+\left[-2+\frac{20}{\gamma^2}\right]G-\frac{13}{\gamma}G^2+2G^3=\frac{15}{4}\frac{1}{\gamma^4}-\frac{45}{4}\frac{1}{\gamma^5}+O\left[\frac{1}{\gamma^6}\right]$$

$$-\frac{C_1^{\pi}}{6m}=1+\left(\frac{2}{15}\alpha_1+\frac{7}{2}\right)\varepsilon^2+O(\varepsilon^4).$$

By using these expressions we easily obtain

$$\lim_{\varepsilon\to 0}\varepsilon^{-2}\left(\frac{C_5}{m}-\sqrt{1+\varepsilon^2}\right)=\frac{2}{3}$$

$$\lim_{\varepsilon\to 0}-\varepsilon^{-2}\left[5\frac{C_3}{m}+2\sqrt{1+\varepsilon^2}\right]=\frac{4}{3}$$

$$\lim_{\varepsilon\to 0} \varepsilon^{-6}\left[\sqrt{1+\varepsilon^2}\left(\frac{3}{\gamma}G-\varepsilon^2-3\overline{K}^{44}\varepsilon^4\right]+\frac{C_1^{\pi}}{6m}\left(\varepsilon^2+\overline{K}^{44}\varepsilon^4-\frac{3}{\gamma}\sqrt{1+\varepsilon^2}\right]=$$

$$=\left(\overline{K}^{44}-\frac{1}{6}\right]\left(\frac{2}{15}\alpha_1+2\right]+\frac{1}{9}\ .$$

Equations $(16)_{2-4}$ with $K^{\alpha\beta\gamma}$ given by (A1), (A2), (A3) give the constitutive function $S'^{\alpha\beta\gamma}$ and consequently $S^{\alpha\beta\gamma}$ from the inverse of (14).

This latter, expressed in covariant formm, is eq. (45).

REFERENCES

[1] R.B.Miller, Intense Charged Particle Beams(Plenum Press, N.Y., N.Y.,1982)

[2] J.G.Siambis, Phys.Fluids 22, 1372 (1979)

[3] W.A.Newcomb, Phys.Fluids 25, 846 (1982)

[4] P.Amendt and H.Weitzner, Phys.Fluids 28, 949 (1985)

[5] K.O.Friedrichs,Comm.Pure Appl.Math.,7,345 (1954)

[6] I-Shih Liu, I.Muller and T.Ruggeri, Annals of Physics 169,191 (1986)

[7] S.R.de Groot, W.A.van Leeuwen and Ch.G.van Weert, Relativistic kinetic theory., (North Holland, Amsterdam, 1980)

[8] W.Israel, Annals of Physics 100,310,1976.

RELATIVISTIC EXTENDED THERMODYNAMICS II

Ingo Müller
FB Phys. Ing. Wissensch., Technische Universität Berlin
D-1000 Berlin

1. Review

First I review briefly the general theory that was explained in the preceding paper by Ruggeri.

The objective of relativistic extended thermodynamics is the determination of the 14 fields

V^α - particle flux
$T^{\alpha\beta}$ - energy-momentum

The necessary field equations are based upon the equations of balance of

particle number $\quad V^\alpha{}_{,\alpha} = 0$
energy-momentum $\quad T^{\alpha\beta}{}_{,\beta} = 0$
fluxes $\quad A^{\alpha\beta\gamma}{}_{,\gamma} = I^{\alpha\beta} \quad$ where $\quad A_\alpha^{\alpha\gamma} = m^2c^2V^\gamma$

The symmetric flux tensor $A^{\alpha\beta\gamma}$ and the flux production $I^{\alpha\beta}$ are given by constitutive relations whose general form for a viscous, heat-conducting gas, is as follows

$$A^{\alpha\beta\gamma} = \hat{A}^{\alpha\beta\gamma}(V^\mu, T^{\mu\nu})$$
$$I^{\alpha\beta} = \hat{I}^{\alpha\beta}(V^\mu, T^{\mu\nu}) .$$

If the functions $\hat{\underset{\sim}{A}}$ and $\hat{\underset{\sim}{I}}$ are known, we may eliminate $\underset{\sim}{A}$ and $\underset{\sim}{I}$ between the balance equations and the constitutive equations and come up with a set of 14 field equations for the determination of the fields V^α, $T^{\alpha\beta}$. Every solution of these field equations is called a thermodynamic process.

The form of the constitutive functions $\hat{\underset{\sim}{A}}$ and $\hat{\underset{\sim}{I}}$ is strongly restricted by the following physical principles and requirements.

i.) Entropy Principle

The entropy inequality

$$h^\alpha{}_{,\alpha} \geq 0 \quad \text{where} \quad h^\alpha = \hat{h}^\alpha(V^\mu, T^{\mu\nu})$$

must be satisfied for all thermodynamic processes.

ii.) Principle of Relativity

The field equations must have the same form in all frames. Thus, in particular, the constitutive functions $\hat{A}, \hat{I}$ and $\hat{h}$ must be invariant under a change of space-time coordinates from x^δ to $\overset{*}{x}{}^\delta$. I.e. we must have

$$\overset{*}{A}{}^{\alpha\beta\gamma} = A^{\alpha\beta\gamma}(\overset{*}{V}{}^\mu, \overset{*}{T}{}^{\mu\nu})$$
$$\overset{*}{I}{}^{\alpha\beta} = I^{\alpha\beta}(\overset{*}{V}{}^\mu, \overset{*}{T}{}^{\mu\nu})$$
$$\overset{*}{h}{}^{\alpha} = h^{\alpha}(\overset{*}{V}{}^\mu, \overset{*}{T}{}^{\mu\nu}),$$

so that the constitutive functions are isotropic functions of their variables.

iii.) Requirement of Hyperbolicity

The field equations must be symmetric hyperbolic. This guarantees the local existence of solutions of Cauchy problems and finite speeds for all propagating waves.

2. Results

Ruggeri has demonstrated the severity of the restrictions that are implied by the entropy principle. They may be summarized in the form

$$V^\alpha = \frac{\partial h'^\alpha}{\partial \xi}, \quad T^{\alpha\beta} = \frac{\partial h'^\alpha}{\partial \Lambda_\beta}, \quad A^{\alpha\beta\gamma} = \frac{\partial h'^\alpha}{\partial \Sigma_{\beta\gamma}} - \frac{1}{4} g^{\beta\gamma} g_{\mu\nu} \frac{\partial h'^\alpha}{\partial \Sigma_{\mu\nu}}$$

which shows that $V^\alpha, T^{\alpha\beta}, A^{\alpha\beta\gamma}$ - expressed in terms of the Lagrange multipliers $\xi, \Lambda_\beta, \Sigma_{\beta\gamma}$ - can be obtained from a single "vectorial potential", viz. h'^α. This statement implies a large number of integrability conditions which, in turn, imply the desired restrictions. Unfortunately, because of the non-suggestive character of the Lagrange multipliers, these restrictions have no suggestive significance.

In order to have suggestive restrictions we must return to the old set ($V^\alpha, T^{\alpha\beta}$) of variables or, even better, to the set ($n, U^\alpha, \pi, t^{\langle\alpha\beta\rangle}, q^\alpha, e$) which is defined in terms of ($V^\alpha, T^{\alpha\beta}$) by the decompositions

$$V^\alpha = nU^\alpha$$
$$T^{\alpha\beta} = t^{\langle\alpha\beta\rangle} + (p(n,e)+\pi)h^{\alpha\beta} + \frac{1}{c^2}(U^\alpha q^\beta + U^\beta q^\alpha) + \frac{1}{c^2} e U^\alpha U^\beta,$$

where $h^{\alpha\beta} = \frac{1}{c^2} U^\alpha U^\beta - g^{\alpha\beta}$ is a projector. n and e are the particle and energy density respectively while U^α is the 4-velocity. $t^{\langle\alpha\beta\rangle}$ is the stress deviator, q^α the heat flux and π is the dynamic pressure, i.e. the non-equilibrium part of the pressure. The latter three quantities vanish in equilibrium. In equilibrium the pressure is a function of n, e, given by the thermal equation of state $p(n, e)$.

We recall that the principle of relativity requires $\hat{\underset{\sim}{A}}, \hat{\underset{\sim}{I}}$ and $\hat{\underset{\sim}{h}}$ to be isotropic functions. If we restrict the attention to linear functions $\hat{\underset{\sim}{A}}, \hat{\underset{\sim}{I}}$ in terms of $\pi, t^{\langle\alpha\beta\rangle}, q^\alpha$ and to quadratic functions $\hat{\underset{\sim}{h}}$ we thus obtain

$$I^{\alpha\beta} = B_1^\pi \pi g^{\alpha\beta} - \frac{4}{c^2} B_1^\pi \pi U^\alpha U^\beta + B_3 t^{\langle\alpha\beta\rangle} + B_4 (q^\alpha U^\beta + q^\beta U^\alpha)$$

$$\begin{aligned} A^{\alpha\beta\gamma} = {} & (C_1^0 + C_1^\pi \pi) U^\alpha U^\beta U^\gamma + \frac{c^2}{6}(nm^2 - C_1^0 - C_1^\pi \pi)(g^{\alpha\beta}U^\gamma + g^{\beta\gamma}U^\alpha + g^{\gamma\alpha}U^\beta) + \\ & + C_3 (g^{\alpha\beta} q^\gamma + g^{\beta\gamma} q^\alpha + g^{\gamma\alpha} q^\beta) - \\ & - \frac{6}{c^2} C_3 (U^\alpha U^\beta q^\gamma + U^\beta U^\gamma q^\alpha + U^\gamma U^\alpha q^\beta) + \\ & + C_5 (t^{\langle\alpha\beta\rangle} U^\gamma + t^{\langle\beta\gamma\rangle} U^\alpha + t^{\langle\gamma\alpha\rangle} U^\beta) . \end{aligned}$$

$$\begin{aligned} h^\alpha = {} & (h|_E + A_1^\pi \pi + A_1^{\pi^2} \pi^2 + A_1^q q^\beta q_\beta + A_1^t t^{\langle\beta\gamma\rangle} t_{\langle\beta\gamma\rangle}) U^\alpha + \\ & + (A_2^0 + A_2^\pi \pi) q^\alpha + A_3^0 t^{\langle\alpha\beta\rangle} q_\beta \end{aligned}$$

The coefficients $h|_E$, A, B and C may all be functions of n and e and the entropy principle will restrict these functions in their generality.

We may group the results of the entropy principle in three parts:

i.) The evaluation of the entropy principle suggests new scalar equilibrium variables. Instead of (n, e) it suggests to choose (T, α), viz. the

absolute temperature T

Fucacity $\qquad \alpha = - \dfrac{e - Th|_E + p}{nT}$

T and α are the "natural" equilibrium variables of thermodynamics and statistical mechanics. If only the thermal equation of state $p(T, \alpha)$ is known we may easily transform from (n, e) to (T, α), because in the course of the calculations one obtains the relations

$$n = -\frac{1}{T}\dot{\mu}$$

$$e = \mu' - \mu .$$

The dot denotes differentiation with respect to α and the prime denotes differentiation with respect to $\ln T$.

ii.) The evaluation of the entropy principle relates all coefficients C and A in the representations of $A^{\alpha\beta\gamma}$ and h^{α} respectively to the thermal equation of state. Indeed one obtains

$$C_1^0 = -\frac{1}{2c^2}\frac{T_1'}{T} \qquad A_1^{\tau} = 0$$

$$C_1^{\tau} = -\frac{3}{c^2}\frac{1}{T}\frac{\left[\!\left[\begin{matrix} -\ddot{\mu} & \dot{\mu}-\dot{\mu}' & \dot{T}_1 \\ \dot{\mu}-\dot{\mu}' & \mu'-\mu'' & T_1'-T_1 \\ \dot{T}_1 & T_1'-T_1 & \frac{5}{3}T_2 \end{matrix}\right]\!\right]}{\left[\!\left[\begin{matrix} -\ddot{\mu} & \dot{\mu}-\dot{\mu}' & \dot{T}_1 \\ \dot{\mu}-\dot{\mu}' & \mu'-\mu'' & T_1'-T_1 \\ -\dot{\mu} & -\mu' & \frac{5}{3}T_1 \end{matrix}\right]\!\right]} \qquad A_1^{\tau 2} = -\frac{c^2}{6}C_1^{\tau}\frac{\left[\!\left[\begin{matrix} \ddot{\mu} & \dot{\mu}-\dot{\mu}' \\ \dot{\mu}-\dot{\mu}' & \mu'-\mu'' \end{matrix}\right]\!\right]}{\left[\!\left[\begin{matrix} -\ddot{\mu} & \dot{\mu}-\dot{\mu}' & \dot{T}_1 \\ \dot{\mu}-\dot{\mu}' & \mu'-\mu'' & T_1'-T_1 \\ -\dot{\mu} & -\mu' & \frac{5}{3}T_1 \end{matrix}\right]\!\right]}$$

$$C_3 = -\frac{1}{2}\frac{1}{T}\frac{\left[\!\left[\begin{matrix} \dot{\mu} & -\dot{T}_1 \\ T_1 & T_2 \end{matrix}\right]\!\right]}{\left[\!\left[\begin{matrix} \dot{\mu} & -\dot{T}_1 \\ \mu' & T_1-T_1' \end{matrix}\right]\!\right]} \qquad A_1^q = -\frac{1}{2c^2}\frac{-\frac{1}{T}\dot{T}_1 - 10C_3\dot{\mu}}{\left[\!\left[\begin{matrix} \dot{\mu} & -\dot{T}_1 \\ \mu' & T_1-T_1' \end{matrix}\right]\!\right]}$$

$$C_5 = -\frac{1}{2}\frac{1}{T}\frac{T_2}{T_1} \qquad A_1^t = -\frac{1}{T}C_5\frac{1}{T_1}$$

$$A_2^0 = \frac{1}{T}$$

$$A_2^{\tau} = -\frac{\frac{1}{T}\dot{T}_1 - \frac{c^2}{3}C_1^{\tau}\dot{\mu}}{\left[\!\left[\begin{matrix} \dot{\mu} & -\dot{T}_1 \\ \mu' & T_1-T_1' \end{matrix}\right]\!\right]}$$

$$T_1 = T^6 \qquad \left[-2m^2c^2\int\frac{\dot{\mu}}{T^7}dT + A_1(\alpha)\right]$$

$$T_2 = T^8\left\{2m^2c^2\int\frac{1}{T^3}\left[-2m^2c^2\int\frac{\ddot{\mu}}{T^7}dT + A_1(\alpha)\right]dT + A_2(\alpha)\right\}$$

$$A_3^0 = -C_3\frac{2}{T_1}$$

Most of the coefficients are given in terms of determinants, because nothing is gained by expanding these expressions. Note that most of the C's and A's contain either the function $T_1(T,\alpha)$ or $T_2(T,\alpha)$ or both. But these two functions are again expressible in terms of $\mu(T,\alpha)$ by integration, as shown in the frame, except that the integrations bring in two functions of α , viz. $A_1(\alpha)$ and $A_2(\alpha)$. This fact qualifies the prior statement that all C's and A's are known from the thermal equation of state $\mu(T,\alpha)$.

iii.) In the evaluation of the entropy principle there is a residual inequality that gives rises to inqualities concerning the coefficients B in $I^{\alpha\beta}$. They read

$$B_1^{\pi}\frac{\begin{vmatrix} -\ddot{\mu} & \dot{\mu}-\dot{\mu}' \\ \dot{\mu}-\dot{\mu}' & \mu'-\mu'' \end{vmatrix}}{\begin{vmatrix} \ddot{\mu} & \dot{\mu}-\dot{\mu}' & \dot{T}_1 \\ \dot{\mu}-\dot{\mu}' & \mu'-\mu'' & T_1'-T_1 \\ -\dot{\mu} & -\mu' & \frac{5}{3}T_1 \end{vmatrix}} \geq 0 \qquad B_4\frac{\dot{\mu}}{\begin{vmatrix} \dot{\mu} & -\dot{T}_1 \\ \mu' & T_1-T_1' \end{vmatrix}} \geq 0 \qquad B_3\frac{1}{T_1} \leq 0 .$$

and we shall come back to their significance.

The requirement of hyperbolicity as well gives rise to inequalities. Among those we have

$$\begin{bmatrix} \frac{1}{T}\ddot{\mu} & -\frac{1}{2T}(\dot{\mu}-\dot{\mu}') \\ -\frac{1}{2T}(\dot{\mu}-\dot{\mu}') & \frac{1}{4T}(\mu''-\mu') \end{bmatrix} - \text{pos. def.} \qquad A_1^{\pi 2}<0 \quad A_1^{q}>0 \quad A_1^{t}<0 .$$

The positive definiteness of the matrix implies the customary thermodynamic stability conditions according to which the specific heats and the compressibilities must be positive. The conditions on the A's guarantee that the entropy has a maximum in equilibrium.

Ordinary thermodynamics as opposed to extended thermodynamics, which is described here, has as basic fields only n, U^α and e. Ordinary thermodynamics must therefore require constitutive equations for π, q^α and $t^{<\alpha\beta>}$. Such equations can be derived in an approximate manner in extended thermodynamics as follows: We rewrite the balance of fluxes with the explicit flux production

$$A^{\alpha\beta\gamma}{}_{,\gamma} = \left(g^{\alpha\beta} - \frac{4}{c^2}U^\alpha U^\beta\right) B_1^\pi \pi + B_3 t^{\langle\alpha\beta\rangle} + B_4\left(q^\alpha U^\beta + q^\beta U^\alpha\right)$$

and insert the equilibrium value

$$A^{\alpha\beta\gamma}\big|_E = C_1^0 U^\alpha U^\beta U^\gamma + \frac{c^2}{6}\left(nm^2 - C_1^0\right)\left(g^{\alpha\beta}U^\gamma + g^{\beta\gamma}U^\alpha + g^{\gamma\alpha}U^\beta\right)$$

on its left hand side. If we then decompose the flux balance into spatial and temperal parts we obtain the equations

$$\pi = \underbrace{-\frac{1}{2T}\frac{1}{B_1^\pi}\frac{\begin{bmatrix} \overset{\circ\circ}{\mu} & \mathring{\mu}-\mathring{\mu}' & \dot{T}_1 \\ \mathring{\mu}-\mathring{\mu}' & \mu'-\mu'' & T_1'-T_1 \\ -\mathring{\mu} & -\mu' & \frac{5}{3}T_1 \end{bmatrix}}{\begin{bmatrix} -\overset{\circ\circ}{\mu} & \mathring{\mu}-\mathring{\mu}' \\ \mathring{\mu}-\mathring{\mu}' & \mu'-\mu'' \end{bmatrix}}}_{\nu + \frac{2}{3}\mu \geq 0} \left\{U^\alpha{}_{,\alpha}\right\} \qquad \text{— } \nu = \text{BULK VISCOSITY}$$

$$q_\mu = \underbrace{-\frac{1}{2T}\frac{1}{B_4}\frac{\begin{bmatrix} \mathring{\mu} & -\dot{T}_1 \\ \mathring{\mu}' & T_1-T_1' \end{bmatrix}}{\mathring{\mu}}}_{\kappa} \left\{h^\alpha_\mu\left(\frac{1}{T}T_{,\alpha} - \frac{1}{c^2}\frac{dU_\alpha}{d\tau}\right)\right\} \qquad \text{— } \kappa = \text{HEAT CONDUCTIVITY}$$

$$t_{\mu\nu} = \underbrace{-\frac{1}{2T}\frac{1}{B_3}T_1}_{\mu} \left\{h^\beta_\mu h^\gamma_\nu U_{\langle\beta,\gamma\rangle}\right\} \qquad \text{— } \mu = \text{VISCOSITY}$$

They are formally identical to the constitutive equations of Navier Stokes and Fourier - the latter in the form appropriate to relativistic theories. We may thus interpret the coefficients indicated by frames as viscosities and heat conductivity. The previously derived inequalities for B_1^π, B_3 and B_4 are thus seen to indicate positive viscosities and positive heat conductivity.

More importantly these interpretations offer the possibility to measure B_1^{*}, B_3 and B_4 by measurements of the viscosities and the heat conductivity.

This finishes the summary of the results of the constitutive theory. We have seen that the thermal equation of state $p = p(T, \alpha)$ plays an important role and that function is not furnished by the general theory. It must either be measured or determined from statistical mechanics of gases in equilibrium. Since measurement is difficult in a relativistic gas, we shall rely on statistical mechanics.

3. Thermal Equation of State for an Ideal Gas

From the kinetic theory of gases we know that the energy momentum tensor is a second moment of the distribution function f of the gas

$$T^{\alpha\beta} = c \int p^{\alpha} p^{\beta} f \, dP .$$

Here p^{α} is the energy-momentum vector of an atom and dP is the invariant element of momentum space. In particular, in equilibrium this relation implies an expression for the equilibrium pressure

$$p = \frac{1}{3} c \, h_{\alpha\beta} \int p^{\alpha} p^{\beta} f|_E \, dP ,$$

where $f|_E$ is given by the Maxwell-Jüttner distribution

$$f|_E = \frac{y}{\exp\left\{\frac{\alpha}{k} + \frac{U_{\alpha} p^{\alpha}}{kT}\right\} \mp 1}$$

k is the Boltzmann constant and y determines the minimal element of the phase space that can accommodate a state. The $\pm$ sign refers to Bosons and Fermions respectively. We see that $f|_E$ depends on the absolute temperature T and on the fugacity α so that insertion into the above formula for p gives the thermal equation of state $p = p(T, \alpha)$.

Before I proceed to exploit special cases I wish to illustrate the spectrum of properties of ideal gases by the following table in which the framed field contains the general Maxwell-Jüttner distribution - albeit in the rest frame of the gas - and where the other fields show special forms of $f|_E$ appropriate to various limiting cases. The relativistic influence becomes more important as the "relativistic coldness" $\gamma = \frac{mc^2}{kT}$ decreases. Note that according to the table the Maxwell-Jüttner distribution contains the Maxwell distribution and the Planck distribution as opposite limiting cases.

$\gamma = \frac{mc^2}{kT}$	NON-RELATIVISTIC	RELATIVISTIC	ULTRA-RELATIVISTIC
NON-DEGENERATE	$g e^{-(\frac{\alpha}{k}+\gamma)} e^{-\frac{p^2}{2mkT}}$ MAXWELL	$g e^{-\frac{\alpha}{k}} e^{-\gamma\sqrt{1+\frac{p^2}{m^2c^2}}}$	$g e^{-\frac{\alpha}{k}} e^{-\frac{cp}{kT}}$
DEGENERATE	$\frac{g}{e^{(\frac{\alpha}{k}+\gamma)} e^{\frac{p^2}{2mkT}} \mp 1}$	$\frac{g}{\exp\left\{\frac{\alpha}{k}+\gamma\sqrt{1+\frac{p^2}{m^2c^2}}\right\} \mp 1}$	$\frac{g}{\exp\left\{\frac{\alpha}{k}+\frac{cp}{kT}\right\} \mp 1}$
STRONGLY DEGENERATE FERMI	$\begin{cases} g & 0 \le p < \sqrt{-2mkT(\frac{\alpha}{k}+\gamma)} \\ 0 & p > \sqrt{-2mkT(\frac{\alpha}{k}+\gamma)} \end{cases}$	$\begin{cases} g & 0 \le p < mc\sqrt{\left(\frac{\alpha/k}{\gamma}\right)^2-1} \\ 0 & p > mc\sqrt{\left(\frac{\alpha/k}{\gamma}\right)^2-1} \end{cases}$	$\begin{cases} g & 0 \le p < -\alpha\frac{T}{c} \\ 0 & p > -\alpha\frac{T}{c} \end{cases}$
STRONGLY DEGENERATE BOSE	$\frac{g}{e^{\frac{p^2}{2mkT}}-1}$ $(p \neq 0)$	$\frac{g}{e^{\gamma(\sqrt{1+\frac{p^2}{m^2c^2}}-1)}-1}$ $(p \neq 0)$	$\frac{g}{e^{\frac{cp}{kT}}-1}$ $(p \neq 0)$ PLANCK

4. Results for Degenerate Relativistic Gases

In no case, except the non-relativistic, non-degenerate limit can the integral for the calculation of $p(T,\alpha)$ be calculated explicitly. These integrals are functions of T and α which we must consider as the relevant special functions for the case under consideration. In the general case these special functions have the form

$$I_m(\alpha,\gamma) = \int_0^\infty \frac{\cosh m\vartheta}{\exp\left\{\frac{\alpha}{k}+\gamma\cosh\vartheta\right\} \mp 1}\, d\vartheta .$$

With $I_{m,n} = I_m - I_n$ we thus obtain for the thermal equation of state

$$p(T,\alpha) = 4\pi g\, m^4 c^5 \frac{1}{24}\left(I_{4,2} - 3 I_{2,0}\right)$$

and for the coefficients C we get

$$C_1^0 = 4\pi g m^3 c^3 \frac{1}{24}\left(4I_{3,1} + I_{5,3}\right)$$

$$C_1^1 = -\frac{3}{2}\frac{m}{c^2}\,\frac{\begin{vmatrix} I_2 & 3I_3+I_1 & 2I_4 \\ 3I_3+I_1 & 8(I_4+I_2) & 5I_5+3I_3 \\ 2I_4 & 5I_5+3I_3 & 3I_6+I_2+\frac{A}{\gamma^6} \end{vmatrix}}{\begin{vmatrix} I_1 & 2I_2 & 3I_3-I_1 \\ 2I_2 & 3I_3+I_1 & 4I_4 \\ 3I_3-I_1 & 4I_4 & 5I_5-3I_3+2I_1 \end{vmatrix}}$$

$$C_3 = -\frac{4}{5} m \,\frac{\begin{vmatrix} 3I_{3,1} & 2I_{4,2} \\ 5I_{5,3}+2I_{3,1} & 3I_{6,2}-\frac{A}{\gamma^6} \end{vmatrix}}{\begin{vmatrix} 12 I_{3,1} & 8I_{4,2} \\ 8I_{4,2} & 5I_{5,3}+2I_{3,1} \end{vmatrix}}$$

$$C_5 = \frac{1}{5} m \,\frac{3I_{6,4} - 5I_{4,2} + \frac{A}{\gamma^6}}{I_{5,3} - 2I_{3,1}}$$

Similar expressions could be given for the A's but these are left out for brevity.

Note that only one function $A(\alpha)$ occurs in these results. This is the function which was formerly called the function $A_2(\alpha)$, because $A_1(\alpha)$ turns out to be zero in statistical mechanics.

5. Results for Non-Degenerate Relativistic Gas

In this case the relevant special functions are modified Bessel functions of order 2, viz.

$$K_n(\gamma) = \int_0^\infty \cosh n\theta \, e^{-\gamma \cosh\theta} \, d\theta .$$

The thermal equation of state reads

$$p(T,\alpha) = nkT \qquad \text{with} \qquad n = e^{-\frac{\alpha}{k}} 4\pi y m^3 c^3 \frac{K_2}{\gamma}$$

and the C's assume the forms

$$C_1^0 = n m^2 \left(1 + 18 \frac{G}{\gamma}\right)$$

$$C_1^1 = -\frac{6m}{c^2} \frac{\left(2 - \frac{5}{\gamma^2}\right) + \left(\frac{19}{\gamma} - \frac{30}{\gamma^3}\right) G - \left(2 - \frac{45}{\gamma^2}\right) G^2 - \frac{9}{\gamma} G^3 + \left(1 - \frac{1}{\gamma^2} + \frac{5}{\gamma} G - G^2\right) \frac{m^3 c^3}{48} \frac{A}{n \gamma^5}}{\frac{3}{\gamma} - \left(2 - \frac{20}{\gamma^2}\right) G - \frac{13}{\gamma} G^2 + 2 G^3}$$

$$C_3 = -\frac{m}{\gamma} \frac{1 + \frac{6}{\gamma} G - G^2 + \frac{m^3 c^3}{240} \frac{A}{n \gamma^5}}{1 + \frac{5}{\gamma} G - G^2}$$

$$C_5 = m \left(\frac{6}{\gamma} + \frac{1}{G}\right) + \frac{m}{G} \frac{m^3 c^3}{240} \frac{A}{n \gamma^5}$$

where G stands for K_3 / K_2.

6. Results for Degenerate Non-Relativistic Gas

In the case of degenerate non-relativistic gases the relevant special functions are

$$i_n\left(\frac{\alpha}{k} + \gamma\right) = \int_0^\infty \frac{x^n}{\exp\left\{\frac{\alpha}{k} + \gamma + x^2\right\} \mp 1} \, dx$$

and the thermal equation of state reads

$$p(T,\alpha) = \frac{2}{3}\left(e - n m c^2\right) \qquad \text{or}$$

$$p(T,\alpha) = 4\pi y \sqrt{2}^3 (kT)^{5/2} m^{3/2} i_4\left(\frac{\alpha}{k} + \gamma\right).$$

The coefficients C assume the forms

$$C_1^0 = n m^2 \left(1 + \frac{6}{\gamma}\right)$$

$$C_1^\pi = -6 \frac{m^2}{kT} \frac{1}{\gamma}$$

$$C_3 = -\frac{2}{5} m - \frac{27 i_2 i_8 - 35 i_4 i_6}{105 i_2 i_6 - 125 i_4^2} \frac{m}{\gamma} .$$

$$C_5 = m + \frac{7}{5} m \frac{i_6}{i_4} \frac{1}{\gamma}$$

7. Transition to Non-Equilibrium Extended Thermodynamics

In the relativistic gas we have quite naturally had 14 fields including the dynamic pressure whereas in a non-relativistic gas there is no such quantity. Accordingly we usually consider non-relativistic extended thermodynamics as a theory with 13 basic fields, viz. densities of mass, momentum, momentum flux and energy flux.

Now the question arises what happens to the 14th variable, i.e. the dynamic pressure, and the corresponding field equation in the non-relativistic limit. In order to answer that question we rely on the kinetic interpretations of the moments, viz.

relativistic

$$A^{\alpha_1\alpha_2 \ldots \alpha_n} = c \int p^{\alpha_1} p^{\alpha_2} \ldots p^{\alpha_n} f \, dP$$

where $dP = \frac{1}{p_0} d\underset{\sim}{p}$ is the invariant element of momentum space and where $p^\alpha p_\alpha = m^2 c^2$. The latter relation implies

$$p_0 = mc \sqrt{1 + \frac{p^2}{m^2 c^2}}$$

$$p_0 \approx mc \left(1 + \frac{1}{2} \frac{p^2}{m^2 c^2} - \frac{1}{8} \frac{p^4}{m^4 c^4} + O\left(\frac{1}{c^6}\right)\right)$$

non-relativistic

$$F^{a_1 a_2 \ldots a_n} = m \int c^{a_1} c^{a_2} \ldots c^{a_n} f_c \, d\underset{\sim}{p},$$

where c^a is the velocity of an atom and f_c is the classical distribution function.

If the relativistic balance equations are decomposed into spatial and temporal parts and if the relativistic moments are approximated according to the above approximate form of μ_0 we obtain with little calculation the corresponding non-relativistic balance equations as shown by the following juxtaposition.

$$
\begin{array}{ll}
V^{\alpha}_{,\alpha}=0: & \frac{1}{c}\frac{\partial A^{0}}{\partial t}+\frac{\partial A^{b}}{\partial x^{b}}=0 \\
T^{\alpha\beta}_{,\beta}=0 & \left[\begin{array}{l} \frac{1}{c}\frac{\partial A^{a0}}{\partial t}+\frac{\partial A^{ab}}{\partial x^{b}}=0. \\ \frac{1}{c}\frac{\partial A^{00}}{\partial t}+\frac{\partial A^{00b}}{\partial x^{b}}=0. \end{array}\right. \\
A^{\alpha\beta\gamma}_{,\gamma}=0 & \left[\begin{array}{l} \frac{1}{c}\frac{\partial A^{\langle ab\rangle 0}}{\partial t}+\frac{\partial A^{\langle ab\rangle c}}{\partial x^{c}}=I^{\langle ab\rangle} \\ \frac{1}{c}\frac{\partial A^{a00}}{\partial t}+\frac{\partial A^{a0c}}{\partial x^{c}}=I^{a0} \\ \frac{1}{c}\frac{\partial A^{000}}{\partial t}+\frac{\partial A^{00c}}{\partial x^{c}}=I^{00} \end{array}\right.
\end{array}
$$

$$
\begin{array}{l}
\frac{\partial F}{\partial t}+\frac{\partial F^{c}}{\partial x^{c}}=0 \\
\frac{\partial F^{a}}{\partial t}+\frac{\partial F^{ac}}{\partial x^{c}}=0 \\
\frac{\partial \frac{1}{2}F^{aa}}{\partial t}+\frac{\partial \frac{1}{2}F^{aac}}{\partial x^{c}}=0 \\
\frac{\partial F^{\langle ab\rangle}}{\partial t}+\frac{\partial F^{\langle ab\rangle c}}{\partial x^{c}}=\frac{1}{m}I^{\langle ab\rangle} \\
\frac{\partial \frac{1}{2}F^{abb}}{\partial t}+\frac{\partial \frac{1}{2}F^{abbc}}{\partial x^{c}}=\frac{c}{m}I^{a0} \\
\frac{\partial F^{aabb}}{\partial t}+\frac{\partial F^{aabbc}}{\partial x^{c}}=\frac{4c^{2}}{m}I^{00}
\end{array}
$$

We conclude that the 14th relativistic equation, the one for the dynamic pressure π (say), corresponds in the non-relativistic case to the doubly contracted 4th moment F^{aabb}. Indeed comparison of the moments themselves shows that we must set

$$\lambda = -\frac{1}{12c^2}\Delta,$$

where Δ is the non-equilibrium part of F^{aabb}. It is thus obvious that the dynamic pressure in a gas is a relativistically small quantity which can be neglected in non-relativistic theories.

References

The preceding paper by Ruggeri and the present one are up-dated versions of the paper

Liu, I-Shi, Müller, I., Ruggeri, T. Relativistic Thermodynamics of Gases. Ann. of Physics 169 (1986)

The interested reader is referred to that paper for more detail and for a discussion of the sense in which the present theory incorporates the older theories by Müller and Israel and makes them more specific.

RELATIVISTIC EXTENDED THERMODYNAMICS:
General Assumptions and Mathematical Procedure.

Tommaso RUGGERI
Dipartimento di Matematica - Università di Bologna (Italy)

Introduction.

As well known the pioneering papers of Müller |1| and Israel |2| are the first tentative to obtain a causal relativistic phenomenological theory that gives a system of equations of hyperbolic type such that wave velocities are finite in agreement with the relativity principle. This approach was based substantially on the idea to modify the Gibbs relation in non equilibrium. The reader will find all of this presentation in the lectures of Israel in this volume. This procedure was fundamental for long time for his simplicity, but a more refined analysis showed that several degrees of freedom are left and as a consequence some assumptions do not seem completely justified from a "rational" point of view and moreover the equations so obtained offered some mathematical inconvenient |3| .
For the previous reason quite recently Liu, Müller and Ruggeri have analysed the possibility to have a new theory that starts with very few natural assumptions and use only universal principles. This tentative was presented in the paper |4| and in this lecture I give with more details the general assumptions and the mathematical tool of this new approach of non equilibrium thermodynamics of relativistic fluids that we call "Extended Thermodynamics" (E.T.).

Assumptions of the Extended Thermodynamics.

- The objective of extended thermodynamics of relativistic fluids is the determination of the 14 fields

$$V^{\alpha}(x^{\mu}) \quad \text{- particle, particle flux vector;} \qquad (1)$$
$$T^{\alpha\beta}(x^{\mu}) \quad \text{- stress, energy momentum tensor.}$$

- For the determination of the 14 state variables one needs the field equations i.e. the conservation laws of particle number and energy - momentum,

$$V^{\alpha}{}_{,\alpha} = 0 \, , \qquad (2)$$

$$T^{\alpha\beta}{}_{,\beta} = 0 \, , \qquad (3)$$

and the (extended) balance law of fluxes

$$A^{\alpha\beta\mu}{}_{,\mu} = I^{\alpha\beta} \quad . \qquad (4)$$

We assume that $T^{\alpha\beta}$, $A^{\alpha\beta\mu}$ and $I^{\alpha\beta}$ are completely symmetric tensors and moreover that

$$I^{\alpha}{}_{\alpha} = 0 \; , \quad A^{\alpha\beta}{}_{\beta} = m^2c^2V^{\alpha} . \qquad (5)$$

The extended balance law and the trace conditions (5) (that guaranteed that the independent field equations are 14)are suggested by kinetic theory of gases (see e.g. Marle |5|).

- As is usual in the continuum approach the field equations contain unknowns that are not in the list of state variables, i.e. the tensors $A^{\alpha\beta\mu}$ and $I^{\alpha\beta}$ and therefore constitutive equations are requested. We assume that the constitutive relations are in a local form, i.e.:

$$A^{\alpha\beta\mu} \equiv \hat{A}^{\alpha\beta\mu}(V^{\gamma},T^{\gamma\rho}); \quad I^{\alpha\beta} \equiv \hat{I}^{\alpha\beta}(V^{\gamma},T^{\gamma\rho}). \tag{6}$$

If the functions (6) are known, the field equations become a full set of quasi linear first order partial differential system, the solutions of which are called "thermodynamic processes".
Now the problem becomes a constitutive problem, i.e. the restrictions of the acceptable constitutive equations that are compatible with universal physical principles.

Universal Principles.

- We assume in this theory the following natural principles:

i) *Entropy Principle* : There exists an entropy-entropy flux vector $-h^{\alpha}$ that is a constitutive quantity:

$$h^{\alpha} \equiv \hat{h}^{\alpha}(V^{\gamma},T^{\gamma\rho}) \tag{7}$$

such that

$$h^{\alpha}{}_{,\alpha} \leq 0 \tag{8}$$

for all thermodynamic processes.

ii) *Relativity Principle* : The field equations have the same form in all frames. This statement implies that the constitutive functions are invariant under a change of frame.

iii) *Convexity and Hyperbolicity* : There exists a time-like covector $\{\xi_{\alpha}\}$ for which the function $h = h^{\alpha}\xi_{\alpha}$ is a strictly convex function $\delta^2 h > 0$. This condition implies that our system is a symmetric hyperbolic one and therefore we have not only causality but also (as we explain later) well-posed Cauchy problem. Moreover this condition is justified also from a physical point of view, because guarantees that the density of entropy $-h$ have a maximum in equilibrium and therefore we obtain also the thermodynamical stability.

Even if the problem is conceptually simple, it is very hard to obtain the constitutive functions (6) and (7) that satisfy the universal principles. The key to resolve the problem consists in an appropriate new mathematical technique for exploiting the entropy principle that holds for a generic quasi linear first order system of balance laws.
First I explain this procedure in the present case and after I give a brief survey on the general problematic.

Sketch of exploiting the entropy principle.

As the inequality (8) is satisfied only for the solutions of the systems (2), (3), (4) it is necessary that there exists a set of multipliers ξ, Λ_{α} and $\Sigma_{\alpha\beta}$ ($\Sigma_{\alpha\beta} = \Sigma_{\beta\alpha}$; $\Sigma^{\alpha}{}_{\alpha} = 0$) such that:

$$\xi V^{\alpha}{}_{,\alpha} + \Lambda_{\alpha} T^{\alpha\beta}{}_{,\beta} + \Sigma_{\alpha\beta}(A^{\alpha\beta\gamma}{}_{,\gamma} - I^{\alpha\beta}) - h^{\alpha}{}_{,\alpha} \geq 0 \tag{9}$$

for all V^{μ}, $T^{\mu\nu}$ (see Friedrichs and Lax |6| for a general hyperbolic system and Liu |7| for second order non hyperbolic system).

The condition (9) implies:

$$dh^{\alpha} = \xi dV^{\alpha} + \Lambda_{\beta} dT^{\alpha\beta} + \Sigma_{\beta\gamma} dA^{\alpha\beta\gamma} \tag{10}$$

and the residual inequality:

$$\Sigma_{\alpha\beta} I^{\alpha\beta} \leq 0 . \tag{11}$$

Following the idea of Boillat |8| and Ruggeri and Strumia |9|, we define the four-vector h'^{α} :

$$h'^{\alpha} = \xi V^{\alpha} + \Lambda_{\beta} T^{\alpha\beta} + \Sigma_{\beta\gamma} A^{\alpha\beta\gamma} - h^{\alpha} \tag{12}$$

and (10) becomes:

$$dh'^{\alpha} = V^{\alpha} d\xi + T^{\alpha\beta} d\Lambda_{\beta} + A^{\alpha\beta\gamma} d\Sigma_{\beta\gamma} . \tag{13}$$

From convexity arguments it is possible to consider the following univalent change of variables

$$(V^{\mu} , T^{\mu\nu}) \to (\xi, \Lambda^{\alpha}, \Sigma^{\alpha\beta})$$

and then (13) implies:

$$V^{\alpha} = \partial h'^{\alpha}/\partial \xi ; \quad T^{\alpha\beta} = \partial h'^{\alpha}/\partial \Lambda_{\beta} \quad ; A^{\alpha<\beta\gamma>} = \partial h'^{\alpha}/\partial \Sigma_{\beta\gamma} \tag{14}$$

(the brackets <> indicates the deviator part).
Therefore the advantage of this unusual procedure is that the entropy principle is <u>fully exploiting</u> by having

$$V^{\alpha}, T^{\alpha\beta}, A^{\alpha\beta\gamma}$$

as derivative of the single "vector potential" h'^{α}. Now by the relativity principle it is possible to show that h'^{α} is an isotropic function of the news variables ξ, Λ_{β} and $\Sigma_{\alpha\beta}$ and therefore by the Hamilton-Cayley theorem must have the form:

$$h'^{\alpha} = \gamma_{o} \Lambda_{\alpha} + \gamma_{1} \Sigma^{\alpha\beta} \Lambda_{\beta} + \gamma_{2} \Sigma^{2\alpha\beta} \Lambda_{\beta} + \gamma_{3} \Sigma^{3\alpha\beta} \Lambda_{\beta} , \tag{15}$$

where the 4 coefficients γ's may depend on the scalars:

$$\xi, \quad G_{A} = \Lambda^{\alpha} \Sigma^{A}_{\alpha\beta} \Lambda^{\beta} , \quad Q_{i} = \Sigma^{i+1\alpha}{}_{\alpha}$$

(A = 0,1,2,3; i = 1,2,3).
Therefore if we insert (15) and (14) into (2)-(4), we have in the field $\underline{u}' \equiv (\xi, \Lambda_{\alpha}, \Sigma_{\alpha\beta})$ ("main field") the most general system that is compatible with entropy and relativity principle. We observe that at this step the theory is <u>completely non linear</u> and contains only <u>four constitutive functions</u> γ_{A} (characterising the material) that are arbitrary except for the restrictions that come for the symmetry conditions for the tensors $T^{\alpha\beta}$ and $A^{\alpha\beta\gamma}$, the trace conditions (5.2) and the convexity condition (we shall discuss this point later).
This powerful procedure (proposed by Ruggeri for a generic system of partial differential equations in the papers |10|, |11|), have the disadvantage that we do not known a priori the relations between the components of the main field $\underline{u}'$, that we use as variables, and the usual physical ones. Therefore the problem is now to obtain the physical meaning of these variables or better the relations from ξ, Λ_{α}, $\Sigma_{\alpha\beta}$ and the usual fields V^{μ} and $T^{\mu\nu}$. To reach this aim, first of all we define the equilibrium state.

Equilibrium state and identification of the new variables.

We define as equilibrium the state for which the production term vanishes $I^{\alpha\beta}|_E = 0$ and the entropy source (11)

$$s = - \Sigma_{\alpha\beta} I^{\alpha\beta} \geq 0$$

have the minimum value zero: $s|_E = 0$.
These conditions implies:

$$\Sigma_{\alpha\beta}|_E = 0,$$

and

$$\{ \partial I^{\alpha\beta}/\partial\Sigma_{\mu\nu} + \partial I^{\mu\nu}/\partial\Sigma_{\alpha\beta}\}_E \text{ negative definite.}$$

At this point, we note that the differential condition (10) is the non equilibrium generalisation of the Gibbs condition (in vectorial form) that in the present theory becomes a consequence of the assumptions of the structure of balance laws for the differential system of the E.T. and not the starting point as in Müller - Israel approach. Therefore if we evaluate (10) in equilibrium we obtain the results:

$$\xi|_E = g/T, \quad \Lambda^\alpha|_E = - u^\alpha/T \tag{17}$$

and from (12):

$$h'^\alpha = p\, u^\alpha/T \tag{18}$$

where g is the Gibbs free energy, u^α is the four-velocity: $u^\alpha u_\alpha = c^2$, p is the pressure and T is the absolute temperature.
Therefore we known the physical interpretations of the main field components in the equilibrium state and these variables are objects of high importance for relativistic fluids. In fact the last two (17) play a very important role in several questions and Israel call these variables "thermal potential" and "inverse-temperature four vector".
We note that the new variable $\Sigma_{\alpha\beta}$ is the only one that vanishes in equilibrium and therefore it substitutes the usual non-equilibrium variables $t^{<\alpha\beta>}$ (stress deviator), q^α (heat flux) and Π (non equilibrium pressure) from which only 9 are independent because of the conditions:

$$t^{<\alpha\beta>} u_\beta = 0, \quad q^\alpha u_\alpha = 0.$$

Therefore the equilibrium properties permit us to understand also the physical meaning of these variables in non-equilibrium:

ξ - is the non equilibrium *Thermal Potential*,

Λ^α - is the non equilibrium *Inverse-Temperature* (or better *Coldness*) *four-vector*,

$\Sigma^{\alpha\beta}$ - is the non equilibrium *Dissipative Tensor*.

In conclusion these variables play a privileged role because they permit to obtain easily a theory completely far of the equilibrium, are independent and all non equilibrium dissipative terms are contained only in a single tensor $\Sigma_{\alpha\beta}$. Moreover these are the only variables for which the original system assumes the form of a symmetric hyperbolic system as we explain in the next section.
For these reasons the more natural variables do not appear to be the

most familiar ones V^{α} and $T^{\alpha\beta}$.
However it is always possible to go back to the old variables proceeding in the following way: solve the first two equations (14) for ξ, Λ^{α} and $\Sigma^{\alpha\beta}$ in terms of V^{μ} and $T^{\mu\nu}$ and insert there values in the last equation (14) and in (12) to obtain $A^{\alpha\beta\gamma}$ and h^{α} as function of V^{μ} and $T^{\mu\nu}$. Practically this can be done in the neighbourhood of equilibrium. In fact near to the equilibrium state it is possible to write:

$$\Sigma^{\alpha\beta} = \sigma^{\alpha\beta} , \quad \xi = g/T + \chi , \quad \Lambda^{\alpha} = - u^{\alpha}/T + \lambda^{\alpha} \tag{19}$$

where $\sigma^{\alpha\beta}$, χ and λ^{α} are small perturbations. Insert (19) into (14.1) and (14.2) we obtain a linear algebraic system of 14 eqs. for the 14 perturbation variables $\sigma^{\alpha\beta}$, χ, λ^{α}.
Using this last procedure in |4| we have obtained the following results: If the thermal and caloric equation of state $p \equiv p(n,T)$; $e \equiv e(n,T)$ are known, the constitutive functions

$A^{\alpha\beta\gamma} \equiv \hat{A}^{\alpha\beta\gamma}(V^{\mu},T^{\mu\nu})$	-	is known to within an arbitrary function of a single variable,
$h^{\alpha} \equiv \hat{h}^{\alpha}(V^{\mu},T^{\mu\nu})$	-	is known,
$I^{\alpha\beta} \equiv \hat{I}^{\alpha\beta}(V^{\mu},T^{\mu\nu})$	-	is known to within three non-negative functions of two variables that are related to measurable quantities, i.e.: viscosity, bulk viscosity and heat conductivity.

This results are completely in agreement with the 14 moments approach of relativistic Boltzmann equation in the case of ideal fluids and the arbitrary function that appear in the triple tensor is zero for the kinetic theory.
We append now some questions about the convexity condition and the general mathematical aspect of the procedure presented here.

General Mathematical Structure of E.T. and Symmetric Hyperbolic Systems.

The mathematical structure of the balance laws of the E.T. and the entropy principle become a particular case of the following general problems related to hyperbolic systems.
Let us consider a quasi-linear first-order system of N balance laws:

$$\partial_{\alpha}\underline{F}^{\alpha}(\underline{u}) = \underline{f}(\underline{u}) \tag{20}$$

for the R^{N} unknown vector $\underline{u} \equiv \underline{u}(x^{\alpha})$, such that all solutions satisfy the supplementary "entropy principle" inequality:

$$\partial_{\alpha} h^{\alpha}(\underline{u}) \leq 0 . \tag{21}$$

This problem was studied, in a classic formalism, first by Friedrichs and Lax in 1971 |6| under the hypothesis that h^{o} is a strictly convex function of the field $\underline{u} \equiv \underline{F}^{o}$. They were able to prove that exists a N x N symmetric positive definite matrix $\underset{\sim}{H}(\underline{u})$, such that the new system obtained by (20) times H:

$$\underset{\sim}{H}(\underline{u})\{\partial_{\alpha}\underline{F}^{\alpha}(\underline{u}) - \underline{f}\} = 0 \tag{22}$$

is a symmetric hyperbolic system.

This proof was extended in a covariant formalism by Friedrichs |12| in 1974. The important result of these authors has however, the following disadvantages: 1) The conservative form (divergence structure in space-time) in the new system (22) is lost with the impossibility to define the usual weak solutions and to study shocks in particular; 2) We have not an explicit expression for the functional dependence of $\underline{F}^\alpha$ with respect to $\underline{u}$, i.e. we do not have a characterisation of the compatible constitutive equations.
To remove these difficulties in 1974 Boillat |8| and in 1981 Ruggeri and Strumia |9| gave (respectively in a classical and relativistic formalism) a new proof, showing that exists a privileged choose of the field variables (main field $\underline{u}'$) such that the original "conservative" system (20) becomes symmetric hyperbolic.
Before giving a brief sketch of the proof, we recall some definitions in a covariant formalism that can be useful also for readers that are not familiar with hyperbolic systems.
The system (20) is a particular case of

$$\underset{\sim}{A}^\alpha(\underline{u})\ \partial_\alpha \underline{u} = \underline{f}(\underline{u}) \tag{23}$$

when the matrices $\underset{\sim}{A}^\alpha$ are gradient of vectors $\underline{F}^\alpha$: $\underset{\sim}{A}^\alpha = \partial \underline{F}^\alpha / \partial \underline{u}$.
Definition of Hyperbolicity: The system (23) is said to be hyperbolic if a time-like covector $\{\xi_\alpha\}$ exists, such that the following two statements hold: i) $\det(\underset{\sim}{A}^\alpha \xi_\alpha) \neq 0$; ii) $\forall$ any covector $\{\zeta_\alpha\}$ of space type, the following eigenvalue problem : $\underset{\sim}{A}^\alpha(\zeta_\alpha - \mu\xi_\alpha)\underline{d} = 0$, has only real proper values μ and N linearly independent eigenvectors $\underline{d}$, i.e. forming a basis of R^N. The covectors $\{\zeta_\alpha - \mu\xi_\alpha\}$ built with any proper value μ are called "characteristic", while the $\{\xi_\alpha\}$ fulfilling i), ii) are said "subcharacteristic".
Definition of Symmetric Hyperbolic Systems: A system (23) is said to be symmetric hyperbolic if : a) $\underset{\sim}{A}^\alpha = (\underset{\sim}{A}^\alpha)^T$; b) a covector $\{\xi_\alpha\}$ exists such that the matrix $A^\alpha \xi_\alpha$ is positive definite $\forall\ \underline{u} \in \mathcal{D}$, $\mathcal{D}$ being a convex open subset of R^N.

By linear algebra the definition of symmetric hyperbolic system implies the hyperbolicity, therefore these last conditions are more restrictive of those that guarantee hyperbolicity. The importance of the symmetric systems consists in the property that for any symmetric hyperbolic system (also in the case of which the proper eigenvalue are not all distinct) there exist theorems that guarantee the well-posedness of the Cauchy problem for smooth initial data: existence, uniqueness and continuous dependence in a neighbourhood of the initial manifold (see e.g. |13|).
Now we give the sketch of the proof of the symmetrisation of (20) under the constraint (21) and a convexity condition. The proof follows substantially the procedure given in |9| , but we present here a little variant that is more general, because now we do not suppose as in |9| that the time congruence $\{\xi_\alpha\}$ is constant.

Entropy Theorem:

"The class of "constitutive equations" $\underline{F}^\alpha(\underline{u})$, $h^\alpha(\underline{u})$, $\underline{f}(\underline{u})$ for a quasi linear first order system of balance laws (20) compatible with an entropy principle (21) are every and only those for which a four-vector h'^α and a privileged field $\underline{u}'$ (main field) exist, such that:

$$\underline{F}^\alpha = \partial h'^\alpha / \partial \underline{u}' \ ; \quad h^\alpha = \underline{u}' . \partial h'^\alpha / \partial \underline{u}' - h'^\alpha \ ; \quad \underline{u}' . \underline{f} \leq 0. \tag{24}$$

Moreover if the quadratic form:

$$Q = \delta \underline{u}' . \delta \underline{F}^{\alpha} \xi_{\alpha} > 0 \qquad (25)$$

is positive definite for any generic time-like covector $\{\xi_{\alpha}\}$ and for all non vanishing variations $\delta\underline{u}'$, then the original system (20) becomes in the field $\underline{u}'$ a symmetric hyperbolic system. Therefore the Cauchy problem turns out to well-posed (locally in time) under suitable smooth initial data. The condition (25) implies for constant congruences that the function $h' = h'^{\alpha}\xi_{\alpha}$ is a convex function of $\underline{u}'$. Moreover as h' is the Legendre transformation of the function $h = h^{\alpha}\xi_{\alpha}$ with respect to the field $\underline{u} = \underline{F}^{\alpha}\xi_{\alpha}$, we have also that h is a convex function of $\underline{u}$.

Sketch of the proof : Proceeding in the same way as in the relativistic fluid, the compatibility from (20) and (21) implies that there exists a set of multipliers $\underline{u}' \in R^N$, such that:

$$\underline{u}' . \{\partial_{\alpha}\underline{F} - \underline{f}\} - \partial_{\alpha} h^{\alpha} \geq 0 \qquad \forall\, \underline{u},\ \partial_{\alpha}\underline{u};$$

that implies:

$$\underline{u}' . d\underline{F}^{\alpha} = dh^{\alpha} , \qquad \underline{u}' . \underline{f} \leq 0 . \qquad (26)$$

Suppose that it is possible to choose as field $\underline{u}'$ (we'll justify after this assumption), then from (26.1) we obtain:

$$\underline{F}^{\alpha} . d\underline{u}' = dh'^{\alpha} , \qquad \text{where} \quad h'^{\alpha} = \underline{u}' . \underline{F}^{\alpha} - h^{\alpha} \qquad (27)$$

and therefore

$$\underline{F}^{\alpha} = \partial h'^{\alpha} / \partial \underline{u}' . \qquad (28)$$

Substituting (28) into (20) the system becomes:

$$\frac{\partial^2 h'^{\alpha}}{\partial \underline{u}' \partial \underline{u}'} \partial_{\alpha}\underline{u}' = \underline{f}(\underline{u}') . \qquad (29)$$

The system (29) is in the form (23) where the matrices in this case are hessian matrices of h'^{α} and therefore are all symmetric.
Then the system (29) is a (very special) symmetric hyperbolic system if the matrix:

$$\xi_{\alpha} \frac{\partial^2 h'^{\alpha}}{\partial \underline{u}' \partial \underline{u}'} \quad \text{is positive definite} \qquad (30)$$

But it is very easy to see, using (28), that (30) is equivalent to the condition (25). We restrict now the attention to the particular case in which ξ_{α} is constant. In this case (30) is equivalent to the convexity of $h' = h'^{\alpha}\xi_{\alpha}$. Moreover putting $\underline{u} = \underline{F}^{\alpha}\xi_{\alpha}$ and $h = h^{\alpha}\xi_{\alpha}$ from (26) and (27) follows for constant ξ_{α} :

$$\underline{u} = \partial h' / \partial \underline{u}' , \qquad \underline{u}' = \partial h / \partial \underline{u} , \qquad h' = \underline{u}' . \underline{u} - h$$

and then the map from $\underline{u}$ and $\underline{u}'$ is globally univalent, h' is the Legendre transformation of h, and h results also a convex function of $\underline{u}$. The globally univalence from $\underline{u}$ and $\underline{u}'$ justifies the previous assumption to take $\underline{u}'$ as field.
Of course these questions remain valid also under the weaker condition

that (25) holds at least for one congruence $\{\xi_\alpha\}$, but the advantage to impose the strong condition that (25) is valid for all time-like covectors $\{\xi_\alpha\}$ consists in the fact that in this case it is possible to prove that characteristic and shocks velocities are bounded automatically with respect to light velocity (see |14|,|9|,|15|,|16|).

Therefore also for these considerations the field u', that we have used in the E.T. of relativistic fluids as field variables, is privileged with respect the most familiar fields. In the paper |4| we have shown the restrinctions coming from (25) in the particular case in which ξ_α is equal to the four-velocity u_α of the fluid.
In a paper in preparation I have intention to study the constraints that arise for a generic time-like covector from (25) and also to discuss some consequences of the theory far from equilibrium state and shocks waves.

A Comment about Einstein Equations.

In all these questions we have assumed the metric as assigned. Of course if the metric is also unknown it is necessary to append the Einstein equations.Now if we suppose that this equations remain unchanged in E.T., we have a situation that is conceptually very different with respect the standard theories in which the balance laws are the usual 5 and the others are considered as constitutive equations.
In fact in the previous approach only one component (for example the internal energy) in the energy-momentum tensor is considered as field variable and the remaining ones are considered constitutive quantities. Therefore the Einstein equations depend through $T^{\alpha\beta}$ explicitly on the material that we consider. In Extended Thermodynamics the situation is different, because we have assumed that all energy-momentum tensor is a field variable and then in E.T. the Einstein equations become universal equations (valid independently from the constitution of the material). Of course the solutions of the total system (2), (3), (4) and the gravitational equation, change when we change the material because in (4) the knowledge of the triple tensor and the production term is ensured only when the constitutive equations (6) are assigned.

References.

|1| I. Müller, Zur Ausbreitungsgeschwindigkeit von Störungen in Kontinuierlichen Medien, Dissertation TH Aachen (1966).

|2| W. Israel, Non stationary Irreversible Thermodynamics. A Causal Relativistic Theory. Ann. of Phys. 100 (1976).

|3| T. Ruggeri, Symmetric Hyperbolic Systems of Conservative Equations for a Viscous-Heat conducting Fluid. Acta Mech. 47 (1983).

|4| I-Shih Liu, I. Müller and T. Ruggeri, Relativistic Thermodynamics of Gases. Ann. of Phys. 169 (1986).

|5| C. Marle, Sur l'Etablissement des Equations de l'Hydrodynamique des fluides relativistes dissipatives. Ann. Inst. H. Poincaré 10 (1969).

|6| K.O. Friedrichs and P.D. Lax, Systems of Conservation Equations with a Convex Extension. Proc. Nat. Acad. Sci. USA 68 (1971).

|7| I-Shih Liu, Method of Lagrange Multipliers for Exploitation of the Entropy Principle. Arch. Rat. Mech. Anal. 46 (1972).

|8| G. Boillat, Sur l'Existence et la Recherce d'Equation de Conservation supplèmentaires pour les Systèmes Hyperboliques. C.R. Acad. Sc. Paris 278-A (1974).

|9| T. Ruggeri and A. Strumia, Main Field and Convex Covariant Density for Quasi-linear Hyperbolic Systems; Relativistic Fluid Dynamics. Ann. Inst. H. Poincaré 34 (1) (1981).

|10| A. Morro and T. Ruggeri, Propagazione del Calore ed Equazioni Costitutive. 8a Scuola Estiva di Fisica Matematica del C.N.R. (Pitagora - Tecnoprint, Bologna) (1984).

|11| T. Ruggeri, Struttura dei Sistemi alle Derivate Parziali compatibili con un Principio di Entropia. Suppl. BUMI del GNFM- Fisica Matematica 4 (5) (1985).

|12| K.O. Friedrichs, On the Laws of Relativistic Electro-Magneto-Fluid Dynamics. Comm. Pure Appl. Math. 27 (1974).

|13| A. Fisher and D.P. Marsden, The Einstein Evolution Equations as a first order quasi linear Symmetric Hyperbolic Systems. Comm. Math. Phys. 28 (1972).

|14| G. Boillat and T. Ruggeri, Limite de la Vitesse des Chocs dans les Champs à Densité d'Energie Convex. C.R. Acad. Sc. Paris 289-A (1979).

|15| A. Strumia, Wave Propagation and Symmetric Hyperbolic Systems of Conservative Laws with Constrained Field variables. Submitted to Nuovo Cimento.

|16| G. Boillat and A. Strumia, in preparation.

RELATIVISTIC HYDRODYNAMICS AND HEAVY ION REACTIONS

D. Strottman
Theoretical Division
Los Alamos National Laboratory
Los Alamos, NM 87545

The use of hydrodynamics to describe the collision of hadronic matter has a long history which dates from work of Fermi [1], Pomeranchuk [2] and Landau [3] in the early fifties. They attempted to describe proton-proton scattering and the concomitant production of pions using statistical and hydrodynamical concepts. Their success encouraged other, later applications to different reactions, both at higher energies as well as for heavier, composite particles. Since this early work, the models have been refined and fresh concepts have been advanced utilizing new ideas from particle physics, quantum chromodynamics and other fields such as astrophysics. This article will very briefly review a few of the varied applications relativistic hydrodynamics has in the area of heavy ion reactions and anti-proton annihilation. A two-fluid model which overcomes certain of the limitations of the usual relativistic hydrodynamics in describing the physical processes and which also avoids the problems with causality associated with the introduction of dissipation into the hydrodynamic equations will be described. We refer the reader to the literature for more detailed descriptions of the application of hydrodynamics to heavy ion reactions [4,5] and hadron-hadron collisions [6,7].

It is not *a priori* apparent that hydrodynamics will be valid for the description of heavy ion reactions. An examination of the conditions necessary for the validity of hydrodynamics indicates that the requirements are only marginally fulfilled. For example, the number of particles which are involved in a heavy ion collision ranges from perhaps only a hundred to a thousand. Hence, the number of degrees of freedom is large compared to one, but relatively small compared to a usual fluid. If one creates a quark-gluon plasma during the collision, then the number of degrees of freedom will increase by at least a factor of three.

There is also the condition that there be sufficient time for the establishment of local, thermal equilibrium; this also is marginally satisfied. A lower limit on the collision time for two heavy ions may be roughly estimated as the nuclear diameter divided by the velocity of light, or about 5×10^{-23} s. Nucleons interact by exchanging pions and it requires about 5×10^{-24} s for two adjacent nucleons to exchange a pion. Since this interaction time is about one-tenth of the total collision time, some degree of local equilibrium will be established. This will be particularly true for central collisions of large nuclei for which the matter in the interior will be confined for longer periods than the above estimate. Further, for moderate bombarding energies, the actual reaction time is around 20×10^{-23} s which is appreciably larger than the simple above argument suggested.

Bondorf and Zimányi [8] have investigated the approach to equilibrium using a time-dependent Boltzmann equation. They concluded that the pion and proton spectra suggest that the momentum distributions are very near their equilibrium values. During sufficiently energetic reactions, new particles such as pions or deltas may be created; such particles are short lived, either being rapidly absorbed in the case of pions, or decaying into a nucleon and a pion. Montvay and Zimányi [9] have investigated whether chemical equilibrium is reached; they conclude that it is not reached although the system is not very far from it.

Finally, it is not unrealistic to treat the nucleons as classical particles for the energy regimes in which we shall be interested. For relativistic nucleons their momenta is greater than 1 GeV/c and their corresponding de Broglie wavelength is 0.4 fm[1] which is less than the radius of a nucleon and much less than that of a nucleus.

The applications of relativistic fluid dynamics to heavy ion reactions have assumed there to be no dissipation. (There is some early work on hadron-hadron reactions by the Russian school which attempted to include the effects of viscosity. This work assumed the validity of the Landau equations and is reviewed by Feinberg [6].) In the Los Alamos effort the three dimensional relativistic Euler equations are solved numerically using the particle-in-cell method developed by Harlow [10,11]. The particle-in-cell method allows calculations in cases of extreme distortion and shear including cases where cavities appear in the fluid. It also allows beautiful graphical representations of the fluid. However, it consumes vast amounts of computer memory.

The equation of state for nuclear matter is unknown; indeed, one of the goals of heavy ion reactions is to investigate the equation of state. Since theoretical calculations of the energy and pressure of nuclear matter as a function of density and

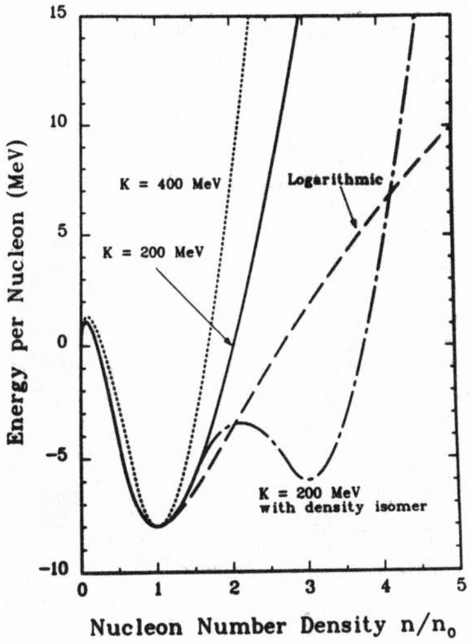

Fig. 1. A plot of the ground state energy per nucleon $E(n)$ as a function of density for four different phenomenological expressions. The quantity K is the compression modulus of nuclear matter, the canonical value of which is 200 MeV.

---

[1]It is customary to give masses in energy units; e.g., the mass of a nucleon is 939 MeV (million electron volts), that of a pion is 139 MeV and a delta is 1232 MeV or 1.232 GeV. One usually disregards the mass difference of the neutron and proton and refers to them generically as nucleons. Since the total energy of a particle is the sum of its rest mass and its kinetic energy, the Lorentz contraction factor is $\gamma = 1 + {}^{T}/_{m}$ where T is the kinetic energy in MeV and m is its rest mass. For the highest energies γ may exceed 100. The unit of length is a fermi (fm) which is 10^{-13} cm.

temperature are quite uncertain, one usually is forced to make assumptions regarding the equation of state. It is normal in nuclear physics to call the energy per nucleon $E(n)$ the equation of state rather than expressing the pressure as a function of temperature and density. The
two are of course equivalent since one may obtain the pressure from the usual thermodynamic relation

$$P = \partial E/\partial V)_S \tag{1}$$

where S is the entropy. Examples of some zero-temperature equations of states which have been used are shown in fig. 1. One further usually assumes that the matter obeys a Fermi gas equation of state for non-zero temperature. For densities greater than five times normal nuclear matter density, a number of possible scenarios have been proposed. Currently, the most plausible one is that at sufficiently high densities or temperatures, the nucleons 'melt' and a quark-gluon plasma is formed in which the identities of the individual nucleons is lost and the constituent quarks and gluons are free to briefly roam about the relatively large collision volume.

An example of the time development of a heavy ion reaction is shown in fig. 2 which shows the collision of ^{20}Ne on ^{238}U at 393 MeV/nucleon and two equal mass nuclei at 800 MeV/nucleon. From a knowledge of the velocity vectors of the fluid in each cell at the end of the calculation, one may calculate the double differential cross section $d^2\sigma/dE\,d\Omega$ which may then be compared with experiment. In general the

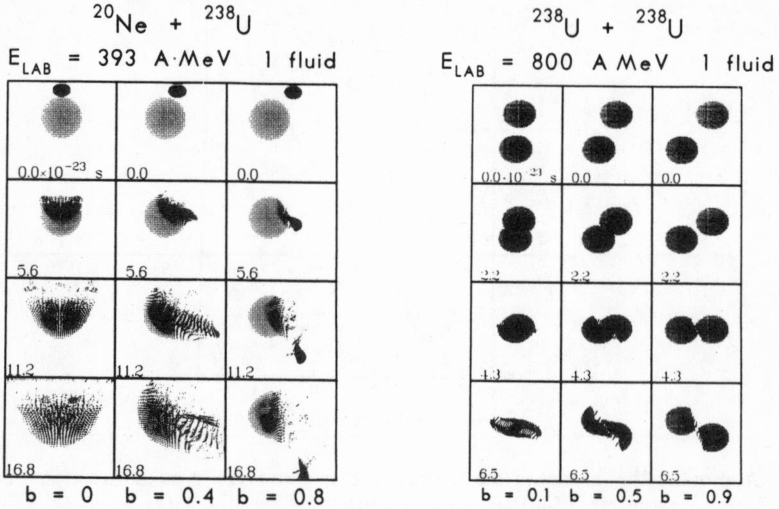

Fig. 2. Matter distributions for 393 MeV/nucleon ^{20}Ne on ^{238}U (left) and two equal-mass nuclei at 162 MeV/nucleon (right) in the center-of-mass (equivalent laboratory energy is 800 Mev/nucleon). Three impact parameters are shown in units of the sum of the radii of the two nuclei. Since the Euler equations are scale invariant, the right figure applies to arbitrary mass nuclei, although the time scale is appropriate only for ^{238}U on ^{238}U.

agreement is satisfactory. In fact hydrodynamic calculations predicted that in certain reactions, the relative incompressibility of nuclear matter would cause the projectile to glance off the target and produce a distinctive signal in the angular distribution. This "sideways" flow was subsequently experimentally observed. For details see ref. 4.

A close examination of fig. 2 will demonstrates that the calculation reproduces the result that for a one-fluid system with no viscosity, the mean-free-path of the matter is zero. (This is more easily seen when the matter from each nucleus is plotted in color as in ref. 5.) This is not a serious problem at low bombarding energies of less than a GeV per nucleon for which the mean free path of a nucleon in the nucleus is much shorter than the nuclear diameter. However, as the energy of the projectile increases, the nucleus becomes more transparent and the assumption of a zero mean-free-path becomes untenable. The effect of non-zero mean-free-paths has been known for some time from high energy proton-nucleus experiments at Fermi Lab and CERN; in these experiments the so-called leading particles punch through the target and carry away a significant amount of the energy of the incident proton. One could simulate to a small extent the effects of a non-zero mean-free-path by introducing viscosity. However, this would introduce all the problems associated with the acausal behaviour as demonstrated by Hiscock and Lindblom [12,13]. In any event, this would be inadequate when the mean-free-paths become so long that some of the nucleons can traverse the entire target and emerge on the far side. Further, as the energies increase it becomes less likely that local thermal equilibrium is instantaneously established at the interface of the two nuclei.

To describe the situation in which large mean-free-paths are involved, a two-fluid model was introduced [14]. To obtain the equations which describe the two-fluid model, each nucleus is assumed to be a fluid which has the identical properties of the fluid representing the other nucleus. When the two fluids collide they are allowed to exchange energy and momentum at a finite rate proportional to the relative velocity of the two nuclei and to the nucleon-nucleon cross section σ_{NN}. Thus, the rate of momentum loss is finite and the two fluids will interpenetrate. The amount of interpenetration is small at low energies for which σ_{NN} is large and increases as σ_{NN} decreases. The Euler equations which ensure particle number conservation remain unchanged, but the equations ensuring energy and momentum conservation must be modified to allow an interchange of these quantities. The changes are in the form of additional terms, the magnitude of which can be estimated from kinetic theory: if one knows the collision rate and the amount of energy and momentum lost in each collision, then the total amount of loss may be found.

The expression for the collision rate is

$$R_{coll} = N_1 N_2 \sigma_{NN} v_{rel}$$

where N_1 and N_2 are the densities of the two fluids and v_{rel} is the relativistic generalization of the relative velocity. The generalized Euler equations for fluid one are

$$\partial_t \mathbf{M}_1 + \nabla(\mathbf{v}_1 \cdot \mathbf{M}_1) = -\nabla P - R_{coll} K (\gamma_1 \mathbf{v}_1 - \gamma_2 \mathbf{v}_2)/Y \quad (2)$$

$$\partial_t E_1 + \nabla(\mathbf{v}_1 E_1) = -\nabla P - R_{coll} K (\gamma_1 - \gamma_2)/Y \quad (3)$$

where M_1 and E_1 are the momentum and energy densities of fluid one and Y is the scalar product of the two four-velocities

$$Y = (u_1 \cdot u_2).$$

The quantity K determines the amount of energy-momentum loss and is fixed by comparing with high energy nucleon-nucleus reactions. The equations for fluid two are obtained by interchanging the indices 1 and 2.

Unlike the Euler equations, eqs (2) and (3) are not scale invariant; the calculated results will depend the masses of the nuclei involved which is entirely reasonable. A similar consequence occurs if one uses the Navier-Stokes equations. However, unlike the case of the Navier-Stokes equations which introduces dissipation through higher order derivatives of the velocity, the two-fluid model partially achieves the same result by eliminating derivatives in the additional terms.

The additional coupling terms in eqs. (2) and (3) describe the friction between the two nuclei entirely in terms of two-body collisions of the constituent nucleons. It is assumed that the nucleon-nucleon cross section is the free NN cross section σ_{NN} and is independent of density and temperature; this assumption is surely poor at high temperatures and densities. It is further assumed that the Fermi velocities of the nucleons may be ignored. For large relative velocities this is a good approximation (the Fermi velocity at normal nuclear density is approximately 0.27 c); for lower bombarding energies, one must worry about the effects due to the Fermi velocity. For methods which partially take into account the effects of the Fermi velocity, the reader is referred to refs. 5 and 14. In addition both the one-fluid and two-fluid models necessarily omit binding energy effects.

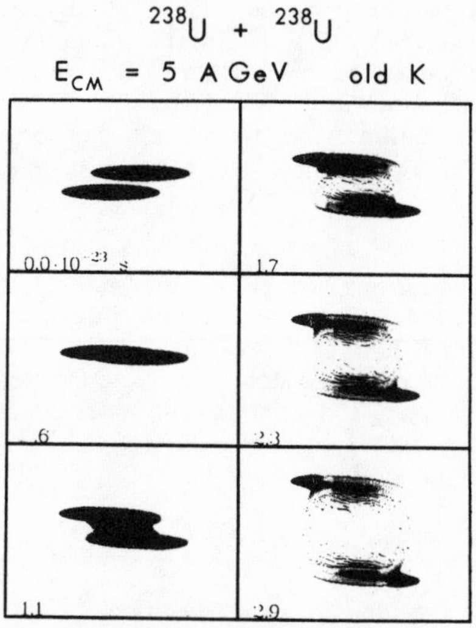

Fig. 3. Matter distributions for ^{238}U on ^{238}U calculated in the center-of-mass system with the two-fluid model. The center-of-mass kinetic energy is 5 GeV/nucleon (equivalent laboratory energy is 73.7 GeV/nucleon) and the impact parameter is 0.3.

In figure 3 results are shown for a collision between two equal mass nuclei, each having an energy of 5 GeV per nucleon in the center of mass frame. (This energy corresponds to a velocity of 0.987c in the center of mass or a velocity of 0.99992c in the laboratory frame.) The effects of a non-zero mean-free-path are immediately evident. The two nuclei essentially pass through each other, although each nucleus exerts a drag upon the other. In the one-fluid model the matter at the interface of the two nuclei would have come to a halt. All the kinetic energy must be converted into thermal energy. Hence, the one-fluid model can expect to exhibit a larger thermal pressure than does the two-fluid model. This will result in the nuclear matter blowing up and disintegrating sooner.

Experiments have recently begun at CERN which collide 200 GeV/nucleon ^{16}O ions on nuclear targets in a search for signals of a quark-gluon plasma. Similar experiments will soon begin at lower energies at Brookhaven National Laboratory. In all these experiments relativistic hydrodynamics will play an essential role in the interpretation of results.

Another interesting hadronic process which can involve the use of hydrodynamics is the annihilation of anti-protons inside a nucleus. The annihilation of an anti-proton and a proton results in 1.87 GeV being localized for a short time in a very small volume. Thus the energy density is very briefly twice the normal value. If we assume the entire energy appears as thermal energy, then a fireball is generated. If the annihilation occurs at rest or for a very slow anti-proton, no shock wave is generated [15]. Rather, the disturbance propagates outward from the annihilation point via a sound wave. If, however, the anti-proton carries a significant amount of kinetic energy, the situation is much different. The additional kinetic energy drives the hadronic matter into the nucleus and a shock wave is generated [16]. In fig. 4 the matter distribution resulting from an anti-proton annihilation is given. The incoming anti-proton had a kinetic energy of 0.4 GeV. From such interactions one can hope to learn about the nature of nuclear matter in regions of small density but very high temperature. This promising field is still in its infancy.

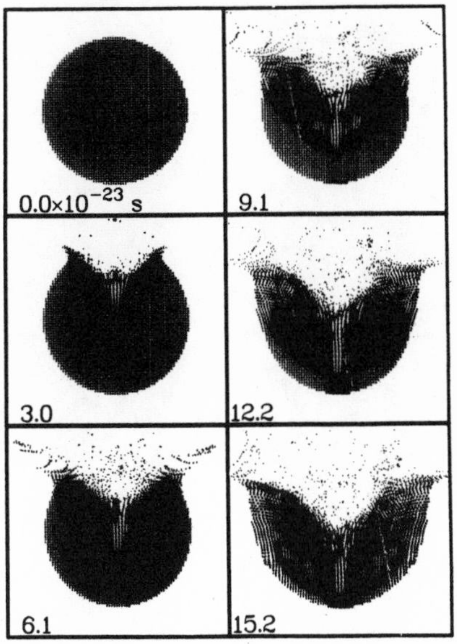

Fig. 4. Nuclear matter distributions resulting from the annihilation of a 400 MeV anti-proton in a nucleus. Only the central 2 fm of the nucleus is plotted.

References

1. E. Fermi, Prog. Theor. Phys. **5**(1950)570.
2. I.U. Pomeranchuk, Dokl. Akad. Nauk. USSR **78**(1951)889.
3. L. D. Landau, Izv. Akad. Nauk. SSSR Sci Fiz. **17**(1953)51.
4. H. Stöcker and W. Greiner, Phys. Rept. **137**(1986)277.
5. R.B. Clare and D. Strottman, Phys. Rept. **141**(1986)177.
6. E.L. Feinberg, Phys. Rept. **5**(1972)237.
7. E.V. Shuryak, Phys. Rept **61**(1980)71.
8. J. Bondorf and J. Zimányi, Phys. Scri. **24**(1981)758.
9. I. Montvay and J. Zimányi, Nucl. Phys. **A316**(1979)490.
10 F.H. Harlow, Los Alamos Scientific Report, LAMS-1956(1955).
11. F.H. Harlow, A.A. Amsden and J.R. Nix, J. Comp. Phys. **20**(1976)119.
12. W.A. Hiscock and L. Lindblom, Ann. Phys. (NY) **151**(1983)466.
13. W.A. Hiscock and L. Lindblom, Phys. Rev. **31D**(1985)725.
14. A.A. Amsden, A.S. Goldhaber, F.H. Harlow and J.R. Nix, Phys. Rev. **C27**(1978)2080.
15. D. Strottman, Phys. Lett. **119B**(1982)39.
16. D. Strottman and W.R. Gibbs, Phys. Lett. **149B**(1984)288.

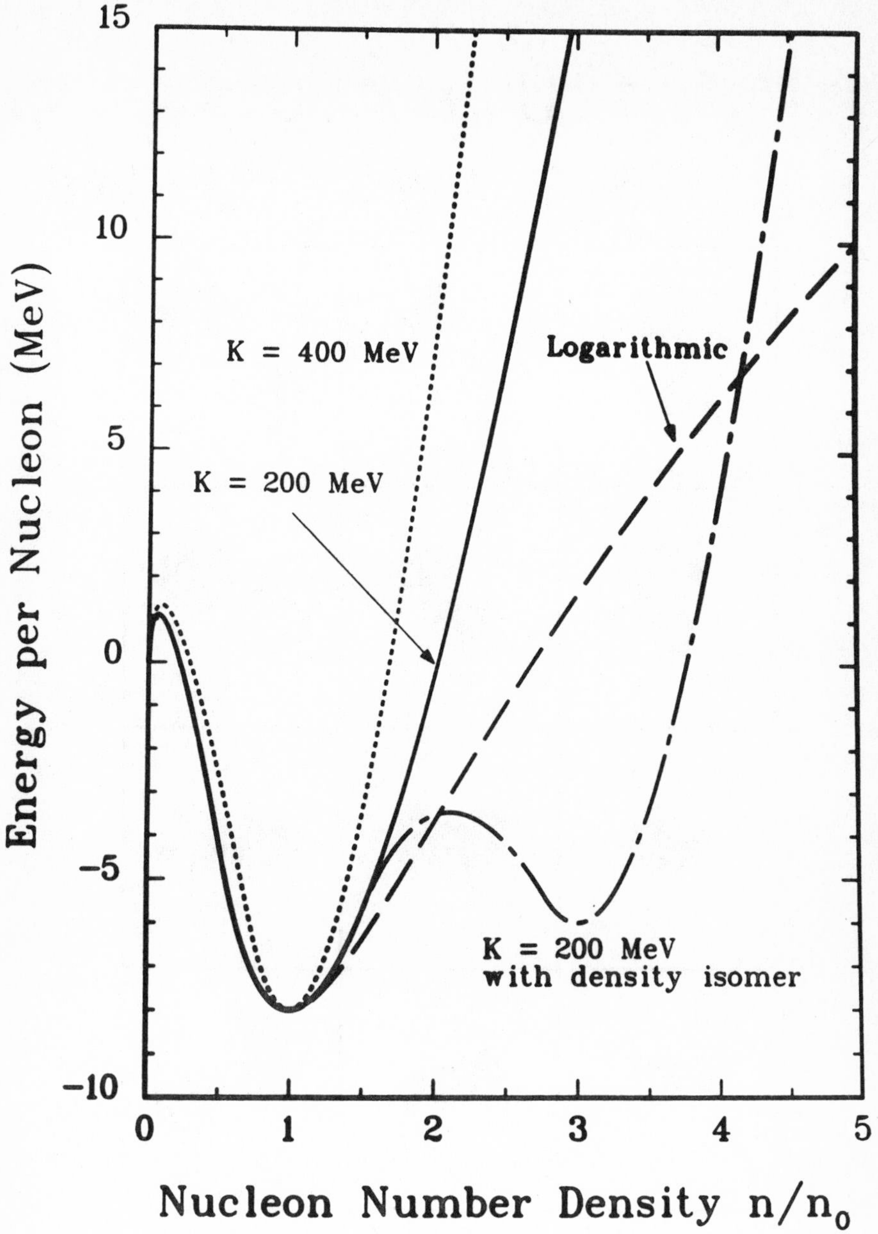
15
10
5
0
−5
−10
0
1
2
3
4
5
Energy per Nucleon (MeV)
Nucleon Number Density n/n₀
K = 400 MeV
K = 200 MeV
Logarithmic
K = 200 MeV
with density isomer

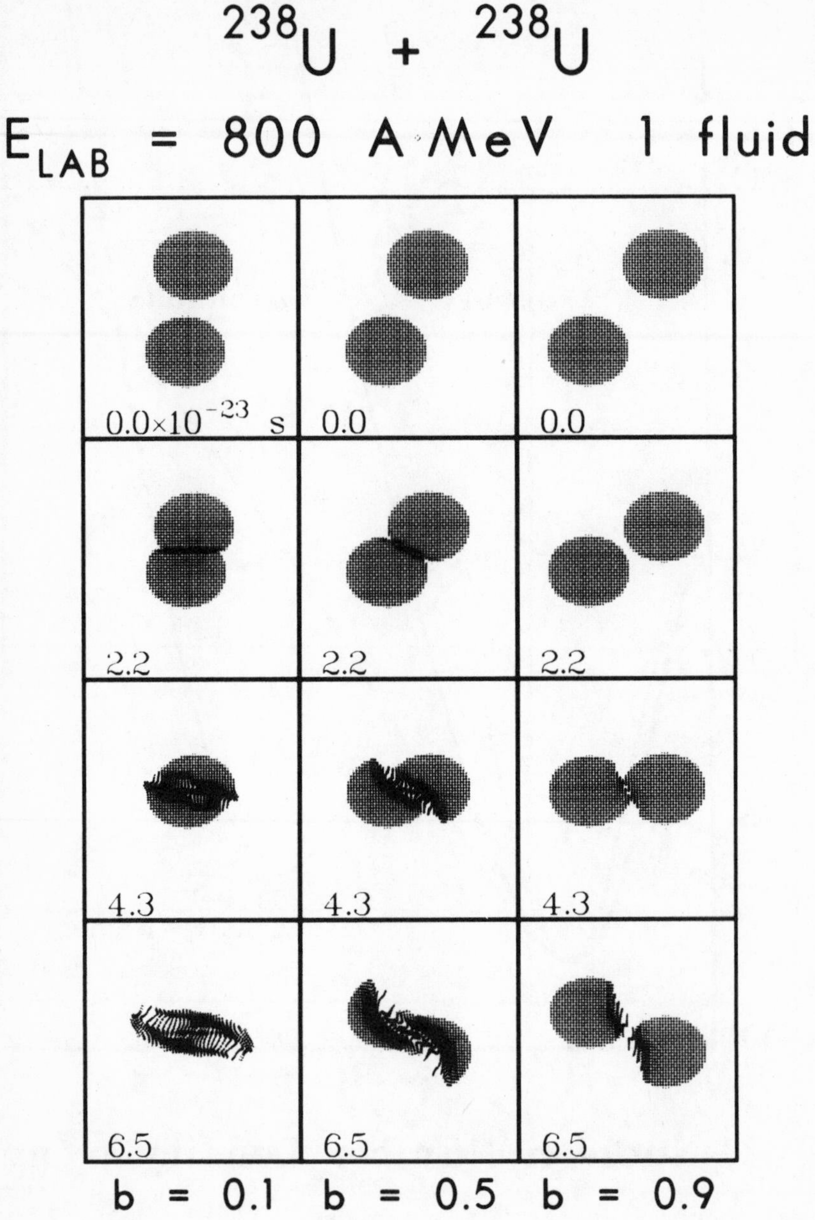

$^{238}U + ^{238}U$
E_{LAB} = 800 A MeV 1 fluid
0.0×10^{-23} s
0.0
0.0
2.2
2.2
2.2
4.3
4.3
4.3
6.5
6.5
6.5
b = 0.1
b = 0.5
b = 0.9

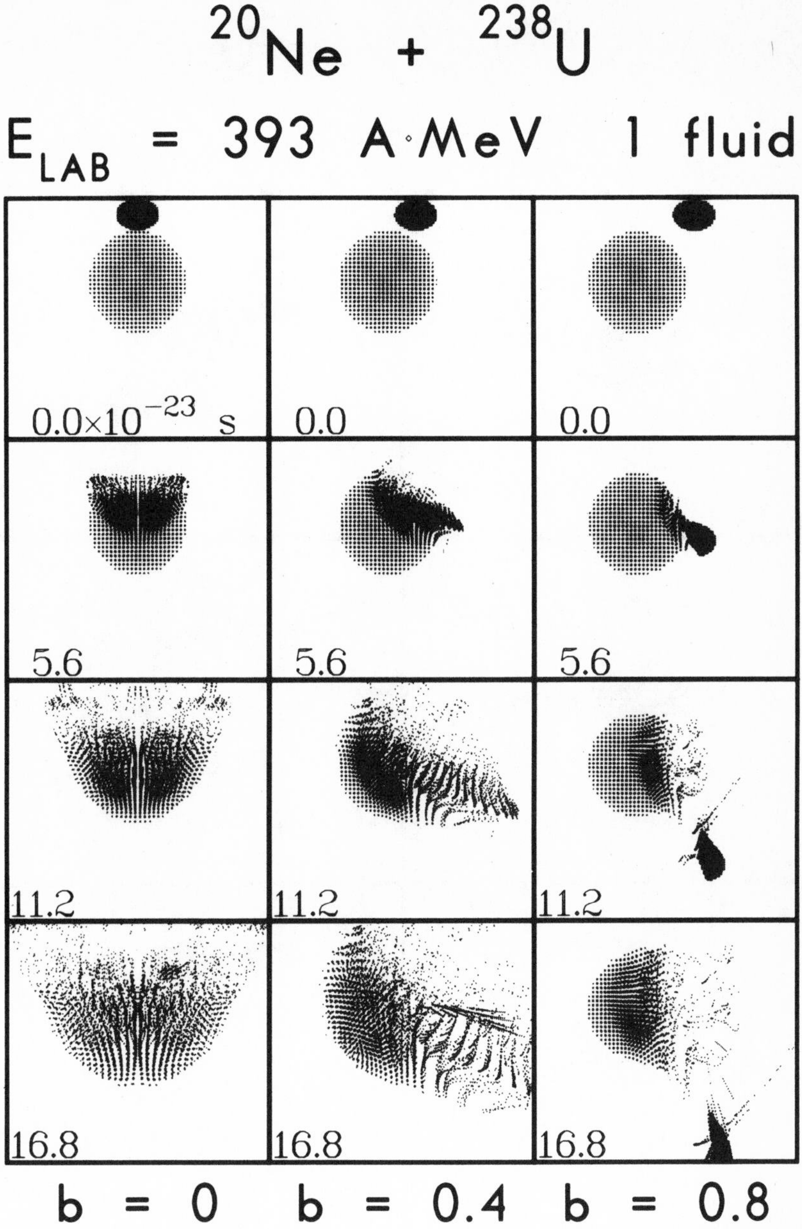
20Ne + 238U
E_LAB = 393 A·MeV 1 fluid
0.0×10^-23 s
0.0
0.0
5.6
5.6
5.6
11.2
11.2
11.2
16.8
16.8
16.8
b = 0
b = 0.4
b = 0.8

$^{238}U + {}^{238}U$

E_{CM} = 5 A·GeV old K

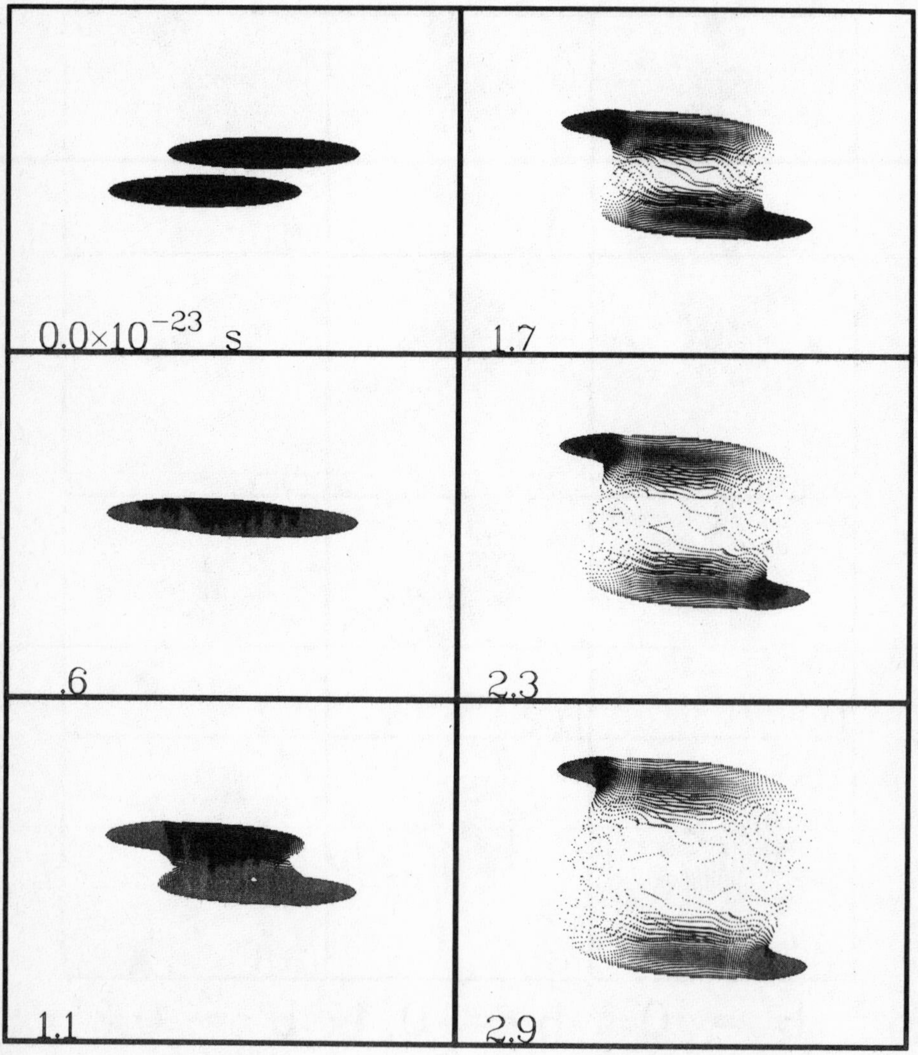

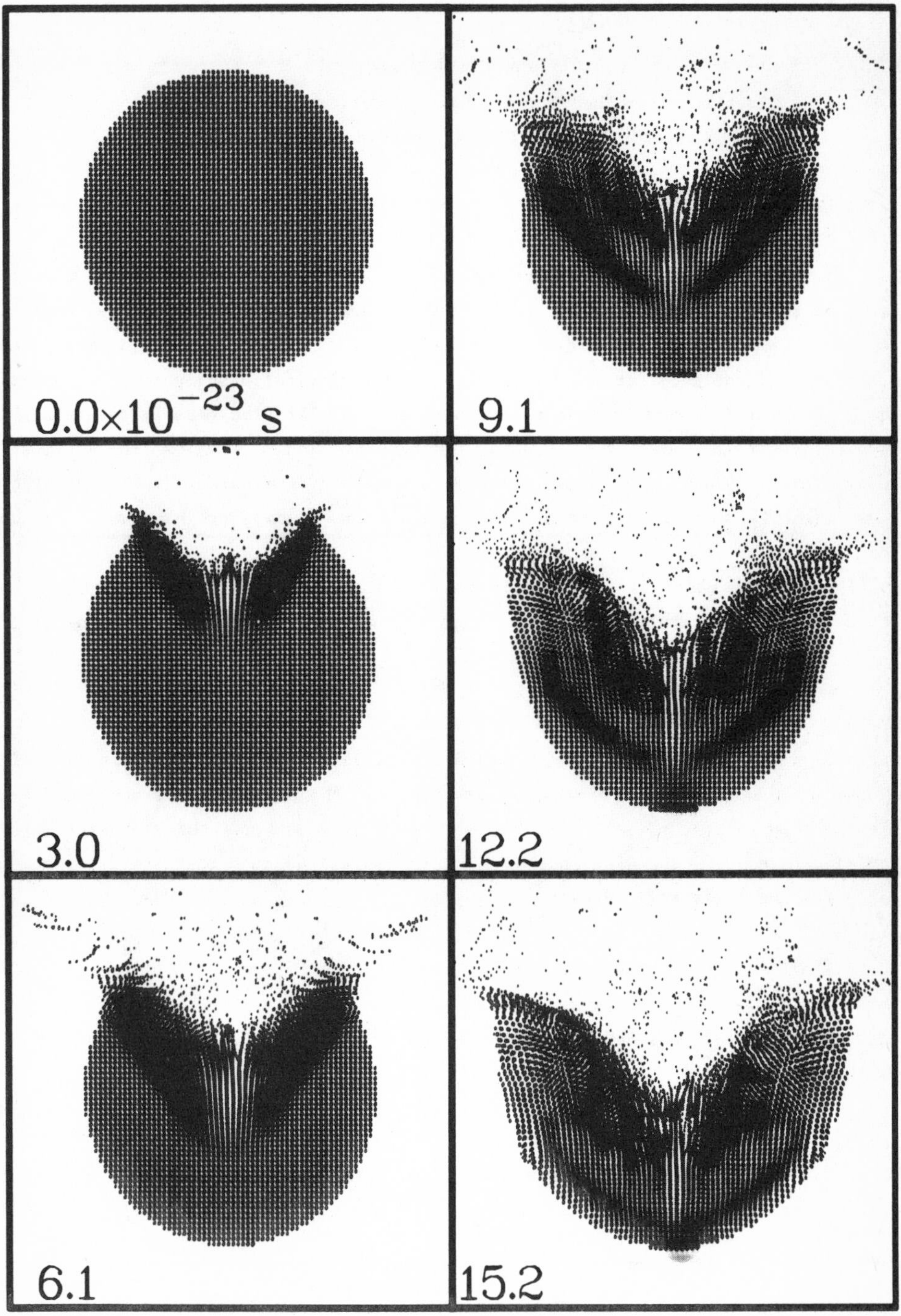
0.0×10^−23 s
9.1
3.0
12.2
6.1
15.2

SOME PROBLEMS IN RELATIVISTIC HYDRODYNAMICS

Ch.G. van Weert
Institute for Theoretical Physics, University of Amsterdam
Valckenierstraat 65, 1018 XE Amsterdam, the Netherlands

1. Introduction

The following material is intended as a contribution to the discussion of the status of relativistic hydrodynamics. In the lectures of Werner Israel [1] there has been great emphasis on the undesirable features of the relativistic Navier-Stokes equations, and the need for a more general phenomenological description. It is hoped that some aspects of this unresolved problem can be clarified here by referring hydrodynamics to an underlying level of description which we take to be kinetic theory. For the purpose of illustration this model suffices since it already allows us to identify some of the major obstacles on the route from dynamics to hydrodynamics.

The kinetic model is very simple in principle. It consists of an equation of motion for the distribution function f(x,p) which in relativity takes the form [1,2]

$$p^{\mu}\partial_{\mu}f(x,p) = C[f] \quad , \tag{1}$$

where the right-hand side is the socalled collision term. The distribution function depends on the space-time point $x = x^{\mu} = (t,\underset{\sim}{x})$ and the on-shell four-momentum variable $p = p^{\mu} = (p^0,\underset{\sim}{p})$, $p^2 = m^2$. (We set $\hbar = c = k_B = 1$, and use the Minkowski metric with signature -2 .) The distribution function is defined such that its first moment yields the local particle current

$$J^{\mu}(x) = \int d\omega\ p^{\mu}\ f(x,p) \quad , \tag{2}$$

where $d\omega = d^3p/(2\pi)^3p_0$, and its second moment the local energy-momentum tensor

$$T^{\mu\nu}(x) = \int d\omega\ p^{\mu}p^{\nu}f(x,p) \quad . \tag{3}$$

The explicit form of the collision term is not important here. However, we require it to be consistent with the local conservation laws

$$\partial_{\mu}\ J^{\mu}(x) = 0 \quad , \tag{4}$$

$$\partial_\nu T^{\mu\nu}(x) = 0 \quad , \tag{5}$$

which reflect the conservation of particle number and energy-momentum at the microscopic level. (In general, there may be more conserved currents corresponding to the various conserved quantum numbers carried by the particles such as baryon number, lepton number, electric charge, etc. [3].)

One further requirement on the collision term derives from the second law which states that the entropy production must be non-negative. In kinetic theory one defines an entropy flow according to

$$S^\mu(x) = - \int d\omega \; p^\mu[f \log f - \theta(1 + \theta f)\log(1 + \theta f)] \tag{6}$$

with $\theta = \pm 1$. Hence, this requirement implies that the inequality

$$\partial_\mu S^\mu(x) = - \int d\omega \; C[f] \log \frac{f}{1 + \theta f} \geqslant 0 \tag{7}$$

must be satisfied for all allowable distribution functions.

As an aside we may add that, strictly speaking, the second law only states that the total entropy cannot decrease. The local second law presumes that the non-uniformities in the system are smooth on the scale of the correlations between the particles. Since the same proviso applies to kinetic theory itself, the local H-theorem (7) is consistent within the confines of this theory. However, there is no foundation for extending the local second law to a general postulate of macroscopic physics.

2. Local-Equilibrium Ansatz

A system left to itself will tend to a state of thermal equilibrium. In this state the entropy reaches its maximum value. A necessary condition for equilibrium is, therefore, that the entropy production (7) vanishes. However, this condition is not sufficient. Indeed, if one substitutes in (7) a distribution function of the form

$$f^{(0)} = [e^\psi - \theta]^{-1} \tag{8}$$

with

$$\psi(x,p) = \alpha(x) + \beta_\mu(x)p^\mu \tag{9}$$

a linear combination of the collision invariants, one finds that the entropy production vanishes identically on account of the microscopic conservation laws respected by the collision term.

One calls $f^{(0)}(x,p)$, with $\alpha(x)$ and $\beta_\mu(x)$ arbitrary functions of space-time, a local equilibrium (LE) distribution function. Its principal feature is that it describes a state of zero entropy production. However, it is important to keep in mind that $f^{(0)}$ is, in general, not a solution of the kinetic equation (1). Therefore, it does not represent any actual state of the system for which we may write

$$f = f^{(0)} + \delta f \quad , \tag{10}$$

with δf the deviation from local equilibrium.

There is no a priori reason for the state of an arbitrary system to be close to local equilibrium; this would depend on the interaction and the external conditions. Nevertheless, much work in kinetic theory rests on the LE ansatz which assumes that δf is small and that only linear deviations need to be retained. If quadratic terms are neglected the entropy flow (6) takes the familiar form [1]

$$S^\mu(x) = \alpha(x)J^\mu(x) + \beta_\nu(x)T^{\mu\nu}(x) + \beta^\mu(x)P^{(0)}(x) \quad . \tag{11}$$

The last term stands for the integral

$$\beta^\mu P^{(0)} = \theta \int d\omega\, p^\mu \log\left(1 + \theta f^{(0)}\right) \quad , \tag{12}$$

where $P^{(0)}$ may be identified with the LE thermostatic pressure. For the first variation one immediately finds

$$\delta\beta_\mu P^{(0)} = - J^{(0)}_\mu \delta\alpha - T^{(0)}_{\mu\nu} \delta\beta^\nu \quad , \tag{13}$$

which may be called the LE Gibbs-Duhem relation.

The last two formulae imply that in the linear theory the entropy production is given by

$$\partial_\mu S^\mu = \left(J-J^{(0)}\right)^\mu \partial_\mu \alpha + \left(T-T^{(0)}\right)^{\mu\nu} \partial_\mu \beta_\nu \quad . \tag{14}$$

This is the well-known bilinear expression in terms of gradients and non-equilibrium fluxes which are linear in the deviation from local equilibrium. One may note that no relationship between the gradients and the fluxes has been imposed. In standard irreversible thermodynamics one assumes that such a relationship exists and that it is linear. This reproduces the whole formalism of Navier-Stokes (NS) hydrodynamics. However, one may also impose non-linear constitutive laws and still be compatible with the second law [4]. The resulting hydrodynamic equations, which contain corrections of order two, three, etc. in the gradients of the macroscopic variables, are usually referred to as Burnett, super-Burnett, etc. Therefore, the LE ansatz does by no means imply a linear constitutive law.

Finally, we like to point out that because of the special role played by the conserved quantities, the LE ansatz is tailored to the description of a fluid in terms of the standard hydrodynamic densities. In fact, $f^{(0)}$ represents the distribution of maximum entropy with respect to these densities. If for whatever reason one wishes to include additional variables in the macroscopic description, as is done in "extended thermodynamics" (see also sect. 5), the LE ansatz, and thereby $f^{(0)}$, should be extended accordingly [5].

3. Conditions of Fit

One of the recurrent questions in relativistic hydrodynamics is how to fit the arbitrary functions $\alpha(x)$ and $\beta_\mu(x)$ to the actual state of the system in such a way that $f^{(0)}$ represents this state as accurately as possible. This problem is already discussed in the book of Chapman and Cowling [6]. In their view any "normal" distribution function must depend on the values of the conserved densities, since their values are the least affected by collisions; conversely, since there are five collision invariants (for a simple gas), one can expect that the densities determined by these invariants are the only local densities on which a normal distribution function can depend.

As already remarked at the end of the preceeding section, $f^{(0)}$ is the unique distribution function that maximizes the entropy with respect to the conserved densities as relevant variables [3,5]. In this construction the five parameters $\alpha(x)$ and $\beta_\mu(x)$ are regarded as Lagrange multipliers which must be fitted such that the local hydrodynamic densities, namely, particle density, energy density and momentum density, are locally given by $f^{(0)}$. This defines a local chemical potential $\mu(x)$, a local temperature $T(x) = \beta^{-1}(x)$, and a local hydrodynamic velocity through

$$\alpha(x) = -\beta(x)\mu(x) \tag{15}$$

$$\beta_\mu(x) = \beta(x)U_\mu(x) \quad , \quad U^2 = 1 \quad . \tag{16}$$

This construction also works in other cases (e.g. photon or neutrino gas), since there are always as many Lagrange multipliers as there are collision invariants and corresponding hydrodynamic densities.

On the grounds of principle, the Eckart (E) prescription [1] has no place in this scheme, because it matches the hydrodynamic velocity to the particle current, which is not a conserved density in relativistic hydrodynamics. Nevertheless, on the practical side, Eckart hydrodynamics enjoys some popularity in astrophysics [7]. There is not really a conflict though: typical astrophysical applications concern transport through a medium of non-relativistic massive particles. It is then natural to use this medium as the preferred reference frame, and numerically there will be no difference to speak of. To this we may add that many formal results of the theory are not dependent on the choice of hydrodynamic velocity, and one can often avoid the issue altogether. by leaving the choice open.

However, one cannot avoid to decide in what frame of reference the chosen conditions of fit are to be imposed. Essentially, there are two possibilities: the one is to impose them in the laboratory 3+1 frame, that is,

$$J_0 - J_0^{(0)} = 0 \quad , \tag{17}$$

$$T_{0\mu} - T_{0\mu}^{(0)} = 0 \quad . \tag{18}$$

This uniquely fixes $\alpha(x)$, $\beta(x)$, and $U_\mu(x)$, but these conditions of fit are clearly not covariant: two observers looking at the same system will fit a different $f^{(0)}$ to it. In a recent paper [8] van Kampen argues that this feature uniquely comes out of a scheme in which hydrodynamics is derived by an elimination of fast variables. Physically the non-covariance means that what is observed as a slow variable by one observer does not appear as a slow variable to another.

Non-covariance is of course, sacrilege to the pure relativist. He certainly will take recourse to the second possibility which is to impose the conditions of fit in the local rest frame determined by the hydrodynamic velocity. This leads to the socalled Landau-Lifshitz conditions [1]

$$U_\mu (J-J^{(0)})^\mu = 0 \quad , \tag{19}$$

$$U_\mu (T-T^{(0)})^{\mu\nu} = 0 \quad . \tag{20}$$

As stressed by van Kampen [8] this fitting of $f^{(0)}$ is nonlinear, because determining the local rest frame is part of the prescription.

The difference between the van Kampen (vK) and Landau-Lifshitz (LL) prescriptions may be illustrated by considering the entropy. In both cases the conditions of fit ensure that the actual entropy density is equal to the LE one. That is, in the case of the vK conditions one has

$$S_0 = S_0^{(0)} \quad , \tag{21}$$

and in the LL case

$$S = S_\mu U^\mu = S_\mu^{(0)} U^\mu \quad . \tag{22}$$

For the first variation of these quantities one gets

$$\delta S_0 = \alpha\delta N_0 + \beta_0\delta E_0 - \underset{\sim}{\beta} \cdot \delta\underset{\sim}{M} \tag{23}$$

in the former case, and

$$\delta S = \alpha\delta N + \beta\delta E \tag{24}$$

in the latter. Here N_0, E_0 and N,E are the particle density and energy density in the laboratory and local reference frames, respectively, and

$$\underset{\sim}{M} = \left(E_0 + P^{(0)}\right)\underset{\sim}{v} \tag{25}$$

is the momentum density in the laboratory frame. Hence, the rest frame entropy density only depends on the particle and energy densities, like in ordinary thermodynamics, whereas the laboratory entropy density also depends on the momentum density. Note, however, that the difference is of second order in the velocity. With regard to linear hydrodynamics, therefore, we can forget the difference, provided we identify the laboratory frame with the global rest frame of the system.

4. Navier-Stokes Hydrodynamics

As we have seen above, doing thermodynamics in the laboratory frame can be rather awkward because of the occurrence of the velocity as a thermodynamic parameter and Lorentz contraction factors. Nevertheless, the practitioners of relativistic hydrodynamics seem to consider this as a lesser evil, that is, they seem to favour the vK conditions of fit; see the lectures of Holm and Weitzner in this volume. When challenged by the pure relativist on this matter, such a practician could put forward a number of arguments in defense. He could point out, for example, that the laboratory frame approach is nearly always the most natural one in a given situation. Indeed, in a typical hydrodynamical problem the boundary conditions usually define a laboratory frame that is particularly well suited for a description of the system. He could also refer to van Kampen [8] and argue that hydrodynamics as an approximate description of nature is intrinsically observer biased; see also the discussion in refs. [9,10]. And, if he wanted to be unpleasant, he could suggest that covariant hydrodynamics may be intrinsically unstable because of the occurrence of non-inertial terms, the socalled Eckart terms.

To discuss this last issue, we recall that under the LE ansatz the entropy production has the form (14). In this expression the gradients may be rearranged by making use of the Gibbs-Duhem relation (13), where the zeroth-order densities may be replaced by the actual ones on the strength of the condition of fit (for arbitrary U_μ). We can then work the entropy production into the form

$$\partial_\mu S^\mu = \beta I^\mu [\beta^{-1} \partial_\mu \beta + (hN)^{-1}\, \partial_\mu P^{(0)}] + \beta\, \Pi^{\mu\nu}\, \overline{\partial_\mu U_\nu} \quad , \tag{26}$$

where the viscous pressure tensor is the space-like part of $T^{\mu\nu}$ with the thermostatic pressure subtracted [1]. Irrespective of whether LL or E conditions are imposed, the heat flow is given by [2]

$$I^\mu = \Delta^{\mu\nu}(T_{\nu\lambda}U^\lambda - hJ_\nu) \tag{27}$$

where h is the enthalpy per particle and $\Delta^{\mu\nu} = g^{\mu\nu} - U^\mu U^\nu$ the space-like projector.

Now, in deriving the result (26), except in the LL "gauge", we had to make use of Euler's equation

$$DU^\mu = (hN)^{-1}\, \Delta^{\mu\nu}\partial_\nu P^{(0)} \tag{28}$$

to rewrite the acceleration (Eckart term) in terms of the gradient of $P^{(0)}$. Within the LE ansatz this is consistent because the terms neglected are of second order. Still this seemingly harmless operation removes a potentially dangerous time derivative. What may happen if this term is left in has been analyzed by Hiscock and Lindblom [11]. They find that this term gives rise to instabilities on a very short time scale, not unlike the well-known run-away solutions of the Lorentz-Dirac equation in classical electrodynamics.

On this basis Hiscock and Lindblom (HL) draw the conclusion that standard first-order covariant hydrodynamics must be abondoned. This seems to me slightly rash especially since the hydrodynamical theory as constructed by HL is rather artificial in the following sense. Their entropy production contains two independent vector fluxes, namely a particle and an energy diffusion current, whereas eq. (26) contains only one independent heat flux (27). As also pointed out by Israel [1], it is only this combined heat flux that has physical significance. Moreover, we have argued above that the spurious instabilities introduced by Eckart terms have to do with second-order corrections to the Euler equation (28). It is therefore highly suspect that in HL's treatment such instabilities already occur at the level of the linearized equations. Still their work is important for calling attention to the fact that the covariant gauges may give rise to (non-linear) instabilities not encountered in the non-covariant vK gauge. This issue certainly deserves further study.

5. Beyond Navier-Stokes

Another objection voiced against standard hydrodynamics concerns the causality of the relativistic Navier-Stokes (NS) equations. In this connection one usually refers to the paradox of the infinite propagation velocity due to the parabolic nature of the equations describing heat conduction, diffusion and other transport processes. However, from the point of view of statistical mechanics there is no paradox because the fundamental hydrodynamic restriction $\kappa\ell \ll 1$, with κ the wave number of the macroscopic disturbance and ℓ the mean free path, precludes propagation velocities faster than the thermal speed of the particles [12]. One can see this reflected in the two NS sound modes [2]

$$z = \pm\, c_s \kappa - i\Lambda_s \kappa^2 \tag{29}$$

which involves the velocity of sound c_s. It may also be noted that the imaginary part is purely dissipative like the heat and viscosity modes

$$z = -\, i\, \frac{\beta\lambda}{Nc_p}\, \kappa^2 \quad , \tag{30}$$

$$z = -\, i\, \frac{\eta}{Nh}\, \kappa^2 \quad , \tag{31}$$

where λ and η are the heat conductivity and viscosity coefficients, respectively. No instabilities show up here.

The NS equations are able to account for a large class of phenomena in fluids and gases. However, their validity only ranges up to order $\kappa^2\ell^2$ which is the proper domain of hydrodynamics. How to account for more general circumstances? Three major approaches may be distinguished:

i. **Burnett expansion.** An expansion in higher orders of $\kappa\ell$ was first derived by Burnett in 1935 from the Chapman-Enskog solution of the Boltzmann equation [6]. We shall not discuss this theory any further because the general opinion is that it does not constitute a reliable extension of hydrodynamics.

ii. **Extended thermodynamics.** The central idea is that the description of a more general state of the system requires additional state variables, in particular the heat flow and the viscous pressure tensor. For example, one writes the Gibbs relation as

$$\delta S = \delta S^{(0)} + \gamma_1\, I^{\mu}\delta I_{\mu} + \gamma_2\, \Pi^{\mu\nu}\delta\Pi_{\mu\nu} \tag{32}$$

with $\gamma_{1,2}$ constants depending on the thermodynamic state. Like the Chapman-Enskog method provides a basis for the Burnett expansion, so the Grad moment method serves to justify extended thermodynamics [13]. There is no need to go in more detail here because the topic of extended thermodynamics is well covered in this volume. However, a few comments will be made later on.

iii. **Generalized hydrodynamics.** The essential feature of this approach is that the transport coefficients are made frequency and wave-number dependent. From a macroscopic point of view this step is just as ad hoc as the introduction of more variables. However, in recent years new advances in non-equilibrium statistical mechanics [14] and neutron scattering experiments [15] seem to have vindicated this memory function formalism, as it is sometimes called.

Technically the derivation of generalized hydrodynamics is rather involved, although the main reasoning is fairly straightforward [9]. The basic assumption is that the five hydrodynamic densities suffice for a macroscopic description of a fluid. The final result is then a set of five linear hydrodynamic equations of the typical form [3]

$$[z + \kappa\Omega(\kappa) - i\kappa^2 K(z,\kappa)]A(z,\kappa) = A(0) \quad , \tag{33}$$

where A stands for the Fourier transform of any of the five hydrodynamic densities as a function of the complex frequency z, and wave number κ . The quantity Ω , only dependent on κ , is called the frequency matrix, and K the memory kernel.

The link with ordinary linear hydrodynamics is easily established. Up to order $\kappa\ell$ one can forget K. One then completely recovers the linearized Euler hydrodynamics. To next order one replaces K by its homogeneous and static limit

$$\eta = \lim_{z \to 0} \lim_{\kappa \to 0} K(z,\kappa) \quad , \tag{34}$$

which may be identified with the static transport coefficients determining the NS hydrodynamics to order $\kappa^2\ell^2$. Higher-order expansions lead to the Burnett $O(\kappa^3\ell^3)$ and the super-Burnett $O(\kappa^4\ell^4)$ equations. As already mentioned these equations are not particularly useful.

Since a straightforward expansion does not work one needs a more serious analysis of the memory kernel. This necessarily involves a large input from microphysics, because a complete specification is tentamount to a full solution of the many-body problem. Now about this problem two attitudes of mind can be perceived. Either one can try to get around the difficulty by guessing what the solution could look like. For example, macroscopic hyperbolic equations of the Maxwell-Cattaneo type [1] can be obtained in this way. Or one accepts that on length and time scales of the order of the mean free path, one cannot avoid but to study the interaction processes between the particles. This implies that one has to solve, at least approximately, the equations of motion, or in kinetic theory, to calculate the spectrum of the collision operator.

Recently, such calculations have been performed for a classical hard sphere gas by Kamgar-Parsi and Cohen [16]. The results are particularly interesting because they illustrate very clearly the limitations of various methods. For instance, the calculations show that, as expected, the NS modes are correct up to $k\ell \sim 0.1$, and that the Burnett corrections do not give a reliable extension to higher values of κ .

Another interesting feature of the results is that they allow a qualitative comparison with the Grad moment method, and the sort of mode calculations that have been done by Kranys and Israel and Stewart for a relativistic gas; see [2] for references. For illustration we reproduce one of the figures of the Kamgar-Parsi-Cohen paper.

The figure shows what is typical of the moment method, namely, a finite absorption cut-off for large wave numbers. The larger the number of moments, the larger the value of $\kappa\ell$ where this horizontal behaviour sets in. This unphysical behaviour is a well-known artifact of the moment method when truncated at any finite number of moments.

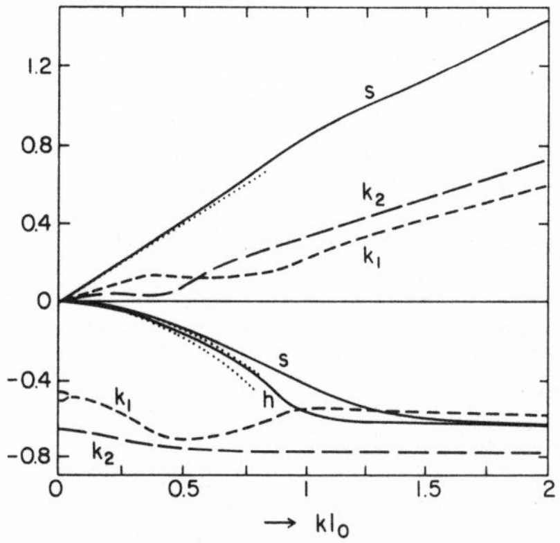

Fig. 1. Extended heat (h), extended sound (s) and nearest kinetic modes obtained with the moment method for hard spheres [16].

This behaviour is not unexpected, since in the transition regime $\kappa\ell \sim 1$ the effect of free streaming begins to contribute. The moment method, like the Chapman-Enskog method, is mainly concerned with collisions and virtually ignores free streaming and the associated continuous eigenvalue spectrum. In the transition regime both discrete and continuum modes are important. In this regime there seems to exist no other approximation method other than simply inverting the collision operator.

One last result I would like to mention is that the five extended hydrodynamic modes as calculated for hard spheres can account for the data obtained from scattering experiments on Xenon up to $\kappa\ell \sim 0.8$. For real fluids, such as liquid Argon, similar results have been reported [16].

The lesson to be learned, I think, is the following. Extensions of hydrodynamics to smaller wave lengths necessarily involve some dynamical input. (In this I share the viewpoint of Zeldovich [1].) Since the dynamics depends on the kind of system in question, it seems very unlikely that any a priori choice of additional variables will be of universal significance. This especially applies to such special approximations as Grad's 14-moment method. As long as it is not made clear that the heat flow and viscous pressure play any special role with respect to either the dynamics or experiment, one can expect no more than a qualitative description for a limited range of experimental conditions. After all, causality and stability are only necessary but not sufficient conditions for the validity of relativistic hydrodynamics.

References

[1] W. Israel, this volume.

[2] S.R. de Groot, W.A. van Leeuwen, and Ch.G. van Weert, Relativistic Kinetic Theory, 1980.

[3] Ch.G. van Weert, Physica 111A(1982)537.

[4] L.S. García-Colin, Physica 118A(1983)341.

[5] I - Shih Liu, I. Muller, and T. Ruggeri, Ann.Phys. 169(1986)191.

[6] S. Chapman and T.G. Cowling, The Mathematical Theory of Non-uniform Gases, 1970.

[7] D. Mihalas and B.W. Mihalas, Foundations of Radiation Hydrodynamics, 1984.

[8] N.G. van Kampen, J.Stat.Phys. 46(1987)709.

[9] Ch.G. van Weert, Ann.Phys. 140(1982)133.

[10] R. Balian et al., Phys.Rep. 131(1986)1.

[11] W.A. Hiscock and L. Lindblom, Phys.Rev. D31(1985)725.

[12] H.D. Weymann, Am.J.Phys. 35(1967)488.

[13] L.S. García-Colin et al., J.Stat.Phys. 29(1982)387; 37(1984)465.

[14] H. Grabert, Projection Operator Techniques in Non-Equilibrium Statistical Mechanics, Springer Tracts 95.

[15] S.W. Lovesey, Theory of neutron scattering from condensed matter, 1984.

[16] B. Kamgar-Parsi and E.G.D. Cohen, Physica 138A(1986)249.

FONDAZIONE C.I.M.E.
CENTRO INTERNAZIONALE MATEMATICO ESTIVO
INTERNATIONAL MATHEMATICAL SUMMER CENTER

"Logic and Computer Science"

is the subject of the First 1988 C.I.M.E. Session.

The Session, sponsored by the Consiglio Nazionale delle Ricerche and the Ministero della Pubblica Istruzione, will take place under the scientific direction of Prof. PIERGIORGIO ODIFREDDI (Università di Torino) at Villa «La Querceta», Montecatini Terme (Pistoia), Italy, ***from June 20 to June 28, 1988.***

Courses

a) ***Overview of Computational Complexity Theory.*** (6 lectures in English).
Prof. Juris HARTMANIS (Cornell University, Ithaca).

Outline

Computational complexity theory is the study of the quantitative laws that govern computing. During the last twenty-five years, complexity theory has grown into a rich mathematical theory and today, it is one of the most active research areas in computer science. Among the most challenging open problems in complexity theory is the problem of understanding what is and is not feasible computable and more generally, a thorough understanding of the structure of the feasible computations. The best known of these problems is the classic P = ? NP problem. It is interesting to note that these problems, which were formulated in computer science, are actually basic problems about fundamental quantitative nature of mathematics. In essence, the P = ? NP problem is a question of how much harder is it to derive (computationally) a proof of a theorem than to check the validity of a proof.

The lectures on Computational Complexity will rewiew the basic concepts and techniques of complexity, summarize the earlier results and then review the more recent results about the structure of feasible computations.

b) ***Non-Traditional Logics for Computation.*** (6 lectures in English).
Prof. Anil NERODE (Cornell University, Ithaca).

Outline

A primer of non-traditional logics for non-experts. Many non-traditional logics are receiving wide attention in computer science because of potential importance in specification, development, and verification of programs and systems.

Prerequisites: some knowledge of undergraduate algebra, topology, computer science, and classical predicate logic. Otherwise self-contained.

I. Classical propositional and first order logic. Its models and Herbrand universes. Proof procedures of Gentzen natural deduction, Beth-Smullyan tableaux and resolution and unification. Discussion of automated classical propositional and first order reasoning.

II. Classical propositional and first order intuitionist logic. Its models such as Kripke models, Beth models, continuous function models, cpo models, Heyting valued models, categorical models. Models with discrete equality for Kroneckerian constructive algebra versus models with apartness for Brouwer's intuitionist analysis and Bishop's constructive analysis. Proof procedures of Gentzen, Fitting tableaux, natural deduction, resolution and unification for intuitionist logics. Discussion of automated intuitionist propositional and first order reasoning. Kleene's realizability for intuitionistic arithmetic as the archetype for lambda calculus realizability and the extraction of programs from proofs.

III. Modal first order logics, classical and intuitionist. The correspondence theory. Modal logics arising in computing such as algorithmic and dynamic logic, and temporal logics for description of concurrent processes and programs.

IV. Many sorted first order logics of all of the above sorts, model and proof theory and automation. Their use in algebraic specification theory, description of communicating processes, etc.

V. Logics for computing based on finite models, such as Gurevich's models for Pascal.

VI. Buchi's monadic second order theory of one successor and Rabin's second order monadic theory of two successors as languages for computing.

VII. Higher order intuitionist logic, de Bruijn and AUTOMATH, Constable and NUPRL, etc. Relation to rewrite rules and lambda calculus via the Curry-Howard isomorphism.

VIII. Other logics, such as logics of knowledge, probabilistic logics and uncertain reasoning logics.

c) ***Program verification.*** (6 lectures in English).
Prof. Richard PLATEK (Odissey Research Association, Ithaca).

Outline

In this lecture series we will review some of the fundamental logical theorems which form the basis of program verification. We will consider proofs of both sequential and concurrent programs. Some of the topics to be considered are:

I. Flowchart Verification: Floyd's verification condition theorem; its relationship to second order logic, infinitary logic, and PROLOG. Partial and total correctness; weakest precondition, strongest postcondition liberal and strict. Extensions of the method of invariants to include concurrency.

II. Hoare Logic: The verification of structured programs. Rules for constructs such as procedure call, recursion, pointers, loop exit statements, etc. Relative completeness results; incompleteness results. Extension of ordinary logic to a logic of partial terms in order to deal with undefined expressions (reading an uninitialized variable, indexing an array out of bounds, etc.).

III. Structured specification languages: An examination of Anna, a specification/assertion language for Ada.

IV. Concurrency: The Owicki-Gries approach; Hoare logic for CSP; the use of temporal logic.

There will also be a review of existing automated program verification systems.

d) ***Logic and Computer Science.*** (6 lectures in English).
Prof. Gerald SACKS (Harvard University, Cambridge, Mass).

Outline

- Logical foundations of prolog.
- Backtracking, cuts and operators.
- Prolog procedures.
- Database manipulation.
- Definite clause grammars and parsing.
- Classical recursion theory.

Basic references

- C. MARCUS, Prolog Programming, Addison-Wesley, 1986.
- I. BRATKO, Prolog Programming for Artificial Intelligence, Addison-Wesley, 1986.

FONDAZIONE C.I.M.E.
CENTRO INTERNAZIONALE MATEMATICO ESTIVO
INTERNATIONAL MATHEMATICAL SUMMER CENTER

"Global Geometry and Mathematical Physics"

is the subject of the Second 1988 C.I.M.E. Session.

The Session, sponsored by the Consiglio Nazionale delle Ricerche and the Ministero della Pubblica Istruzione, will take place under the scientific direction of Prof. MAURO FRANCAVIGLIA (Università di Torino), and Prof. FRANCESCO GHERARDELLI (Università di Firenze) at Villa «La Querceta», Montecatini (Pistoia), Italy, ***from July 4 to July 12, 1988.***

Courses

a) ***String Theory and Riemann Surfaces.*** (6 lectures in English).
Prof. L. ALVAREZ-GAUME (University of Boston, USA).

Contents

In these lectures we will review the new developments of string theory and its connections with the theory of Riemann surfaces, super-Riemann surfaces and their moduli spaces. The point of view taken will be to start with the conformal field theory formulation of string theory and then develop in detail the operator formalism for strings on higher genus surfaces. A tentative plane of the lectures is:

Lecture 1 - Introduction to string theory and conformal field theory (Part 1)
Lecture 2 - Introduction to string theory and conformal field theory (Part 2)
Lecture 3 - Perturbation theory for Bosonic strings. Belavin-Knizhnik theorem, Mumford forms, string infinities and the boundary of moduli spaces.
Lecture 4 - The operator formulation of string theory (Part 1)
Lecture 5 - The operator formulation of string theory (Part 2)
Lecture 6 - Virasoro action on moduli space, more general conformal theories, nonpertubative ideas and recent developments.

b) ***Riemann Surfaces and Infinite Grassmannians.*** (6 lectures in English).
Prof. E. ARBARELLO (Università di Roma "La Sapienza", Roma, Italy).

Contents

— Compact Riemann surfaces and their moduli (stable and semi-stable curves). Picard's group of the moduli space Mg. The Riemann-Roch-Grothendieck theorem.
— Kodaira-Spencer theory. Schiffer variations. Calculus of cohomology à la Mayer-Vietoris. The cotangent bundle of moduli space.
— Lie algebras d (Virasoro) and D (Virasoro-Heisenberg) and their relations with moduli space Mg.
— The Lie algebra gl_∞ and the "Boson-Fermion correspondence". Relations with the Lie algebras d and D. Calculation of the relevant cohomology.
— The infinite Grassmannian Gr and its geometry. The central extension of GL_∞. Tautological and determinant bundles; the function τ. Relation with Plücker's coordinates. The K.P. hierarchy.
— Krichever application. Relations between the functions τ and θ. The correlation function. The trisecting formula. Novikov conjecture.
— The sheaf of differential operators of order less or equal to one, acting on sections of the determinant bundle over Gr. Its restriction to moduli space.
— Some known results about moduli spaces of Riemann surfaces.

References

- L. ALVAREZ-GAUME, C. GOMEZ, C. REINA, Loop Groups, Grassmannians and String Theory, Phys. Lett. 190B, 55-62 (1987).
- E. ARBARELLO, M. CORNALBA, The Picard Groups of the Moduli Spaces of Curves, Topology 26(2), 153-171 (1987).
- E. ARBARELLO, C. DE CONCINI, V. KAC, C. PROCESI, Moduli Space of Curves and Representation Theory, preprint, 1987.
- A.A. BEILINSON, YU. I. MANIN, The Mumford Form and the Polyakov Measure in String Theory, Comm. Math. Phys. 107, 359-376 (1986).
- A.A. BEILINSON, YU. I. MANIN, V.V. SCHECHTMAN, Sheaves of the Virasoro and Neveu Schwarz Algebras, Moscow Univ. preprint, 1987.
- E. DATE, M. JIMBO, M. KASHIWARA, T. MIWA, Transformation Groups for Soliton Equations, in "Nonlinear Integrable Systems. Classical Theory and Quantum Theory", World Sci. (Singapore, 1983), pp. 39-119.
- J. HARER, The Second Homology Group of the Mapping Class Group of an Orientable Surface, Inv. Math. 72, 221-239 (1983).
- J. HÁRER, Stability of the Homology of the Mapping Class Group of Orientable Surfaces, Ann. of Math. 121, 215-249 (1985).
- V.G. KAC, D.H. PETERSON, Spin and Wedge Representations of Infinite Dimensional Lie Algebras and Groups, Proc. Nat. Acad. Sci. U.S.A. 78, 3308-3312 (1981).
- V.G. KAC, Highest Weight Representations of Conformal Current Algebras, in "Geometrical Methods in Field Theory", World Sci. (Singapore, 1986), pp. 3-15.
- N. KAWAMOTO, Y. NAMIKAWA, A. TSUCHIYA, Y. YAMADA, Geometric Realization of Conformal Field Theory on Riemann Surfaces, Nagoya Univ. preprint, 1987.
- YU. I. MANIN, Quantum String Theory and Algebraic Curves, Berkeley I.C.M. talk, 1986.
- E.Y. MILLER, The Homology of the Mapping Class Group, Journ. Diff. Geom. 24, 1-14 (1986).
- D. MUMFORD, Stability of Projective Varieties, L'Enseignement Mathém. 23, 39-110 (1977).
- A. PRESSLEY, G. SEGAL, Loop Groups, Oxford Univ. Press (Oxford, 1986).
- G. SEGAL, G. WILSON, Loop Groups and Equations of KdV Type, Publ. Math. I.H.E.S. 61, 3-64 (1985).
- C. VAFA, Conformal Theories and Punctured Surfaces, preprint, 1987.
- E. WITTEN, Quantum Field Theory, Grassmannians and Algebraic Curves, preprint, 1987.

c) ***The Topology and Geometry of Moduli Spaces.*** (6 lectures in English).
Prof. N.J. HITCHIN (Oxford University, Oxford, UK).

Contents

Topics will include:

(i) Moduli space instantons
(ii) Moduli space of monopoles
(iii) Moduli space of vortices
(iv) Teichmüller space
(v) Moduli spaces related to Riemann surfaces

References

- M.F. ATIYAH, Instantons in 2 and 4 Dimensions, Comm. Math. Phys. 93, 437-451 (1984).
- D. FREED, K. UHLENBECK, Instantons and 4 - Manifolds, Springer Verlag (Berlin, 1984).
- N.J. HITCHIN, A. KALBADE, U. LINDSTROM, M. ROCEK, Hyperkähler Metrics and Supersymmetry, Comm. Math. Phys. 108, 535-589 (1987).
- M.F. ATIYAH, N.J. HITCHIN, Geometry and Dynamics of Magnetic Monopoles, Princeton University Press (Princeton, 1988).
- A. JAPPE, C. TOMBES, Vortices and Monopoles, Birkhäuser (1980).
- N.J. HITCHIN, The Self-Duality Equation on a Riemann Surface, Proc. London Math. Soc. 55, 59-126 (1987).
- M.F. ATIYAH, R. BOTT, The Yang-Mills Equations over Riemann Surfaces, Phil. Trans. Roy. Soc. London, sec. A. 308, 523-615 (1982).

d) ***Differential Algebras in Field Theory.*** (6 lectures in English).
Prof. R. STORA (LAPP, Annecy-le-Vieux, France).

Contents

Lecture 1 - Introduction. The role of locality in perturbative quantum field theory ([1]).

LIST OF C.I.M.E. SEMINARS

		Publisher
1954 -	1. Analisi funzionale	C.I.M.E.
	2. Quadratura delle superficie e questioni connesse	"
	3. Equazioni differenziali non lineari	"
1955 -	4. Teorema di Riemann-Roch e questioni connesse	"
	5. Teoria dei numeri	"
	6. Topologia	"
	7. Teorie non linearizzate in elasticità, idrodinamica, aerodinamica	"
	8. Geometria proiettivo-differenziale	"
1956 -	9. Equazioni alle derivate parziali a caratteristiche reali	"
	10. Propagazione delle onde elettromagnetiche	"
	11. Teoria della funzioni di più variabili complesse e delle funzioni automorfe	"
1957 -	12. Geometria aritmetica e algebrica (2 vol.)	"
	13. Integrali singolari e questioni connesse	"
	14. Teoria della turbolenza (2 vol.)	"
1958 -	15. Vedute e problemi attuali in relatività generale	"
	16. Problemi di geometria differenziale in grande	"
	17. Il principio di minimo e le sue applicazioni alle equazioni funzionali	"
1959 -	18. Induzione e statistica	"
	19. Teoria algebrica dei meccanismi automatici (2 vol.)	"
	20. Gruppi, anelli di Lie e teoria della coomologia	"
1960 -	21. Sistemi dinamici e teoremi ergodici	"
	22. Forme differenziali e loro integrali	"
1961 -	23. Geometria del calcolo delle variazioni (2 vol.)	"
	24. Teoria delle distribuzioni	"
	25. Onde superficiali	"
1962 -	26. Topologia differenziale	"
	27. Autovalori e autosoluzioni	"
	28. Magnetofluidodinamica	"
1963 -	29. Equazioni differenziali astratte	"
	30. Funzioni e varietà complesse	"
	31. Proprietà di media e teoremi di confronto in Fisica Matematica	"

1964 -	32. Relatività generale	C.I.M.E.
	33. Dinamica dei gas rarefatti	"
	34. Alcune questioni di analisi numerica	"
	35. Equazioni differenziali non lineari	"
1965 -	36. Non-linear continuum theories	"
	37. Some aspects of ring theory	"
	38. Mathematical optimization in economics	"
1966 -	39. Calculus of variations	Ed. Cremonese, Firenze
	40. Economia matematica	"
	41. Classi caratteristiche e questioni connesse	"
	42. Some aspects of diffusion theory	"
1967 -	43. Modern questions of celestial mechanics	"
	44. Numerical analysis of partial differential equations	"
	45. Geometry of homogeneous bounded domains	"
1968 -	46. Controllability and observability	"
	47. Pseudo-differential operators	"
	48. Aspects of mathematical logic	"
1969 -	49. Potential theory	"
	50. Non-linear continuum theories in mechanics and physics and their applications	"
	51. Questions of algebraic varieties	"
1970 -	52. Relativistic fluid dynamics	"
	53. Theory of group representations and Fourier analysis	"
	54. Functional equations and inequalities	"
	55. Problems in non-linear analysis	"
1971 -	56. Stereodynamics	"
	57. Constructive aspects of functional analysis (2 vol.)	"
	58. Categories and commutative algebra	"
1972 -	59. Non-linear mechanics	"
	60. Finite geometric structures and their applications	"
	61. Geometric measure theory and minimal surfaces	"
1973 -	62. Complex analysis	"
	63. New variational techniques in mathematical physics	"
	64. Spectral analysis	"

1974 -	65. Stability problems	Ed. Cremonese, Firenze
	66. Singularities of analytic spaces	"
	67. Eigenvalues of non linear problems	"
1975 -	68. Theoretical computer sciences	"
	69. Model theory and applications	"
	70. Differential operators and manifolds	"
1976 -	71. Statistical Mechanics	Ed. Liguori, Napoli
	72. Hyperbolicity	"
	73. Differential topology	"
1977 -	74. Materials with memory	"
	75. Pseudodifferential operators with applications	"
	76. Algebraic surfaces	"
1978 -	77. Stochastic differential equations	"
	78. Dynamical systems	Ed. Liguori, Napoli and Birkhäuser Verlag
1979 -	79. Recursion theory and computational complexity	Ed. Liguori, Napoli
	80. Mathematics of biology	"
1980 -	81. Wave propagation	"
	82. Harmonic analysis and group representations	"
	83. Matroid theory and its applications	"
1981 -	84. Kinetic Theories and the Boltzmann Equation	(LNM 1048)Springer-Verlag
	85. Algebraic Threefolds	(LNM 947) "
	86. Nonlinear Filtering and Stochastic Control	(LNM 972) "
1982 -	87. Invariant Theory	(LNM 996) "
	88. Thermodynamics and Constitutive Equations	(LN Physics 228) "
	89. Fluid Dynamics	(LNM 1047) "
1983 -	90. Complete Intersections	(LNM 1092) "
	91. Bifurcation Theory and Applications	(LNM 1057) "
	92. Numerical Methods in Fluid Dynamics	(LNM 1127) "
1984	93. Harmonic Mappings and Minimal Immersions	(LNM 1161) "
	94. Schrödinger Operators	(LNM 1159) "
	95. Buildings and the Geometry of Diagrams	(LNM 1181) "
1985 -	96. Probability and Analysis	(LNM 1206) "
	97. Some Problems in Nonlinear Diffusion	(LNM 1224) "
	98. Theory of Moduli	(LNM 1337) "

Note: Volumes 1 to 38 are out of print. A few copies of volumes 23,28,31,32,33,34,36,38 are available on request from C.I.M.E.

1986 -	99. Inverse Problems	(LNM 1225)	Springer-Verlag
	100. Mathematical Economics	(LNM 1330)	"
	101. Combinatorial Optimization	to appear	"
1987 -	102. Relativistic Fluid Dynamics	(LNM 1385)	"
	103. Topics in Calculus of Variations	(LNM 1365)	"
1988 -	104. Logic and Computer Science	to appear	"
	105. Global Geometry and Mathematical Physics	to appear	"
1989 -	106. Methods of nonconvex analysis	to appear	"
	107. Microlocal Analysis and Applications	to appear	"

LECTURE NOTES IN MATHEMATICS

Edited by A. Dold and B. Eckmann

Some general remarks on the publication of monographs and seminars

In what follows all references to monographs, are applicable also to multiauthorship volumes such as seminar notes.

§1. Lecture Notes aim to report new developments - quickly, informally, and at a high level. Monograph manuscripts should be reasonably self-contained and rounded off. Thus they may, and often will, present not only results of the author but also related work by other people. Furthermore, the manuscripts should provide sufficient motivation, examples and applications. This clearly distinguishes Lecture Notes manuscripts from journal articles which normally are very concise. Articles intended for a journal but too long to be accepted by most journals, usually do not have this "lecture notes" character. For similar reasons it is unusual for Ph.D. theses to be accepted for the Lecture Notes series.

Experience has shown that English language manuscripts achieve a much wider distribution.

§2. Manuscripts or plans for Lecture Notes volumes should be submitted either to one of the series editors or to Springer-Verlag, Heidelberg. These proposals are then refereed. A final decision concerning publication can only be made on the basis of the complete manuscripts, but a preliminary decision can usually be based on partial information: a fairly detailed outline describing the planned contents of each chapter, and an indication of the estimated length, a bibliography, and one or two sample chapters - or a first draft of the manuscript. The editors will try to make the preliminary decision as definite as they can on the basis of the available information.

§3. Lecture Notes are printed by photo-offset from typed copy delivered in camera-ready form by the authors. Springer-Verlag provides technical instructions for the preparation of manuscripts, and will also, on request, supply special staionery on which the prescribed typing area is outlined. Careful preparation of the manuscripts will help keep production time short and ensure satisfactory appearance of the finished book. Running titles are not required; if however they are considered necessary, they should be uniform in appearance. We generally advise authors not to start having their final manuscripts specially tpyed beforehand. For professionally typed manuscripts, prepared on the special stationery according to our instructions, Springer-Verlag will, if necessary, contribute towards the typing costs at a fixed rate.

The actual production of a Lecture Notes volume takes 6-8 weeks.

.../...

§4. Final manuscripts should contain at least 100 pages of mathematical text and should include

- a table of contents
- an informative introduction, perhaps with some historical remarks. It should be accessible to a reader not particularly familiar with the topic treated.
- a subject index; this is almost always genuinely helpful for the reader.

§5. Authors receive a total of 50 free copies of their volume, but no royalties. They are entitled to purchase further copies of their book for their personal use at a discount of 33.3 %, other Springer mathematics books at a discount of 20 % directly from Springer-Verlag.

Commitment to publish is made by letter of intent rather than by signing a formal contract. Springer-Verlag secures the copyright for each volume.

LECTURE NOTES

ESSENTIALS FOR THE PREPARATION
OF CAMERA-READY MANUSCRIPTS

Springer

Springer-Verlag
Berlin Heidelberg New York
London Paris Tokyo Hong Kong

The preparation of manuscripts which are to be reproduced by photo-offset require special care. Manuscripts which are submitted in technically unsuitable form will be returned to the author for retyping. There is normally no possibility of carrying out further corrections after a manuscript is given to production. Hence it is crucial that the following instructions be adhered to closely. If in doubt, please send us 1 - 2 sample pages for examination.

General. The characters must be uniformly black both within a single character and down the page. Original manuscripts are required: photocopies are acceptable only if they are sharp and without smudges.

On request, Springer-Verlag will supply special paper with the text area outlined. The standard TEXT AREA (OUTPUT SIZE if you are using a 14 point font) is 18 x 26.5 cm (7.5 x 11 inches). This will be scale-reduced to 75% in the printing process. If you are using computer typesetting, please see also the following page.

Make sure the TEXT AREA IS COMPLETELY FILLED. Set the margins so that they precisely match the outline and type right from the top to the bottom line. (Note that the page number will lie outside this area). Lines of text should not end more than three spaces inside or outside the right margin (see example on page 4).

Type on one side of the paper only.

Spacing and Headings (Monographs). Use ONE-AND-A-HALF line spacing in the text. Please leave sufficient space for the title to stand out clearly and do NOT use a new page for the beginning of subdivisons of chapters. Leave THREE LINES blank above and TWO below headings of such subdivisions.

Spacing and Headings (Proceedings). Use ONE-AND-A-HALF line spacing in the text. Do not use a new page for the beginning of subdivisons of a single paper. Leave THREE LINES blank above and TWO below headings of such subdivisions. Make sure headings of equal importance are in the same form.

The first page of each contribution should be prepared in the same way. The title should stand out clearly. We therefore recommend that the editor prepare a sample page and pass it on to the authors together with these instructions. Please take the following as an example. Begin heading 2 cm below upper edge of text area.

MATHEMATICAL STRUCTURE IN QUANTUM FIELD THEORY

John E. Robert
Mathematisches Institut, Universität Heidelberg
Im Neuenheimer Feld 288, D-6900 Heidelberg

Please leave THREE LINES blank below heading and address of the author, then continue with the actual text on the same page.

Footnotes. These should preferable be avoided. If necessary, type them in SINGLE LINE SPACING to finish exactly on the outline, and separate them from the preceding main text by a line.

Symbols. Anything which cannot be typed may be entered by hand in BLACK AND ONLY BLACK ink. (A fine-tipped rapidograph is suitable for this purpose; a good black ball-point will do, but a pencil will not). Do not draw straight lines by hand without a ruler (not even in fractions).

Literature References. These should be placed at the end of each paper or chapter, or at the end of the work, as desired. Type them with single line spacing and start each reference on a new line. Follow "Zentralblatt für Mathematik"/"Mathematical Reviews" for abbreviated titles of mathematical journals and "Bibliographic Guide for Editors and Authors (BGEA)" for chemical, biological, and physics journals. Please ensure that all references are COMPLETE and ACCURATE.

IMPORTANT

Pagination. For typescript, number pages in the upper right-hand corner in LIGHT BLUE OR GREEN PENCIL ONLY. The printers will insert the final page numbers. For computer type, you may insert page numbers (1 cm above outer edge of text area).

It is safer to number pages AFTER the text has been typed and corrected. Page 1 (Arabic) should be THE FIRST PAGE OF THE ACTUAL TEXT. The Roman pagination (table of contents, preface, abstract, acknowledgements, brief introductions, etc.) will be done by Springer-Verlag.

If including running heads, these should be aligned with the inside edge of the text area while the page number is aligned with the outside edge noting that right-hand pages are odd-numbered. Running heads and page numbers appear on the same line. Normally, the running head on the left-hand page is the chapter heading and that on the right-hand page is the section heading. Running heads should not be included in proceedings contributions unless this is being done consistently by all authors.

Corrections. When corrections have to be made, cut the new text to fit and paste it over the old. White correction fluid may also be used.

Never make corrections or insertions in the text by hand.

If the typescript has to be marked for any reason, e.g. for provisional page numbers or to mark corrections for the typist, this can be done VERY FAINTLY with BLUE or GREEN PENCIL but NO OTHER COLOR: these colors do not appear after reproduction.

COMPUTER-TYPESETTING. Further, to the above instructions, please note with respect to your printout that
- the characters should be sharp and sufficiently black;
- it is not strictly necessary to use Springer's special typing paper. Any white paper of reasonable quality is acceptable.

If you are using a significantly different font size, you should modify the output size correspondingly, keeping length to breadth ratio 1 : 0.68, so that scaling down to 10 point font size, yields a text area of 13.5 x 20 cm (5 3/8 x 8 in), e.g.

Differential equations.: use output size 13.5 x 20 cm.

Differential equations.: use output size 16 x 23.5 cm.

Differential equations.: use output size 18 x 26.5 cm.

Interline spacing: 5.5 mm base-to-base for 14 point characters (standard format of 18 x 26.5 cm).
If in any doubt, please send us 1 - 2 sample pages for examination. We will be glad to give advice.

Vol. 1232: P.C. Schuur, Asymptotic Analysis of Soliton Problems. VIII, 180 pages. 1986.

Vol. 1233: Stability Problems for Stochastic Models. Proceedings, 1985. Edited by V.V. Kalashnikov, B. Penkov and V.M. Zolotarev. VI, 223 pages. 1986.

Vol. 1234: Combinatoire énumérative. Proceedings, 1985. Edité par G. Labelle et P. Leroux. XIV, 387 pages. 1986.

Vol. 1235: Séminaire de Théorie du Potentiel, Paris, No. 8. Directeurs: M. Brelot, G. Choquet et J. Deny. Rédacteurs: F. Hirsch et G. Mokobodzki. III, 209 pages. 1987.

Vol. 1236: Stochastic Partial Differential Equations and Applications. Proceedings, 1985. Edited by G. Da Prato and L. Tubaro. V, 257 pages. 1987.

Vol. 1237: Rational Approximation and its Applications in Mathematics and Physics. Proceedings, 1985. Edited by J. Gilewicz, M. Pindor and W. Siemaszko. XII, 350 pages. 1987.

Vol. 1238: M. Holz, K.-P. Podewski and K. Steffens, Injective Choice Functions. VI, 183 pages. 1987.

Vol. 1239: P. Vojta, Diophantine Approximations and Value Distribution Theory. X, 132 pages. 1987.

Vol. 1240: Number Theory, New York 1984–85. Seminar. Edited by D.V. Chudnovsky, G.V. Chudnovsky, H. Cohn and M.B. Nathanson. V, 324 pages. 1987.

Vol. 1241: L. Gårding, Singularities in Linear Wave Propagation. III, 125 pages. 1987.

Vol. 1242: Functional Analysis II, with Contributions by J. Hoffmann-Jørgensen et al. Edited by S. Kurepa, H. Kraljević and D. Butković. VII, 432 pages. 1987.

Vol. 1243: Non Commutative Harmonic Analysis and Lie Groups. Proceedings, 1985. Edited by J. Carmona, P. Delorme and M. Vergne. V, 309 pages. 1987.

Vol. 1244: W. Müller, Manifolds with Cusps of Rank One. XI, 158 pages. 1987.

Vol. 1245: S. Rallis, L-Functions and the Oscillator Representation. XVI, 239 pages. 1987.

Vol. 1246: Hodge Theory. Proceedings, 1985. Edited by E. Cattani, F. Guillén, A. Kaplan and F. Puerta. VII, 175 pages. 1987.

Vol. 1247: Séminaire de Probabilités XXI. Proceedings. Edité par J. Azéma, P.A. Meyer et M. Yor. IV, 579 pages. 1987.

Vol. 1248: Nonlinear Semigroups, Partial Differential Equations and Attractors. Proceedings, 1985. Edited by T.L. Gill and W.W. Zachary. IX, 185 pages. 1987.

Vol. 1249: I. van den Berg, Nonstandard Asymptotic Analysis. IX, 187 pages. 1987.

Vol. 1250: Stochastic Processes – Mathematics and Physics II. Proceedings 1985. Edited by S. Albeverio, Ph. Blanchard and L. Streit. VI, 359 pages. 1987.

Vol. 1251: Differential Geometric Methods in Mathematical Physics. Proceedings, 1985. Edited by P.L. García and A. Pérez-Rendón. VII, 300 pages. 1987.

Vol. 1252: T. Kaise, Représentations de Weil et GL_2 Algèbres de division et GL_n. VII, 203 pages. 1987.

Vol. 1253: J. Fischer, An Approach to the Selberg Trace Formula via the Selberg Zeta-Function. III, 184 pages. 1987.

Vol. 1254: S. Gelbart, I. Piatetski-Shapiro, S. Rallis. Explicit Constructions of Automorphic L-Functions. VI, 152 pages. 1987.

Vol. 1255: Differential Geometry and Differential Equations. Proceedings, 1985. Edited by C. Gu, M. Berger and R.L. Bryant. XII, 243 pages. 1987.

Vol. 1256: Pseudo-Differential Operators. Proceedings, 1986. Edited by H.O. Cordes, B. Gramsch and H. Widom. X, 479 pages. 1987.

Vol. 1257: X. Wang, On the C*-Algebras of Foliations in the Plane. V, 165 pages. 1987.

Vol. 1258: J. Weidmann, Spectral Theory of Ordinary Differential Operators. VI, 303 pages. 1987.

Vol. 1259: F. Cano Torres, Desingularization Strategies for Three-Dimensional Vector Fields. IX, 189 pages. 1987.

Vol. 1260: N.H. Pavel, Nonlinear Evolution Operators and Semigroups. VI, 285 pages. 1987.

Vol. 1261: H. Abels, Finite Presentability of S-Arithmetic Groups. Compact Presentability of Solvable Groups. VI, 178 pages. 1987.

Vol. 1262: E. Hlawka (Hrsg.), Zahlentheoretische Analysis II. Seminar, 1984–86. V, 158 Seiten. 1987.

Vol. 1263: V.L. Hansen (Ed.), Differential Geometry. Proceedings, 1985. XI, 288 pages. 1987.

Vol. 1264: Wu Wen-tsün, Rational Homotopy Type. VIII, 219 pages. 1987.

Vol. 1265: W. Van Assche, Asymptotics for Orthogonal Polynomials. VI, 201 pages. 1987.

Vol. 1266: F. Ghione, C. Peskine, E. Sernesi (Eds.), Space Curves. Proceedings, 1985. VI, 272 pages. 1987.

Vol. 1267: J. Lindenstrauss, V.D. Milman (Eds.), Geometrical Aspects of Functional Analysis. Seminar. VII, 212 pages. 1987.

Vol. 1268: S.G. Krantz (Ed.), Complex Analysis. Seminar, 1986. VII, 195 pages. 1987.

Vol. 1269: M. Shiota, Nash Manifolds. VI, 223 pages. 1987.

Vol. 1270: C. Carasso, P.-A. Raviart, D. Serre (Eds.), Nonlinear Hyperbolic Problems. Proceedings, 1986. XV, 341 pages. 1987.

Vol. 1271: A.M. Cohen, W.H. Hesselink, W.L.J. van der Kallen, J.R. Strooker (Eds.), Algebraic Groups Utrecht 1986. Proceedings. XII, 284 pages. 1987.

Vol. 1272: M.S. Livšic, L.L. Waksman, Commuting Nonselfadjoint Operators in Hilbert Space. III, 115 pages. 1987.

Vol. 1273: G.-M. Greuel, G. Trautmann (Eds.), Singularities, Representation of Algebras, and Vector Bundles. Proceedings, 1985. XIV, 383 pages. 1987.

Vol. 1274: N. C. Phillips, Equivariant K-Theory and Freeness of Group Actions on C*-Algebras. VIII, 371 pages. 1987.

Vol. 1275: C.A. Berenstein (Ed.), Complex Analysis I. Proceedings, 1985–86. XV, 331 pages. 1987.

Vol. 1276: C.A. Berenstein (Ed.), Complex Analysis II. Proceedings, 1985–86. IX, 320 pages. 1987.

Vol. 1277: C.A. Berenstein (Ed.), Complex Analysis III. Proceedings, 1985–86. X, 350 pages. 1987.

Vol. 1278: S.S. Koh (Ed.), Invariant Theory. Proceedings, 1985. V, 102 pages. 1987.

Vol. 1279: D. Ieşan, Saint-Venant's Problem. VIII, 162 Seiten. 1987.

Vol. 1280: E. Neher, Jordan Triple Systems by the Grid Approach. XII, 193 pages. 1987.

Vol. 1281: O.H. Kegel, F. Menegazzo, G. Zacher (Eds.), Group Theory. Proceedings, 1986. VII, 179 pages. 1987.

Vol. 1282: D.E. Handelman, Positive Polynomials, Convex Integral Polytopes, and a Random Walk Problem. XI, 136 pages. 1987.

Vol. 1283: S. Mardešić, J. Segal (Eds.), Geometric Topology and Shape Theory. Proceedings, 1986. V, 261 pages. 1987.

Vol. 1284: B.H. Matzat, Konstruktive Galoistheorie. X, 286 pages. 1987.

Vol. 1285: I.W. Knowles, Y. Saitō (Eds.), Differential Equations and Mathematical Physics. Proceedings, 1986. XVI, 499 pages. 1987.

Vol. 1286: H.R. Miller, D.C. Ravenel (Eds.), Algebraic Topology. Proceedings, 1986. VII, 341 pages. 1987.

Vol. 1287: E.B. Saff (Ed.), Approximation Theory, Tampa. Proceedings, 1985–1986. V, 228 pages. 1987.

Vol. 1288: Yu. L. Rodin, Generalized Analytic Functions on Riemann Surfaces. V, 128 pages, 1987.

Vol. 1289: Yu. I. Manin (Ed.), K-Theory, Arithmetic and Geometry. Seminar, 1984–1986. V, 399 pages. 1987.

Vol. 1290: G. Wüstholz (Ed.), Diophantine Approximation and Transcendence Theory. Seminar, 1985. V, 243 pages. 1987.

Vol. 1291: C. Mœglin, M.-F. Vignéras, J.-L. Waldspurger, Correspondances de Howe sur un Corps p-adique. VII, 163 pages. 1987

Vol. 1292: J.T. Baldwin (Ed.), Classification Theory. Proceedings, 1985. VI, 500 pages. 1987.

Vol. 1293: W. Ebeling, The Monodromy Groups of Isolated Singularities of Complete Intersections. XIV, 153 pages. 1987.

Vol. 1294: M. Queffélec, Substitution Dynamical Systems – Spectral Analysis. XIII, 240 pages. 1987.

Vol. 1295: P. Lelong, P. Dolbeault, H. Skoda (Réd.), Séminaire d'Analyse P. Lelong – P. Dolbeault – H. Skoda. Seminar, 1985/1986. VII, 283 pages. 1987.

Vol. 1296: M.-P. Malliavin (Ed.), Séminaire d'Algèbre Paul Dubreil et Marie-Paule Malliavin. Proceedings, 1986. IV, 324 pages. 1987.

Vol. 1297: Zhu Y.-l., Guo B.-y. (Eds.), Numerical Methods for Partial Differential Equations. Proceedings. XI, 244 pages. 1987.

Vol. 1298: J. Aguadé, R. Kane (Eds.), Algebraic Topology, Barcelona 1986. Proceedings. X, 255 pages. 1987.

Vol. 1299: S. Watanabe, Yu.V. Prokhorov (Eds.), Probability Theory and Mathematical Statistics. Proceedings, 1986. VIII, 589 pages. 1988.

Vol. 1300: G.B. Seligman, Constructions of Lie Algebras and their Modules. VI, 190 pages. 1988.

Vol. 1301: N. Schappacher, Periods of Hecke Characters. XV, 160 pages. 1988.

Vol. 1302: M. Cwikel, J. Peetre, Y. Sagher, H. Wallin (Eds.), Function Spaces and Applications. Proceedings, 1986. VI, 445 pages. 1988.

Vol. 1303: L. Accardi, W. von Waldenfels (Eds.), Quantum Probability and Applications III. Proceedings, 1987. VI, 373 pages. 1988.

Vol. 1304: F.Q. Gouvêa, Arithmetic of *p*-adic Modular Forms. VIII, 121 pages. 1988.

Vol. 1305: D.S. Lubinsky, E.B. Saff, Strong Asymptotics for Extremal Polynomials Associated with Weights on $\mathbb{R}$. VII, 153 pages. 1988.

Vol. 1306: S.S. Chern (Ed.), Partial Differential Equations. Proceedings, 1986. VI, 294 pages. 1988.

Vol. 1307: T. Murai, A Real Variable Method for the Cauchy Transform, and Analytic Capacity. VIII, 133 pages. 1988.

Vol. 1308: P. Imkeller, Two-Parameter Martingales and Their Quadratic Variation. IV, 177 pages. 1988.

Vol. 1309: B. Fiedler, Global Bifurcation of Periodic Solutions with Symmetry. VIII, 144 pages. 1988.

Vol. 1310: O.A. Laudal, G. Pfister, Local Moduli and Singularities. V, 117 pages. 1988.

Vol. 1311: A. Holme, R. Speiser (Eds.), Algebraic Geometry, Sundance 1986. Proceedings. VI, 320 pages. 1988.

Vol. 1312: N.A. Shirokov, Analytic Functions Smooth up to the Boundary. III, 213 pages. 1988.

Vol. 1313: F. Colonius, Optimal Periodic Control. VI, 177 pages. 1988.

Vol. 1314: A. Futaki, Kähler-Einstein Metrics and Integral Invariants. IV, 140 pages. 1988.

Vol. 1315: R.A. McCoy, I. Ntantu, Topological Properties of Spaces of Continuous Functions. IV, 124 pages. 1988.

Vol. 1316: H. Korezlioglu, A.S. Ustunel (Eds.), Stochastic Analysis and Related Topics. Proceedings, 1986. V, 371 pages. 1988.

Vol. 1317: J. Lindenstrauss, V.D. Milman (Eds.), Geometric Aspects of Functional Analysis. Seminar, 1986–87. VII, 289 pages. 1988.

Vol. 1318: Y. Felix (Ed.), Algebraic Topology – Rational Homotopy. Proceedings, 1986. VIII, 245 pages. 1988

Vol. 1319: M. Vuorinen, Conformal Geometry and Quasiregular Mappings. XIX, 209 pages. 1988.

Vol. 1320: H. Jürgensen, G. Lallement, H.J. Weinert (Eds.), Sem groups, Theory and Applications. Proceedings, 1986. X, 416 page 1988.

Vol. 1321: J. Azéma, P.A. Meyer, M. Yor (Eds.), Séminaire Probabilités XXII. Proceedings. IV, 600 pages. 1988.

Vol. 1322: M. Métivier, S. Watanabe (Eds.), Stochastic Analysi Proceedings, 1987. VII, 197 pages. 1988.

Vol. 1323: D.R. Anderson, H.J. Munkholm, Boundedly Controll Topology. XII, 309 pages. 1988.

Vol. 1324: F. Cardoso, D.G. de Figueiredo, R. Iório, O. Lopes (Eds Partial Differential Equations. Proceedings, 1986. VIII, 433 page 1988.

Vol. 1325: A. Truman, I.M. Davies (Eds.), Stochastic Mechanics ar Stochastic Processes. Proceedings, 1986. V, 220 pages. 1988.

Vol. 1326: P.S. Landweber (Ed.), Elliptic Curves and Modular Forms Algebraic Topology. Proceedings, 1986. V, 224 pages. 1988.

Vol. 1327: W. Bruns, U. Vetter, Determinantal Rings. VII,236 page 1988.

Vol. 1328: J.L. Bueso, P. Jara, B. Torrecillas (Eds.), Ring Theor Proceedings, 1986. IX, 331 pages. 1988.

Vol. 1329: M. Alfaro, J.S. Dehesa, F.J. Marcellan, J.L. Rubio d Francia, J. Vinuesa (Eds.): Orthogonal Polynomials and their Applic tions. Proceedings, 1986. XV, 334 pages. 1988.

Vol. 1330: A. Ambrosetti, F. Gori, R. Lucchetti (Eds.), Mathematic Economics. Montecatini Terme 1986. Seminar. VII, 137 pages. 1988

Vol. 1331: R. Bamón, R. Labarca, J. Palis Jr. (Eds.), Dynamic Systems, Valparaiso 1986. Proceedings. VI, 250 pages. 1988.

Vol. 1332: E. Odell, H. Rosenthal (Eds.), Functional Analysis. Pro ceedings, 1986–87. V, 202 pages. 1988.

Vol. 1333: A.S. Kechris, D.A. Martin, J.R. Steel (Eds.), Cabal Semina 81–85. Proceedings, 1981–85. V, 224 pages. 1988.

Vol. 1334: Yu.G. Borisovich, Yu. E. Gliklikh (Eds.), Global Analysi – Studies and Applications III. V, 331 pages. 1988.

Vol. 1335: F. Guillén, V. Navarro Aznar, P. Pascual-Gainza, F. Puerta Hyperrésolutions cubiques et descente cohomologique. XII, 19 pages. 1988.

Vol. 1336: B. Helffer, Semi-Classical Analysis for the Schrödinge Operator and Applications. V, 107 pages. 1988.

Vol. 1337: E. Sernesi (Ed.), Theory of Moduli. Seminar, 1985. VIII, 23 pages. 1988.

Vol. 1338: A.B. Mingarelli, S.G. Halvorsen, Non-Oscillation Domain of Differential Equations with Two Parameters. XI, 109 pages. 1988.

Vol. 1339: T. Sunada (Ed.), Geometry and Analysis of Manifolds Procedings, 1987. IX, 277 pages. 1988.

Vol. 1340: S. Hildebrandt, D.S. Kinderlehrer, M. Miranda (Eds.) Calculus of Variations and Partial Differential Equations. Proceedings 1986. IX, 301 pages. 1988.

Vol. 1341: M. Dauge, Elliptic Boundary Value Problems on Corner Domains. VIII, 259 pages. 1988.

Vol. 1342: J.C. Alexander (Ed.), Dynamical Systems. Proceedings, 1986–87. VIII, 726 pages. 1988.

Vol. 1343: H. Ulrich, Fixed Point Theory of Parametrized Equivarian Maps. VII, 147 pages. 1988.

Vol. 1344: J. Král, J. Lukeš, J. Netuka, J. Veselý (Eds.), Potentia Theory – Surveys and Problems. Proceedings, 1987. VIII, 271 pages 1988.

Vol. 1345: X. Gomez-Mont, J. Seade, A. Verjovski (Eds.), Holomorphic Dynamics. Proceedings, 1986. VII, 321 pages. 1988.

Vol. 1346: O. Ya. Viro (Ed.), Topology and Geometry – Rohlir Seminar. XI, 581 pages. 1988.

Vol. 1347: C. Preston, Iterates of Piecewise Monotone Mappings or an Interval. V, 166 pages. 1988.

Vol. 1348: F. Borceux (Ed.), Categorical Algebra and its Applications. Proceedings, 1987. VIII, 375 pages. 1988.

Vol. 1349: E. Novak, Deterministic and Stochastic Error Bounds in Numerical Analysis. V, 113 pages. 1988.